Advances in Anti-Access and Area Denial (A2/AD)

ADVANCES IN ANTI-ACCESS AND AREA DENIAL (A2/AD)

From Defensive Technologies To A Whole-of-Society Approach

Edited by

Edwin Dado, René Janssen, Roy Hamans,

Marc Gouw and Kerry Malone

LEIDEN UNIVERSITY PRESS

The Open Access edition was made possible by a contribution from The Netherlands Defence Academy

Cover design: Andre Klijsen
Cover illustration: Peter de Vries
Lay-out: Crius Group, Hulshout
Printer: Antilope De Bie Printing, Duffel

ISBN 9789087285036
e-ISBN 9789400605626 (e-PDF)
e-ISBN 9789400605633 (e-PUB)
https://doi.org/10.24415/9789087285036
NUR 950

Contact EU General Product Safety Regulation (GPSR): productsafety@lup.nl

Table of Contents

List of Figures

List of Tables

Foreword

By the Dean of the Faculty of Military Sciences / Netherlands Defence Academy

Traditionally the Faculty of Military Sciences at the Netherlands Defence Academy publishes the Netherlands Annual Review of Military Sciences, a volume dedicated to emerging topics in the field of military research. With this volume our faculty gathers scientific contributions from its' various fields of research in order to give the reader a broad overview on a specialised topic. This year's research topic is on Anti-Access (A2) and Area Denial (Denial). Both are military strategies, where A2 is designed to prevent an adversary from entering a specific area and AD is intended to hinder an opponent already present in an area regarding their freedom of movement and operations.

The urgency of A2/AD (Anti-Access/Area Denial) has increased significantly in recent years due to growing tensions between major powers and rapid military technological developments. A2/AD capabilities make it possible to keep opponents at a distance (anti-access) or severely restrict their freedom of movement within an area (area denial). This undermines traditional military power projection, especially that of Western countries.

An important example is China's establishment of A2/AD zones in the South China Sea and around Taiwan. Through long-range missiles, air defence systems, and maritime capabilities, China is attempting to hinder foreign intervention, particularly by the United States. This increases the risk of escalation in the Indo-Pacific and has direct consequences for global trade routes and chip production.

The urgency is also evident in Europe. Since the war in Ukraine, Russia has strengthened its military presence and air defence in strategic areas. This could limit the freedom of movement of NATO troops and make it more difficult to send reinforcements in crisis situations. In addition, technological innovations such as hypersonic weapons, precision missiles, and drones are accelerating the effectiveness of A2/AD systems. This makes military dominance less self-evident and increases the risk of miscalculation.

In this volume A2/AD is studied from different perspectives, mainly technological, but also chapters on logistics, legal issues and artificial intelligence are included. I am convinced that combining these chapters in this volume will give the reader more insight into the topic. From this place I'd like to thank all authors and editors for bringing these interesting perspectives together in the book.

Prof. dr. Patrick Oonincx
Dean Faculty of Military Sciences
Breda/Den Helder, The Netherlands

Foreword

By the Netherlands Military Representative to the NATO and EU Military Committees

The coming years will be decisive for the future of Europe. One key question is how Russia will position itself towards the free democracies bordering Russia and Belarus once the war in Ukraine has ended. In its Strategic Concept and Family of Plans, NATO has defined how the Alliance will defend its eastern Allies. At the heart of these plans lies "Deterrence and Defence": to convince Putin and his proxies that an attack on any Ally will never yield strategic benefit and, should deterrence fail, to ensure that such an attack is effectively repelled.

As during the Cold War, the reinforcement of NATO's eastern flank is an integral part of these plans. Such reinforcement is critically dependent on Freedom of Movement. The timely movement of units, equipment, and supplies – both immediately prior to and following the outbreak of a potential conflict – is essential. However, Russia has the capability to disrupt the required Freedom of Movement. Around Murmansk, St Petersburg, and Kaliningrad, modern air defence and strike systems are deployed that can engage targets in the air, on land and at sea. From Kaliningrad in particular, movements by land and air through the Suwałki Gap, and by sea and air across the Baltic Sea, are highly vulnerable. The capabilities that Russia has concentrated in these areas are commonly described as Anti-Access/Area Denial (A2/AD): the ability to prevent an adversary from accessing and operating freely within a specific region.

For NATO, Russian A2/AD capabilities constitute a serious threat. They have the potential to isolate the Baltic States and pose risks to the Swedish island of Gotland as well as to south-eastern Finland. At the same time, Western conceptual thinking and practical implementation of A2/AD remain underdeveloped. Yet it is precisely in this domain that Deterrence and Defence could be further strengthened.

This book Advances in Anti-Access Area Denial (A2/AD): From Defensive Technologies to a Whole-of-Society Approach, brings together 16 scientific contributions that explore what A2/AD is, how an effective A2/AD system can be constructed, and how such a system can be countered. This volume is essential reading for military planners, doctrine developers, and all those with an interest in modern warfare. I wish the reader an engaging and thought-provoking journey through these pages, and trust that this volume will deepen understanding and stimulate further reflection on the challenges and opportunities of A2/AD.

Lieutenant-General Dick C. van Ingen MA
Netherlands Military Representative to the NATO and EU Military Committees

Introduction

Roy Hamans, Marc Gouw, Edwin Dado, René Janssen and Kerry Malone

NATO's original definition of Anti-Access Area Denial (A2/AD) as "warfighting strategies focused on preventing an opponent from operating military forces near, into, or within a contested region"[1] no longer captures the realities of contemporary conflict. This volume argues that A2/AD has evolved from a predominantly defensive military deterrent into a technologically advanced, integrated, multidomain system for denial as well as coercion, whose effects increasingly extend into social, political and economic domains or the 'whole-of-society'. This evolution demands fundamental conceptual and practical reappraisal.

The threat of A2/AD is not new, nor does it alter military capabilities per se.[2] What is new is that recent developments show A2/AD is no longer solely about blocking an enemy's military access to a country or operational theatre. Instead, capabilities – military and otherwise – that are traditionally associated with A2/AD are increasingly used beyond the battlefield, beyond military targets, and outside of a declared war. These capabilities are now actively employed to target critical infrastructure, digital systems, and societal cohesion that support open democratic societies. This volume emphasises this shift and addresses it by combining academic and practitioner perspectives on how A2/AD capabilities are utilised today and how they should be countered: not solely as a military issue, but as a technological advancement and strategic concern for entire societies.

Reassessing A2/AD in the Light of Hybrid Warfare

A stark illustration of this shift is the sabotage of Norway's Bremanger dam in April 2025, which marks one of the first formally attributed attacks against a NATO ally's critical infrastructure. The deliberate opening of a dam's floodgate, releasing 500 litres of water per second to flow unnoticed for hours,[3] led Norwegian authorities to formally attribute this state-sponsored sabotage to a hybrid warfare strategy[4] that blends cyber, physical and informational means. According to Beate Gangås, head of the Norwegian Police Security Service PST, such acts aim not only to disrupt infrastructure and demonstrate capability but "to exert influence and create fear or unrest in the population"[5]. These operations are part of adversaries' non-kinetic and kinetic multidomain A2/AD systems, now employed offensively as elements of hybrid warfare campaigns.

Similar dynamics are observed in the ongoing war against Ukraine, where the strategy of systematically targeting critical civilian infrastructure is 'unmistakable' and has been condemned as 'unlawful.[6] Human rights organisations have highlighted how deliberate attacks on energy facilities, water systems and other essential services aim to make daily life impossible and the "sole purpose of terrorizing civilians is a war crime"[7]. In response, the International Criminal Court in The Hague has issued warrants for the arrest of individuals "responsible for the war crime" of "directing attacks at civilian objects" and "causing excessive incidental harm to civilians" as well as "the crime against humanity of inhumane treatment"[8]. These developments show that the use of A2/AD capabilities now extends beyond military conflicts and the battlefield, reaching into society at large, especially in legal, moral, and humanitarian areas, as enemies deliberately target the civilian domain as part of their operational strategy.

A2/AD, Hybrid Wars and the Grey Zone

The examples of Norway and Ukraine show how A2/AD capabilities have advanced beyond traditional targeting of military forces in the theatre to include civilians, critical infrastructure, and digital systems. These attacks are not isolated incidents, but as part of a broader pattern of 'hybrid wars'[9] that blur the distinctions between war and peace, state and non-state actors, combatants and non-combatants, with adversaries employing physical, virtual and cognitive tools in a synchronised way. As the analysis of hybrid warfare already noted, such campaigns combine military and non-military instruments, all forms of warfare and tactics, to exploit vulnerabilities across multiple domains 'simultaneously'[10].

Political leaders increasingly acknowledge that European democracies are operating in a 'grey zone'. As a Dutch minister of Defence observed "we are not living in a state of war [...] but we are not living in peace any more either. In this grey zone, we need to defend against hybrid attacks"[11]. Intelligence officials have similarly begun to warn that adversaries are "testing us in the grey zone with tactics that are just below the threshold of war," according to Blaise Metreweli, the head of MI6, on December 15th 2025,[12] and probing resolve and resilience while avoiding clear triggers for collective defence. In this environment, A2/AD is no longer seen as a way to keep military forces out of a region but as a tool to restrict political decisions, influence public opinion, and weaken confidence through sustained, ambiguous pressure.

A2/AD and the Vulnerability of Open Democratic Societies

Open democratic societies such as Norway and the Netherlands are particularly exposed to these developments because their prosperity and legitimacy depend on the transparent, reliable operation of complex infrastructures and institutions. Medium-sized, trade-dependent economies like the Netherlands rely on open seas, free skies, and the unrestricted movement of people, goods, and data. Such unimpeded movement is facilitated by accessible ports, communication cables and satellites, digital systems, energy grids and sectors that are increasingly targeted by adversaries.[13] Their openness is further reinforced by international agreements, the rule of law and robust regulatory and oversight mechanisms. Yet this very openness makes them vulnerable[14] when data is widely available in the public domain, or can be illicitly accessed through infiltration and hacking.

The interconnectedness of critical infrastructures and related systems amplify the ripple effect of disruptions,[15] magnifying the societal impact of attacks that may initially appear local or technical. Recent global cybersecurity assessments show that geopolitical instability is reshaping the threat landscape, with critical infrastructure increasingly targeted as an entry point for broader coercive strategies. Following such acts, the erosion of confidence is measurable. A 2025 survey by the World Economic Forum, involving hundreds of leaders and academics from 92 countries, reports that 31% now lack confidence in their nation's ability to manage major cyber incidents, up from 26% in 2024.[16] The prominence of these concerns in mainstream publications like those used here, rather than niche military or technical ones, indicates that uncertainty and vulnerability are spreading throughout society at large.

From Military Deterrence to Multi-Domain Coercion

Traditionally A2/AD was understood as a defensive military posture aimed at raising the costs of intervention by a stronger adversary. Earlier definitions of A2/AD distinguished anti-access (A2) measures, designed to prevent an opponent from deploying its forces into a given area, from area-denial (AD) measures intended to keep them from operating freely in that area.[17] In this context, A2/AD aimed to challenge military dominance by establishing 'contested zones'[18] where damage could be inflicted on approaching forces, thereby deterring or complicating intervention, particularly for actors like the United States operating in forward theatres and western force projection in general.[19] Such contested zones would typically be within a few hundred kilometres of the coast, on land and below 15,000 feet in the air. Adversaries have demonstrated that it is possible to engage U.S. forces in this manner, and although few have achieved victory, others have managed to impose costs and retain some advantage for their forces.[20] This illustrated the logic of A2/AD as a defensive deterrent aimed at neutralising physical threats by opposing military forces within defined geographic areas.

Strategic intent aside, technological advancements and cross-domain integration have fundamentally transformed the nature and application of A2/AD systems. These advancements encompass intelligence, surveillance, and reconnaissance (ISR) technologies, long-range and precision strike capabilities, hypersonic systems, cyber warfare (EW), and the formal inclusion of the space domain within military forces. These technologies, which have converged into increasingly integrated multidomain capabilities, also seem to have enhanced their use for more offensive purposes. In recent years, adversaries have also potentially surpassed NATO's defensive systems[21] thereby threatening the European homeland, and reducing the technological advantage of Western forces, including US precision strike capabilities and global force projection.[22]

Threats are thus no longer only physical in theatre or deterrent, as virtual intrusions into cyber and information space pose increasing challenges to public and private systems alike. Because modern economies, administrations and communities depend on networked digital infrastructures, the offensive use of cyber A2/AD alone can jeopardise the functioning of entire societies,[23] with cascading effects that would make effective crisis management or military response difficult. Given their substantial impact on the 'whole of society', it is likely that A2/AD systems like these will be employed during military conflicts. Yet even the perceived threat to economies and societies in a pre-conflict environment risks escalating crises into armed conflict[24] as actors respond to coercive pressure that remains intentionally ambiguous in origin.

Societal Systems, Coherence and Cognitive Targets

As societal systems become more digital, open, and transparent under democratic oversight and market efficiency, their vulnerabilities have become more visible and exploitable. Critical assessments emphasise how geopolitical instability is reshaping cyber risk and how adversaries increasingly target critical infrastructure[25] as part of broader strategies to coerce societies. While high-profile acts of sabotage against pipelines, cables, or dams attract public attention, and given their physical nature can be repaired albeit at significant cost, more subtle activities may be even more consequential.

Less publicly visible threats include orchestrated disinformation campaigns, support for radical or fringe movements and the deliberate transformation of demonstrations into serious security challenges.[26] These activities undermine social trust, polarise political debate, and erode confidence in institutions and international law and order, without producing clearly attributable 'events'. Together with more visible physical and cyber-attacks, they challenge the cognitive and societal aspects of resilience,[27] making deterrence, crisis management, and collective defence more complicated. As a result, A2/AD must be understood as operating along physical, virtual, and cognitive axes, with effects that are both cumulative and immediate. This calls for a re-evaluation of the concept of anti-access/area denial (A2/AD).

Re-appraising A2/AD: Through Technological Advancements to a Whole-of-Society Approach

The reality outlined above underscores the need for a paradigm shift in how A2/AD is conceptualised and addressed. A2/ AD can no longer be regarded merely as a bounded, theatre-specific military challenge focused on access to, and control of, defined operational spaces. Instead it must be understood as a strategic component of hybrid warfare, that targets physical, virtual and cognitive dimensions, with the aim of undermining society as a whole and putting societal resilience under sustained pressure.[28] In this context, modern A2/AD can be viewed as a collection of multidomain capabilities which can be used for damaging both military and civilian systems, eroding cohesion within societies and alliances, and constraining political decision-making under pressure.

The analysis set forth in this introduction thus reframes A2/AD not solely as a military challenge to be confronted during the initial stages of an overt conflict, but as a critical component within a broader, active hybrid confrontation that remains undeclared. In principle, the threat of A2/AD is not new, nor is the way military capabilities are used per se.[29] What is new is that advancements have substantially extended and integrated these capabilities into multidomain systems, which have shifted from a defensive to an offensive nature and extend far beyond the battlefield, ever more into society, as part of active hybrid warfare.

This means, and this is a shift, that while military assets such as tanks, HIMARS, and well-trained and equipped military forces prepared to conduct kinetic operations remain crucial for deterrence, they are no longer sufficient on their own. The effective deployment of these forces increasingly depends on non-military technologies, infrastructure, systems, and personnel, which, as demonstrated here, are vulnerable to being targeted due to the inherent weaknesses of open democratic systems societies. Consequently, military forces can no longer be regarded as shields to safeguard societal security, since such shields are inherently susceptible to circumvention. Contemporary and active confrontation involves synchronised and integrated activities across civilian, military, and informational capabilities to actively coerce open democratic societies.

Since coercion through A2/AD capabilities effectively targets a society, the response should also begin as a social response[30] on the whole-of-society level. Addressing this societal challenge thus calls for a new response[31] based on integrated multidomain A2/AD capabilities that include non-military capabilities[32] and actors, rather than relying solely on traditional military deterrent assets. Countries, such as the Netherlands, would "benefit from a proactive, anticipatory and comprehensive approach to national security" as their "response to an acute threat is too sector-based and often focused on damage limitation"[33]. And such a whole-of-society approach is necessary sooner rather than later, as the longer that visible and 'invisible' risks of A2/AD are denied, "the harder it will be to recover" not unlike the "slow boil of the frog"[34].

Technological Advancements and Overview of the Volume

The proposed re-evaluation of A2/AD forms the core of this edited volume, which explores how to address A2/AD by shifting from a 'defence-only' perspective, via technological advancements, to a 'whole-of-society' approach, as emphasised in the subtitle. The chapters collectively illustrate how A2/AD systems are developing, how they are utilised as part of hybrid warfare strategies, and how they threaten the infrastructure and institutions that underpin open societies. They also consider how states and societies can adapt by enhancing resilience, improving situational awareness, and developing legal, technological, and organisational strategies capable of countering the coercive use of A2/AD capabilities.

This volume is organised to move from conceptual and strategic blind spots to tangible technological, operational, and legal challenges. Early contributions assess the development of A2/AD thinking and emphasise how analysis has often underestimated technological progress and its societal and hybrid aspects. Subsequent chapters explore how to counteract A2/AD as it evolves from static shields to flexible, technologically advanced hybrid warfare systems. They also examine how to overcome emerging threats such as increasingly advanced technological challenges, from hypersonic glide vehicles and drone swarms to missile systems. Additional contributions address specific domains and technological innovations, including multi-band neutralisation of drones using high-power microwave radar, the electromagnetic domain as an unseen foundation of modern A2/AD, navigation warfare targeting positioning, navigation, and timing information, and hydraulic warfare as a low-tech method of area denial in high-tech conflicts, along with the critical condition of cyberspace and decision-making for resilient logistics in A2/AD environments.

The volume also explores the legal aspects of A2/AD technology developments, including self-defence zones in outer space under international law, and the implications of missile tests for national security and international standings. It discusses how to address A2/AD, shifting from a solely defensive, technological advancement approach to a 'Whole-of-Society' perspective, as indicated by the subtitle. Answering this question requires not only a renewed appreciation of A2/AD as a military challenge but also recognising it as a technological, strategic, and society-wide issue. More importantly, it involves bridging the gap between those who simply hope that situations will not worsen significantly and those who actively defend existing infrastructure and prepare to counter further A2/AD-enabled transgressions by adversaries.

Collectively, these chapters explain how A2/AD transforms from mainly defensive military measures aimed at neutralising more capable adversaries, into technologically advanced, integrated, multidomain systems that increasingly extend beyond the battlefield to involve the 'whole-of-society' as part of hybrid warfare. The final chapter synthesises these insights into a policy-oriented call to action for policymakers, military strategists, industry stakeholders, and civil society, emphasising the need to recalibrate existing thinking and adapt strategies and practices to better counter ongoing and future A2/AD threats. The speed and level of adversaries' technological advancement require NATO to follow suit, reviewing doctrine and practice to achieve the integrated joint level necessary to counter A2/AD.[35] This also emphasises the importance of developing capabilities for effective deterrence and intervention. Today's security landscape no longer depends solely on the quantity of tanks or missiles, but also on the effective use of technological advancements and society's ability to adapt in new ways for a different kind of war and peace: a paradigm in which A2//AD occupies an intersectional space that includes military operations, technological developments, infrastructure resilience, and the functioning of democratic institutions within the 'whole-of-society'.

Notes

1 Lasconjarias, *"How to Respond to Anti-Access"*, based on Tangredi, "Anti-access Warfare: Countering A2/AD Strategies".
2 Schmidt, "Countering Anti-Access / Area Denial".
3 World Economic Forum, *Global Cybersecurity Outlook 2026.*
4 Ribeiro, *"Oslo Warns of Escalating Russian Cyber Threat".*

5 Ibid.

6 Amnesty International, *"Ukraine: Russian Attacks on Critical Energy"*.

7 Ibid.

8 International Criminal Court, *"Situation in Ukraine"*.

9 Hoffman, *Conflict in the 21st Century.*

10 Ibid.

11 Government of The Netherlands, *"Government Works to Increase Resilience"*.

12 *The Economist*, *"A Rash of Baltic Cable-Cutting"*.

13 Advisory Council on International Affairs, *Hybrid Threats and Societal Resilience.*

14 Ibid.

15 World Economic Forum, *Global Cybersecurity Outlook 2026.*

16 Ibid.

17 "Anti-access/area Denial (A2/AD)," *Wikipedia*, 2026; Lasconjarias, "How to Respond to Anti-Access/Area Denial (A2/AD)? Towards a NATO Counter-A2/AD Strategy".

18 Posen, "Command of the Commons".

19 Schmidt, "Countering Anti-Access / Area Denial".

20 Ibid.

21 Kemp, "Strategic Security in Northern Europe".

22 Schmidt, "Countering Anti-Access / Area Denial".

23 Russell, *"Strategic Anti-Access/Area Denial in Cyberspace"*.

24 Ibid.

25 World Economic Forum, *Global Cybersecurity Outlook 2026.*

26 Advisory Council on International Affairs, *Hybrid Threats and Societal Resilience.*

27 Ibid.

28 Ibid.

29 Schmidt, "Countering Anti-Access / Area Denial".

30 Zeman, "Social Anti-Access/Area-Denial".

31 Ibid.

32 Sukhankin, "David vs. Goliath".

33 Advisory Council on International Affairs, *Hybrid Threats and Societal Resilience.*

34 Zeman, "Social Anti-Access/Area-Denial".

35 Schmidt, "Countering Anti-Access / Area Denial".

References

Advisory Council on International Affairs (AIV). 2024. *Hybrid Threats and Societal Resilience*. The Hague: AIV, 4 June. https://www.adviesraadinternationalevraagstukken.nl/site/binaries/site-content/collections/documents/2024/06/04/hybride-dreigingen-en-maatschappelijke-weerbaarheid/aiv-hybrid-threats-and-societal-resilience-4jun2024.pdf.

Amnesty International. 2022. *"Ukraine: Russian Attacks on Critical Energy Infrastructure Amount to War Crimes."* News release, October 20. Amnesty International. https://www.amnesty.org/en/latest/news/2022/10/ukraine-russian-attacks-on-critical-energy-infrastructure-amount-to-war-crimes/

Government of the Netherlands. 2024. *"Government Works to Increase Resilience Against Military and Hybrid Threats." Government.nl*, December 6. https://www.government.nl/latest/news/2024/12/06/government-works-to-increase-resilience-against-military-and-hybrid-threats

Hoffman, Frank G. 2007. *Conflict in the 21st Century: The Rise of Hybrid Wars*. Potomac Institute for Policy Studies. https://www.potomacinstitute.org/images/stories/publications/potomac_hybridwar_0108.pdf

International Criminal Court. 2024. *"Situation in Ukraine: ICC Judges Issue Arrest Warrants Against Sergei Ivanovich Kobylash and Viktor Nikolayevich Sokolov." International Criminal Court*, March 5. https://www.icc-cpi.int/news/situation-ukraine-icc-judges-issue-arrest-warrants-against-sergei-ivanovich-kobylash-and

Kemp, Herb. 2020. "Strategic Security in Northern Europe." *Journal of Strategic Security* 14 (1), 78–91. Published by University of South Florida Board of Trustees. Stable URL: https://www.jstor.org/stable/10.2307/26999978

Lasconjarias, Guillaume, and Alessandro Marrone. 2016. *How to Respond to Anti-Access/Area Denial (A2/AD)? Towards a NATO Counter-A2/AD Strategy*. NDC Conference Report No. 01/16. Rome: NATO Defense College, February. Accessed. https://www.ulib.sk/files/english/nato-library/collections/monographs/ndc-conference-report/crep_01_2016_a2_ad.pdf

Posen, Barry R. 2003. "Command of the Commons: The Military Foundation of U.S. Hegemony." *International Security* 28 (1), Summer 5–46. https://www.belfercenter.org/sites/default/files/pantheon_files/files/publication/posen_summer_2003.pdf

Ribeiro, Anna. 2025. *"Oslo Warns of Escalating Russian Cyber Threat after Dam Breach, Citing Moscow as Biggest Risk to National Security." Industrial Cyber*, August 15. https://industrialcyber.co/industrial-cyber-attacks/oslo-warns-of-escalating-russian-cyber-threat-after-dam-breach-citing-moscow-as-biggest-risk-to-national-security/

Russell, Alison Lawlor. 2015. *"Strategic Anti-Access/Area Denial in Cyberspace."* In *Proceedings of the 7th International Conference on Cyber Conflict: Architectures in Cyberspace*, edited by M. Maybaum, A.-M. Osula, and L. Lindström. Tallinn: NATO Cooperative Cyber Defence Centre of Excellence. https://ccdcoe.org/uploads/2018/10/Art-11-Strategic-Anti-Access-Area-Denial-in-Cyberspace.pdf

Schmidt, Andreas. 2017. "Countering Anti-Access / Area Denial: Future Capability Requirements in NATO." *Journal Edition 23*, Joint Air Power Competence Centre, January 2017. https://www.japcc.org/articles/countering-anti-access-area-denial/

Sukhankin, S. 2019. "David vs. Goliath: Kaliningrad Oblast as Russia's A2/AD 'Bubble". *Scandinavian Journal of Military Studies*, 2 (1), 95–110.

Tangredi, Sam J. 2013. "Anti-access Warfare: Countering A2/AD Strategies". Naval Institute Press.

The Economist. 2026. *"A Rash of Baltic Cable-Cutting Raises Fears of Sabotage."* January 6. https://www.economist.com/europe/2026/01/06/a-rash-of-baltic-cable-cutting-raises-fears-of-sabotage

World Economic Forum. 2026. *Global Cybersecurity Outlook 2026: Insight Report*. World Economic Forum, January 12. https://reports.weforum.org/docs/WEF_Global_Cybersecurity_Outlook_2026.pdf

Zeman, Phil. 2021. "Social Anti-Access/Area-Denial (Social A2/AD)." *Journal of Advanced Military Studies* 12 (1), 149–164. Marine Corps University Press. https://doi.org/10.21140/mcuj.20211201007

CHAPTER 1

Strategic Blind Spots in A2/AD Analysis

Peter van Maurik

Abstract

This chapter discusses a conceptual gap between strategic-level perceptions and operational or tactical challenges of Anti-Access/Area Denial (A2/AD). It is written with the intent to shift the debate from a political and strategic conceptual discourse towards a more scientific military operational analysis.

Far too often, the A2/AD concept is portrayed in strategic discourse as a revolutionary development capable of transforming warfare. This chapter, however, argues that the A2/AD logic is deeply rooted in existing defensive principles from military history. By analysing both historical and contemporary cases, it intends to demonstrate that A2/AD remains subject to the same limitations – predictability, reactivity, and susceptibility to adaptation – that have undermined historical denial strategies.

Methodologically, the chapter is based on a combination of theory by strategic thinkers and a comparative historical analysis of three cases: the Maginot Line, Japan's island perimeter defence, and the Soviet Integrated Air Defence System. These represent land, maritime, and air denial strategies, respectively. It then applies the insights to three modern examples – China, Russia, and Iran –to assess how technology and complexity have modified, but not transcended, the fundamental limitations.

The findings suggest that policymakers operating at the strategic level frequently misinterpret the operational feasibility of A2/AD systems. This strategic–operational disconnect risks generating alarmist narratives, distorted procurement priorities, and unbalanced force structures. The chapter concludes by recognising that this gap is essential to avoid overestimating denial capabilities and to develop effective adaptive counterstrategies.

Keywords: A2/AD transformation; Historical cases; Conceptual framework; Blind spots

Introduction

In the last decades, Anti-Access/Area Denial (A2/AD) has emerged as a central concept in Western strategic debates concerning Russia, China and other adversaries. Terms such as 'missile walls' or 'drone barriers' have entered policy discussions, implying that large regions could become 'no-go zones' for advanced militaries.

These interpretations, while politically compelling, risk overstating the strategic effectiveness of denial systems and underestimating their operational fragility.

A potential gap exists between strategic and operational-tactical thinking, leading policymakers to view A2/AD as a stable and decisive form of deterrence. At the tactical and operational levels, however, the effectiveness of denial is highly conditional, transient, and vulnerable to adaptation. This misalignment will lead to flaws in force planning, deterrence policy, and alliance strategy. This chapter, therefore, addresses the following question:

> *"To what extent do historical and operational patterns of denial strategies explain the actual effectiveness of modern A2/AD systems, and what are the implications for strategic-level policymaking?"*

To answer this question, the chapter follows three analytical steps. It starts with a historical analysis to identify recurring structural features and vulnerabilities of denial strategies across history in all three domains (land, sea, air). A contemporary application follows, by assessing how modern technology and integration modify these dynamics in the cases of China, Russia, and Iran. To conclude and to challenge future research, a summary of the strategic implications is provided by evaluating how misinterpretation of operational realities at the strategic level can distort defence planning and policy.

The analysis builds upon three strands of strategic theory: The work of Clausewitz, stating[1] that defence is inherently temporary and reactive; its strength lies in resistance, but its weakness lies in passivity. The paradoxical logic of strategy, explained by Luttwak[2] that every successful strategy generates a counter-adaptation, transforming an advantage into vulnerability. And, finally, Boyd's OODA-Loop theory that tells us that operational success depends on the speed of observation, orientation, decision, and action, and leads to the conclusion that complex systems, of which A2/AD is one, are often slow to adapt and therefore fragile.

This chapter argues that these theoretical perspectives provide the central argument that A2/AD, even when technologically enhanced, remains subject to structural limitations embedded in the logic of denial.

Theoretical and Conceptual Framework

Anti-Access/Area Denial (A2/AD) refers to a set of capabilities and operational concepts designed to prevent an adversary from entering or operating freely within a designated area of military interest. Anti-Access is defined as measures that seek to deter or prevent entry into a theatre through long-range strikes, missile salvos, or cyber disruption. The Area-denial measures aim to restrict movement once the adversary is inside the theatre.[3] Although these categories are analytically distinct, in practice they overlap. A layered missile and sensor network positioned along a coastline simultaneously performs both functions: deterring entry and constraining manoeuvre.[4]

Originally coined in U.S. doctrinal circles during the 1990s to describe potential Chinese and Russian capabilities, the term has since evolved into a catch-all label for integrated defensive postures combining precision strike, surveillance, and electronic warfare.[5] However, the underlying logic – using positional advantage and technology to compensate for inferiority in offensive power – is ancient. This continuity places A2/AD within a long tradition of defensive strategy rather than a revolutionary break.

The Enduring Logic of Defence

Carl von Clausewitz described defence as the 'stronger form of war,' but one that is in essence a 'negative' strength; defence resists but cannot, by itself, achieve a decisive outcome. A defensive posture is based on exploiting terrain and thorough preparation, but, to a large extent, the same is true for the attacker, who has the added option of taking the initiative. This asymmetry of initiative forms the first theoretical pillar of the analysis.

Edward N. Luttwak expanded on this insight with his concept of the *paradoxical logic of strategy*. Every successful measure invites countermeasures, thereby step by step converting an initial advantage into vulnerability. Fortifications provoke bypass; radar networks trigger electronic warfare; missile zones inspire stealth and dispersion. This recursive dynamic – action → reaction → adaptation ensures a dynamic challenge in which a denial strategy cannot adapt fast enough.

The force employment theory of Stephen Biddle adds to this paradox that technological superiority rarely guarantees success; outcomes depend far more on doctrine, training, and coordination.[6] A2/AD's effectiveness does not depend solely on hardware but also on a complex organisation that integrates sensors, shooters, and command systems. Defensive complexity, however, amplifies fragility: the more tightly coupled the system, the more catastrophic small failures become.[7]

Finally, John Boyd's Observe–Orient–Decide–Act (OODA) lecture highlights the temporal dimension of adaptation. A combatant who cycles faster through observation and decision will outpace and, more importantly, out-adapt a slower opponent. Highly centralised architectures, reliant on continuous data fusion and hierarchical approval, risk lagging behind more agile, decentralised adversaries. The logic of A2/AD, which is often very centralised, thus mirrors the paradox: the stronger the defence appears, the more its rigidity exposes it to faster-moving offence.

Denial versus Control

A second pillar of arguments concerns the relationship between *denial* and *control*. Strategic theorists have long differentiated between preventing an opponent's use of space and exercising one's own positive control over it. Biddle and Freedman[8] both note that denial is essentially reactive and temporary: it can raise costs but seldom shapes political outcomes unless paired with offensive action.

Where control of an area requires actual power projection or even occupation, denial of an area depends on more deterrence through threat. This difference creates what Luttwak terms a *cost-imposition illusion*: denial appears economical, but maintaining credible deterrent coverage over time often proves more expensive than a limited offensive manoeuvre that destroys the enemy's means of operating.

Adaptation and the Offence–Defence Cycle

The historical record reveals a persistent *offence–defence cycle* in which every new layer of protection provokes innovation in penetration. Mahnken and Watts argued[9] that military innovation proceeds through iterative adaptation rather than technological revolution. Within this cycle, advantages will come to the actor that adapts faster, not merely to the one with superior technology.

For A2/AD, this means that any temporary advantage conferred by long-range precision or integrated networks will be eroded as opponents exploit weaknesses in sensing, data processing, and command structures. The contemporary proliferation of unmanned systems, cyber tools, and artificial-intelligence-assisted targeting exemplifies this shift. Offensive adaptation targets the *decision process* rather than the physical barrier: disrupting situational awareness yields the same operational effect as destroying assets.

An Analytical Model of the Denial Paradox

Synthesising these theoretical strands, this study employs a simple model to analyse both historical and modern cases. This model rests on four interrelated variables, which together serve as the primary lens for both historical and contemporary cases:

- **Predictability** – The visibility and fixed nature of denial systems expose them to mapping and pre-planning.
- **Reactivity** – Defensive posture yields the initiative to the attacker, who selects the time, place and methods of engagement.
- **Complexity** – Integrated, multi-domain systems enhance effectiveness but simultaneously increase vulnerability to cascading failure.

- **Adaptability** – The decisive factor determining whether systems persist or collapse under pressure.

When combined, these variables explain why denial strategies often appear formidable in peacetime assessments but degrade rapidly when confronted by adaptive opponents in the heat of battle. They also provide a bridge between theory and policy. Strategic-level perceptions tend to overemphasise *complexity* as a capability, while missing its vulnerability: adaptability. This misalignment constitutes the central blind spot examined in this chapter.

This model drives the analysis in the parts that follow: the historical case studies, an examination of contemporary A2/AD systems, and an exploration of operational vulnerabilities and their strategic implications.

Historical Patterns of Denial Strategy

Historical cases serve two analytical purposes. They will demonstrate that the logic behind modern Anti-Access/Area Denial (A2/AD) systems is not new but has been observed throughout history; they reveal structural vulnerabilities that recur across eras, technologies, and operational domains. The three examples, one from each warfare domain, reveal a remarkably similar pattern: perceived strength followed by rapid erosion as adversaries adapt.

Land Denial: The Problem of Predictability

The Maginot Line, developed by France in the interwar period, is one of the most obvious land-based denial strategies in military history. It consisted of an extensive network of fortifications, subterranean infrastructure, artillery casemates, and interlocking fields of fire. At the tactical level, the system was exceptionally robust: French forces operating within the fortified sectors could not be easily dislodged by direct assault.

The system's weakness lay not in its tactical performance but in its strategic rigidity. By creating an extremely strong defensive belt, France simultaneously revealed its vulnerabilities and, crucially, where the German Army would not be deterred from attacking. German planners exploited this predictability by diverting their main effort through the Ardennes, a region the French erroneously deemed impassable to mechanised forces.[10] The German breakthrough in May 1940 rendered the enormous French investment strategically obsolete within days.

The Maginot case highlights three enduring features of denial strategies: First, predictability invites bypass. By signalling its strongest sectors, France inadvertently revealed its weakest defences. Next, the initiative lies with the attacker. The defender must respond to the choices the attacker makes: timing, direction, and attack methods.[11] And finally, the concentration of the budget towards the fortress created a risk for other sectors. It meant that the extensive investment in a single dominant solution hindered broader operational flexibility.

The Maginot Line did not fail because fortifications were ineffective in a narrow tactical sense, but because the strategic assumptions behind it proved incorrect. This paradox – tactically strong yet strategically fragile – reappears throughout the history of denial and is relevant to the analysis of modern A2/AD.

Maritime Denial: How Initiative Was Lost

During the Second World War, Japan pursued an expansive maritime denial strategy in the Pacific, known as the 'outer perimeter defence.' This approach relied on heavily fortified islands equipped with airfields, coastal artillery, garrisons, and naval units. The objective was to impose high costs on U.S. forces for each island seized, thereby hoping to discourage, or at least slow, the American advance and reduce strategic resolve.[12]

Initially, the Japanese system produced notable tactical successes. Prepared positions with overlapping areas of denial and nearby naval forces were intended to inflict significant losses, as seen in the early battles around Guadalcanal. However, bypassing and isolating strongpoints proved the decisive countermeasure. The American 'island-hopping' strategy avoided the strongest Japanese positions, instead capturing less fortified islands and rapidly establishing airbases from which the bypassed outposts could be isolated and neutralised. Once isolated, even the strongest Japanese island fortresses became strategically irrelevant, turning from force multipliers into logistic liabilities.[13]

This case illustrates three structural features of maritime denial: foremost, geographical fixation is exploitable. Islands cannot manoeuvre; once bypassed, their operational value collapses. Then, sustaining these 'fortresses' is the critical vulnerability. Without resupply, even formidable positions decay rapidly. This, in the end, leads to the ineffectiveness of denial, as it declines sharply once the attacker innovates. The rapid U.S. construction of airfields and mobile logistics outpaced Japanese planning.[14]

The Japanese perimeter defence closely mirrors patterns seen in contemporary maritime A2/AD systems, such as those based in the South China Sea. Fixed positions can be exceptionally dangerous at first contact but become strategically dangerous when an opponent identifies an indirect or asymmetric approach.

Air Denial: The Effect of Technological Adaptation

The Soviet Integrated Air Defence System (IADS), developed during the Cold War, represents an example of a modern institutionalised denial concept. It integrated radar networks, interceptor aircraft, command centres, and multiple layers of surface-to-air missile (SAM) sites. The system's goal was to create an impenetrable defensive shield that would deny U.S. bombers and reconnaissance aircraft access to Soviet airspace.[15]

Initially, the system was highly effective. Early incidents – such as the downing of U-2 reconnaissance aircraft – demonstrated that the Soviet network posed a credible threat. However, as the paradoxical logic of Luttwak predicts, success generated sustained countermeasures. The United States and NATO developed a dedicated Suppression of Enemy Air Defences (SEAD) doctrine that combined stand-off jamming, anti-radiation missiles, precision strikes, and stealthy platforms such as the F-117.

By the 1980s, Soviet and Soviet-style IADS structures had become increasingly vulnerable.[16] Their weaknesses lay in a combination of three interlocking problems, beginning with sensor dependence. Blinding, deceiving, or saturating sensors degraded the entire network. The centralised command then becomes a liability, as John Boyd's theory shows. Hierarchical coordination slows reactions and creates vulnerable critical nodes. These two vulnerabilities were exploited through a rapid technological counter-adaptation. Western innovations, especially in stealth and electronic warfare, subsequently eroded the IADS' effectiveness.

The rapid collapse of the Iraqi IADS during Operation Desert Storm in 1991 painfully illustrated that dense, multi-layered denial architectures could be suppressed within days when faced with a technologically advanced and adaptive opponent. This example shows that air-based denial depends on a delicate balance between sensor performance, command-and-control resilience, and adaptation speed. Once adversaries innovate in any of these dimensions, the denial advantage rapidly deteriorates.

Cross-Case Synthesis: The Historical Patterns

Despite their differences in geography, technology, and domain, the historical cases show us a consistent set of five patterns that all link back to the model used:

- Fixed defences are inherently predictable, enabling adversaries to find attack options that the defender has overlooked or was unable to strengthen enough.
- Denial cedes initiative, allowing attackers to define when and how to impose pressure or examine the strength of certain pathways.
- Early success generates counter-innovation. The stronger the initial denial effect, the greater the incentive for the attacker to innovate.

- The ability to sustain systemic resilience, not tactical lethality, determines the long-term viability of denial systems.
- Denial yields only a temporary advantage. Its value is contingent and erodes as adaptation accelerates.

These historical insights form the analytical foundation for the next section, which examines whether modern A2/AD architectures – those of China, Russia, and Iran – represent a genuine departure from historical patterns or reproduce the same paradoxes in updated technological form.

Contemporary A2/AD Systems and Their Limits

Contemporary Anti-Access/Area Denial (A2/AD) systems are frequently portrayed in strategic discourse as unprecedented challenges that can render large regions 'no-go zones' for advanced militaries. China's missile-centric 'counter-intervention' posture, Russia's layered defences in Europe and the Black Sea, and Iran's asymmetric denial strategy in the Persian Gulf are often cited as evidence of a transformation in warfare.[17]

However, when examined through the historical and theoretical lenses developed earlier, these systems display a familiar pattern: technological enhancements have increased the lethality and reach of denial but have not altered its fundamental vulnerabilities. This section analyses each of the major contemporary cases. It shows how they align with the recurring logic of denial – predictability, reactivity, dependence on fixed or centralised infrastructure, and deteriorating effectiveness under sustained adaptation.

China and the Western Pacific 'Missile Wall'

China's A2/AD architecture in the Western Pacific is the most sophisticated and comprehensive denial system in existence today. Its purpose is explicit: deter, delay, or disrupt U.S. and potential Western interventions regarding Taiwan or the maritime disputes in the South China Sea. The system integrates multiple layers:

- Long-range precision strike systems, including DF-21D and DF-26 anti-ship ballistic missiles (ASBMs), are designed to target naval forces at ranges extending several hundred kilometres.
- Surface-to-air missile networks, including HQ-9 variants and S-300/S-400 systems purchased from Russia, are positioned to deny air operations across wide swathes of the first island chain.

- Extensive surveillance and reconnaissance, including over-the-horizon (OTH) radar, electronic intelligence platforms, and an expanding constellation of satellites feed targeted data into the kill chain.
- Maritime militia and coast guard forces provide grey-zone pressure and complicate escalation dynamics.
- Integrated command-and-control (C2) structures link sensors and shooters across services.

At first glance, this appears to address many of the shortcomings of historical denial systems: it is multi-domain, mobile, and technologically advanced. Moreover, the scale of China's arsenal suggests an ability to impose substantial costs on U.S. intervention – perhaps enough to deter or delay a major operation. However, closer analysis reveals that China's A2/AD complex exhibits familiar weaknesses.

First, reliance on vast sensor networks creates systemic vulnerability. Chinese ASBMs require accurate, real-time tracking data on moving naval targets, a task demanding high-fidelity sensing and rapid data fusion. This makes the system highly dependent on the integrity of its surveillance architecture. Cyber disruption, jamming, spoofing, and kinetic strikes against space-based and terrestrial sensors could undermine the kill chain. As with the Soviet IADS, once key nodes are degraded, the whole system becomes less effective.

Second, the size of the denial zone is both a liability and an asset. Covering all the island chains and waters in between requires tracking and interpreting enormous volumes of data, creating OODA-loop overload or interference risks. Such vast areas favour more decentralised and adaptive adversaries.

Third, missile stockpiles are finite. Although China has large inventories, the expenditure rates required to sustain denial against distributed U.S. forces would be high. This mirrors the Japanese perimeter problem: a system designed to deter initial attack may not withstand prolonged attrition.

Fourth, the political effectiveness of A2/AD depends on perception. China could be deliberately exaggerating the range of the DF-21D and DF-26 to influence decision-making. However, this perception advantage may erode quickly once the system is tested, as historical denial cases have so often demonstrated.

In sum, China's denial posture is formidable but not unprecedented. It magnifies denial's strengths – cost-imposition and deterrence – yet inherits its structural fragilities: sensor dependence, systemic complexity, and vulnerability to adaptation.

Russia's Layered Defences in the Baltic

Russia's A2/AD systems are often cited as key challenges for NATO in the Baltic region. The Kaliningrad exclave is described as a dense and formidable A2/AD 'bubble', integrating several modern capacities:

- Iskander-M ballistic missiles, capable of striking regional airfields, ports and other logistical nodes.
- S-300 and S-400 SAM systems, providing high-altitude coverage across parts of Poland and Lithuania and the southern part of the Baltic.
- Coastal defence cruise missiles, extending denial effects into the Baltic Sea.
- Electronic warfare units, capable of disrupting GPS, communications, and radar.
- And probably, in times of tension, mining actions near the naval entry points around Denmark.

On paper, this architecture intends to complicate NATO's freedom of manoeuvre in early phases of a crisis. However, recent empirical evidence, especially from the war in Ukraine, reveals profound shortcomings that echo historical failures of denial.

First, Russian denial systems are highly centralised and poorly integrated across services. Whereas China has invested heavily in joint fire control and data integration, Russian systems suffer from compartmentalisation, inconsistent information-sharing, and rigid command structures. This slows reaction times and limits adaptability.

Second, the system is geographically exposed. Kaliningrad, though heavily defended, has a weak landward flank bordered by NATO territory. Finland's and Sweden's accession to NATO further compresses Russia's maritime manoeuvre space and places its Baltic naval and air assets at a serious disadvantage. Here, the logic mirrors the Maginot Line: a concentrated defensive investment that unintentionally highlights its own critical vulnerabilities.

Third, Ukraine has repeatedly penetrated and degraded Russian defence systems. Through innovative use of drones, precision artillery, and anti-radiation missiles, Ukraine has struck Russian air defence sites, destroyed naval vessels in Sevastopol, and forced Russia's Black Sea Fleet to relocate. This demonstrates that even a seemingly more capable adversary can be defeated by adaptation and asymmetric tactics.

Fourth, Russia faces escalating cost burdens. To maintain credibility, A2/AD systems require continual modernisation – an economic challenge exacerbated by sanctions, industrial constraints, and wartime attrition. As France did with the Maginot Line and the Soviets did with IADS modernisation, denial systems risk consuming disproportionate resources without guaranteeing strategic effectiveness.

Russia's A2/AD posture is, therefore, less a revolutionary threat than a traditional defensive system with the same classical vulnerabilities associated with the Soviet legacy: rigidity, exposure, and likely operational underperformance when confronted by innovation and sustained pressure.[18]

Iran's Asymmetric Denial in the Persian Gulf

Iran's A2/AD approach differs fundamentally from those of China and Russia. Lacking the resources to build a large integrated system, Iran relies on asymmetric denial, designed to impose costs, create uncertainty, and influence global markets rather than achieve decisive control. Key components include:

- Mine warfare in the Strait of Hormuz, one of the world's most strategically significant chokepoints.
- Fast-attack craft and swarm tactics, exploiting geographic proximity and surprise.
- Unmanned aerial vehicles (UAVs) and drones for reconnaissance and harassment.
- Ballistic and cruise missiles capable of striking regional bases and naval assets.
- Proxy forces add political and operational complexity in the grey zone.

Iran's system is not designed to stop a major power outright, but to threaten short-term disruption with disproportionate political and economic consequences. In this sense, Iran explicitly embraces the psychological component of A2/AD, seeking to leverage its position rather than achieve battlefield dominance. However, Iran's denial strategy is also the most fragile of the three.

First, it is unsustainable under concentrated military pressure. Minefields can be cleared; fast-attack craft and drones can be neutralised; fixed missile sites are vulnerable to pre-emption. Iran's 2024 experience with precision Israeli strikes illustrated how quickly key assets can be suppressed when targeted by a technologically superior adversary. Second, Iran depends heavily on surprise and opacity. Once adversaries understand Iranian tactics, the element of unpredictability diminishes, and the effects of denial decline rapidly. Third, its denial capabilities are localised. Iran can impose a temporary disruption in the Strait of Hormuz but cannot sustain a denial across the region or extend it over time. Thus, Iran represents a paradigmatic case of denial's tactical potency but is unsustainable at the strategic level.

Cross-Case Assessment: Continuity Beneath Technological Change

Across the three contemporary cases, several cross-cutting themes emerge:

- Complexity increases vulnerability. Integrated systems are powerful but fragile; disruption of sensors or C2 can collapse entire layers of defence.
- Denial systems remain reactive. The attacker chooses when and how to test, saturate, spoof, bypass, or overmatch the defensive bubble.
- Modern adaptation cycles erode initial advantages faster. From distributed force concepts to Ukrainian drone innovation, counter-denial measures can evolve much more rapidly.
- Sustainment challenges persist as weapon depletion increases faster than the ability to replenish stocks. Russia's industrial base, Iran's asymmetric forces and even China's missile inventories all face limits under prolonged conflict.
- Denial provides political leverage but not durable control. Strategic outcomes depend on offensive initiative and adaptability, not defensive lethality alone.

Taken together, these cases demonstrate that modern A2/AD systems differ in form from historical denial strategies but not in function. Their modern technologies magnify both their strengths and their weaknesses and, crucially, do not resolve the fundamental paradox of denial.

Exposed Vulnerabilities in Modern Denial

Although modern Anti-Access/Area Denial (A2/AD) architectures exhibit greater lethality, range, and multidomain integration than their historical predecessors, they also exhibit the same structural vulnerabilities that become more pronounced as complexity increases. This section identifies four interconnected areas of fragility – systemic complexity, unmanned and distributed threats, the tempo of adaptation, and the sustainability of denial operations. Together, these factors demonstrate that even the most advanced A2/AD systems are unlikely to deliver decisive or durable strategic outcomes once confronted by innovative and adaptive adversaries.

Systemic Complexity as a Source of Vulnerability

Perhaps the most consequential vulnerability of modern A2/AD systems is their reliance on large, interconnected surveillance and command networks. While these networks are designed to enhance sensing, targeting, and fire control, they introduce an increasing number of critical nodes whose failures can compromise the system. Modern missile-based A2/AD depends on:

- persistent surveillance through satellites, over-the-horizon radar, unmanned sensors,
- data fusion across multiple services and domains,
- real-time communication between sensors and launchers,
- centralised or semi-centralised command-and-control, and
- precision guidance dependent on GPS, targeting algorithms, or active radar illumination.

These dependencies create multiple options for serious disruption. Cyber intrusion, electronic warfare, GPS spoofing, jamming, decoy targets, and kinetic strikes against critical radars or satellites can degrade or disable the kill chain. Historical examples demonstrate that once a denial system loses situational awareness, its effectiveness declines sharply – much as the Soviet IADS did as NATO refined SEAD operations.

Furthermore, data overload presents an increasingly significant challenge. As sensors proliferate, the volume of information that must be processed and interpreted grows exponentially. Large denial architectures – such as China's in the Western Pacific – must manage multi-domain data across vast geographic areas. This strains command systems and increases the likelihood of misclassification, delayed reactions, or paralysis. The availability of increasing processing power does not resolve this, as it is primarily a human decision-making problem. In this sense, the very complexity intended to strengthen denial simultaneously magnifies its fragility.

Unmanned and Distributed Systems Undermine Deterrence Logic

Another source of vulnerability arises from the proliferation of unmanned and distributed systems. Modern A2/AD systems are optimised to threaten high-value, manned platforms – naval surface combatants, aircraft, and fixed bases. However, unmanned systems fundamentally alter the attacker's cost-benefit calculus.

First, unmanned systems reduce political risk. Losing a drone swarm or unmanned surface vessel is politically insignificant compared with losing a manned aircraft or ship. This undermines the core psychological leverage that A2/AD seeks to impose.

Second, unmanned systems enable continuous probing. Low-cost drones, USVs, and UUVs can test a denial system's responses, identify emission patterns, and trigger defender reactions – gradually mapping the architecture without requiring high-end assets.

Third, distributed lethality dilutes the defender's targeting efficiency. Instead of concentrating offensive power in a few large platforms, adversaries can disperse firepower across dozens or hundreds of smaller units. This forces A2/AD systems to expend valuable interceptors on cheap targets or risk saturation.

The Adaptation Race: Tempo Determines Success

A third vulnerability derives from the temporal dynamics of conflict. As the theoretical framework is established, denial is intrinsically reactive and thus cedes the initiative to the attacker. The advantage in modern warfare goes to the actor who cycles faster through Boyd's OODA loop:[19] observing, orienting, deciding, and acting more rapidly than the opponent. Modern A2/AD systems often struggle with this challenge for several reasons:

- centralised C2 structures slow decision-making,
- Large sensor networks increase processing demands,
- Rigid doctrine reduces flexibility, and
- fixed or semi-fixed assets limit manoeuvre options.

By contrast, attackers can adopt techniques such as swarming, multi-axis saturation attacks, cyber and electronic shaping, false-target generation, long-range precision strikes, and decentralised mission command.

These approaches reduce predictability, overload defender sensors, and force defenders to expend resources on uncertain threats. Critically, adaptation does not occur only in wartime. Modern militaries increasingly rely on synthetic environments, operational simulations, and wargaming to pre-map denial systems and rehearse countermeasures long before conflict begins.

Advanced commercial and military simulators– such as *Command: Modern Operations* or classified joint-force models – enable planners to identify weaknesses in A2/AD architectures, stress-test kill chains, and experiment with various attack vectors. This process systematically erodes A2/AD's initial advantage, even before a shot is fired.

Thus, while A2/AD may appear static and formidable in peacetime assessments, adversaries often enter conflict with a deep understanding of how to penetrate, bypass, or turn off the system.

Sustainment, Depletion, and the Cost Burden of Denial

Finally, denial systems face enduring challenges of sustainment and cost. While A2/AD is often depicted as a 'cost-effective' strategy for weaker powers, historical and contemporary evidence suggests that maintaining a credible denial posture is expensive and resource-intensive. Four aspects should be rigorously analysed to understand this vulnerability:

Missile depletion is a critical concern

China, Russia, and Iran all rely heavily on missile salvos to impose costs on adversaries. However, missile stockpiles are finite, and replenishment is slow and costly. Once initial salvos are expended, denial effectiveness declines rapidly.

Constant modernisation is required.

Adversaries innovate at high speed. To remain credible, A2/AD systems must continually upgrade sensors, interceptors, software, and C2 networks. This dynamic mirrors the escalating maintenance costs of the Maginot Line and the Soviet IADS.

Logistical vulnerabilities persist.

Sustaining forward-deployed batteries, radars, and missiles requires secure supply lines, energy, spare parts, and maintenance access – each of which can be targeted for disruption.

Economic asymmetry favours the attacker.

Offensive adaptation (e.g., new drone designs, decoys, EW modes) is often cheaper and quicker than developing new defensive layers. This reverses the common assumption that denial is inherently cost-efficient.

Collectively, these factors suggest that denial systems, although potent in the early phases of conflict, struggle to maintain their effectiveness in a prolonged confrontation. As historical cases demonstrated, strategic outcomes hinge not on the initial strength of the defensive system but on the defender's ability to adapt and endure – a challenge that modern denial systems have not fundamentally solved.

Summary: The Paradox Remains

The vulnerabilities identified – systemic complexity, susceptibility to unmanned and distributed threats, slower adaptation cycles, and sustainment challenges – reveal a persistent paradox. Modern A2/AD systems may appear stronger than their historical counterparts, but they replicate the same structural weaknesses that have long undermined the ability to deny. Their sophistication magnifies not only their lethality but also their brittleness.

Strategic and Policy Implications

When considering the vulnerabilities of denial strategies and having argued that these are, in essence, consistent even today, several strategic implications become apparent.

Priority One: Avoid Strategic Alarmism

The most important policy implication is the need to avoid alarmist interpretations of A2/AD. Strategic discourse often presents denial systems, particularly those of China and Russia, as the defining features of a transformed battlespace. Terms such as 'no-go zones,' 'missile walls,' or 'A2/AD bubbles' suggest impermeability and imply that Western forces would incur unacceptable costs if they entered these areas. This rhetoric, even if it were true, has three negative consequences:

First, it exaggerates the strategic effectiveness of denial. As demonstrated, even highly sophisticated systems exhibit predictable and exploitable vulnerabilities, such as reliance on sensors, centralised command structures, and finite missile inventories.

Second, alarmism distorts defence planning. Overemphasis on A2/AD can drive procurement focused primarily on counter-denial solutions, potentially narrowing the force's adaptability. Alarmism can also divert resources from sustained investment in training, logistics, and multi-domain integration.

Third, it may encourage escalation or bad political decision-making. If A2/AD is framed as an existential threat, policymakers might perceive pre-emption or an aggressive forward posture as necessary, thereby increasing the instability of crises. Alternatively, they may give in to aggressive policies that push the boundaries of hybrid warfare ever closer to conflict as vital interests are challenged.

Historical and contemporary evidence support a more nuanced view: A2/AD presents a challenge that can be managed through adaptation and innovation, rather than a decisive barrier that necessitates radical restructuring or fatalistic strategic thinking.

Doctrinal and Organisational Lessons: Reforming Force Structure

A central finding of this study is that adaptation speed, rather than initial capability, is the decisive factor in overcoming denial. This has significant implications for military organisations, which must structure their forces to exploit A2/AD vulnerabilities effectively.

Embrace Distributed and Decentralised Force Design

Concentrated, high-value assets, such as carrier strike groups, large airbases, or fixed C2 nodes, are more vulnerable to missile-based denial and incur a greater negative cost when lost. Distributed force structures, by contrast, reduce the attacker's targeting efficiency, force adversaries to disperse munitions across multiple nodes, and increase operational resilience.

Re-evaluate Command and Control of Multi-Domain Operations

Modern denial spans cyber, space, air, sea, land, and the electromagnetic spectrum. Countering it requires fully integrated multi-domain operations that combine cyber intrusion to degrade sensors, electronic warfare to disrupt targeting, unmanned swarms to saturate defences, long-range fires to attack key nodes, and manoeuvre forces to exploit created openings. This almost implies a centralised command, but this will create the same problems as an A2/AD concept. Several authors have recently published their concerns about this tendency.[20]

Institutionalise Iterative Training and Simulation

As demonstrated, denial systems are most effective when adversaries are unprepared. Synthetic environments, digital twins, and high-fidelity simulation tools allow militaries to map, model, and rehearse operations inside denial zones without risk. This reduces uncertainty, shortens adaptation cycles, and erodes denial's advantage before conflict begins. Of all recommendations, this must be considered the most important one: adaptability can, and must be, trained.

Technological Investment: A Fast-paced Approach is Required

Technological innovation plays a critical yet supportive role in countering the A2/AD strategy, provided it is developed rapidly enough. Three categories of capabilities stand out as essential:

Expand Unmanned Systems and Autonomous Platforms

Unmanned capabilities may, and probably will, reshape the offence–defence balance. They can probe denial systems with minimal political cost, saturate defences through large-scale swarm attacks, conduct persistent surveillance to identify gaps, deliver precision munitions in contested airspace, and exploit detection seams in sensor networks.

Integration of Electronic Warfare and Cyber Operations

EW and cyber tools provide some of the most cost-effective means to degrade denial, as jamming can blind radars or disrupt communications, Cyber intrusion can corrupt targeting data or turn off sensors, GPS spoofing can misdirect missiles or confuse navigation, and deception operations can overload the enemy's decision cycle. Modern cyber-enabled operations will allow non-kinetic means to have an outsized impact.

Investment, Both Money and Time, in Analysis and Training

The ability of forces to adapt faster than their adversaries will be the defining capacity. This requires rapid analysis of large volumes of information, the development of analytical tools, and the ability to train forces, particularly staff, to adapt and capitalise on opportunities. Modern simulation and analysis software, particularly when supported by tailored AI, will be crucial in this case and will significantly enhance the two other essential capabilities.[21]

The Political and Deterrence Dimensions of A2/AD

A2/AD strategies are inherently political: they are designed not only to impose military costs but also to shape perceptions and deter intervention. Policymakers therefore face the dual task of managing both the operational reality and the psychological impact of denial systems.

Avoid Being Manipulated by Perception Campaigns

States such as China will deliberately publicise the reach and capabilities of their ASBMs to influence foreign decision-making. Strategic communication is thus part of the denial strategy. Overestimating these capabilities risks granting adversaries political leverage they may not be able to enforce in wartime.

Recognise the Temporal Nature of Denial

A2/AD is primarily effective before or in the early phases of conflict. Policymakers should therefore avoid strategic decisions that assume denial will remain intact over time. The historical record shows that denial erodes quickly once countermeasures are used.

Improve Force Credibility by Scientific Analysis and Military Planning

A balanced understanding of A2/AD helps preserve deterrence by assuring allies, through detailed operational analysis, that contested regions remain accessible. The insights obtained should be used to ensure proper planning and development of capabilities. In this way, confidence is built among Allies to withstand overreaction to perceived threats.

Prevent Miscalculation in Crisis

Clear communication about capabilities, both one's own and the adversary's, reduces the risk of escalation based on false assumptions. Policymakers should avoid deterministic language that frames denial as insurmountable, as this can create pressure for pre-emptive action.

Summary: Strategic Adaptation over Strategic Alarmism

Ultimately, the strategic and policy implications of A2/AD converge on a simple principle: adaptation, not avoidance, is the key to countering denial. Policymakers should neither underestimate the risks posed by modern denial systems nor accept exaggerated claims about their invulnerability. By focusing on distributed forces, improving C2 in multi-domain operations, leveraging unmanned systems, and sustaining the development of adaptive doctrine and training, states can ensure that A2/AD remains a manageable operational challenge.

Conclusion

This study aimed to examine the gap between strategic perception and operational realities of modern A2/AD systems by analysing them in a historical and theoretical context. While often presented as unprecedented challenges capable of reshaping the strategic environment, the analysis demonstrates that they follow historic and theoretical patterns in the logic of denial. Beneath their advanced technologies lie structural vulnerabilities that undermine denial strategies.

Historical cases illustrate a recurring paradox: denial systems are often tactically formidable yet strategically fragile. They aim to impose higher localised costs but struggle to remain effective once adversaries identify, probe, and exploit their weaknesses. Modern A2/AD architectures inherit this dynamic. Their complexity enhances their lethality but also magnifies their exposure to disruption, saturation, deception and bypass. Moreover, they carry the added risk of potentially catastrophic cascading failures due to their complexity.

The assessment of Chinese, Russian, and Iranian systems reinforces this conclusion. China's 'Missile Wall' offers impressive reach and density, but it relies on vulnerable sensor networks, complex kill chains, and finite missile inventories, even for the Chinese state-led economy. Russia's layered defences in the Baltic and Black Sea regions appear formidable in static assessments but have proven susceptible to disruption, poor integration, and attrition. Iran's asymmetric denial posture shows some tactical potency but lacks the durability, resilience, and depth required in sustained operations.

Across all cases, four key vulnerabilities emerge. First, systemic complexity creates single points of failure; disruption of sensors or C2 nodes can collapse entire layers of defence. Second, the rise of unmanned and distributed systems undermines the foundational cost-imposition logic of A2/AD. Third, denial systems adapt more slowly than their adversaries, ceding initiative in the race to innovate. Fourth, denial postures impose significant sustainment burdens that will erode under prolonged pressure.

The analysis in this chapter supports two important warnings for policymakers and defence planners. First and foremost, any alarmist interpretations of A2/AD distort strategic thinking, risk overestimating adversary capabilities, and can lead to wrong responses. A more measured assessment recognises that denial is neither a revolutionary threat nor a negligible concern. It is a manageable operational challenge that requires thorough analysis, adaptive countermeasures and consistent training rather than radical force restructuring or fatalistic political restraint.

Secondly, at a strategic level, the priority should be on agility rather than avoidance, developing distributed force structures (including C2), integrating cyber and electronic warfare into joint operations, expanding unmanned capabilities, and institutionalising rapid adaptation through simulation-driven training and planning.

To summarise, modern A2/AD systems do not revolutionise warfare in any domain. They are, at best, an updated expression of a consistent defensive logic, initially potent but limited in strategic endurance. Recognising these limits allows policymakers and military leaders to respond effectively, ensuring that denial remains a tactical or operational obstacle rather than a strategic constraint.

Notes

1 von Clausewitz, *On War*, book IV.
2 Luttwak, *The Logic of War and Peace.*
3 NATO, *Allied Joint Doctrine.*
4 Hughes, *Fleet Tactics and Naval Operations.*
5 Erickson and Yang, "Using the Land to Control the Sea".
6 *Biddle, Military Power.*
7 Perrow, *Normal Accidents.*
8 Freedman, *Strategy: A History.*
9 Mahnken and Watts, *What the Evidence Shows.*
10 Eliot and Gooch. *Military Misfortunes.*
11 van Creveld, *The Art of War.*
12 Toll, *Pacific Crucible.*
13 *Toll, The Conquering Tide.*
14 Toll, *Twilight of the Gods.*
15 Zaloga, *Soviet Air Defence Missiles.*
16 Keaney and Cohen. *Gulf War Air Power Survey.*
17 *Chase et al. China's Incomplete Military Transformation;* Zysk, "Russia's Evolving Anti-access/Area-denial Capabilities"; Nadimi, *Iran's Anti-access Strategy.*
18 Biddle, "Theory of Victory".
19 It must be mentioned that John Boyd has never described OODA as a loop, but more as a continuous activity. A loop suggests a sequential activity; I have kept the term because it is commonly used and allows one to picture it as becoming slower and more laborious.
20 Storr, *Something Rotten;* McChrystal, *Team of Teams.*
21 It is why I have used AI to help me write this chapter. If you believe something has potential, you should challenge yourself to test it. I have done so and found it useful and even enjoyable. However, it required me to be focused and adjust my instructions regularly to be more specific. The current AI, in this case ChatGPT, mirrors a very diligent and hardworking, but not above-average, staff officer. Sometimes it is inconsistent and prone to cognitive tunnel vision. However, I must admit it is better than expected and has a good learning curve. That ChatGPT tends to be far too polite and overly complimentary creates an additional challenge: not getting overconfident. In this way, it is a good tool for learning to identify the type of career-oriented staff officer who praises too much but is of no bloody use in the heat of battle.

References

Biddle, S. 2004. *Military Power: Explaining Victory and Defeat in Modern Battle.* Princeton University Press.

Biddle, S. 2022. "Theory of Victory and the Logic of Adaptation in Ukraine: Denial, Fragility, and Sustained Adaptation". *Foreign Affairs*, 101 (6).

Chase, M. S., J. Engstrom, E. Heginbotham, M. O'Hanlon, and A. Ward. 2015. *China's Incomplete Military Transformation: Assessing the Weaknesses of the People's Liberation Army (PLA)*, RAND Corporation.

Clausewitz, C. von. 1976. *On War*, M. Howard & P. Paret, Eds. & Trans, Book VI, Princeton University Press.

Creveld, M. van. 2000. *The Art of War: War and Military Thought.*

Eliot, A. C., and J. Gooch. 2006. *Military Misfortunes: The Anatomy of Failure in War,* Vintage Books.

Erickson, A. S., and D. Yang. 2012. "Using the Land to Control the Sea: China's Anti-Access/Area-Denial Challenge". *Naval War College Review*, 65 (3), 1–29.

Freedman, L. 2013. *Strategy: A History,* Oxford University Press.

Hughes, W. P. 2018. *Fleet Tactics and Naval Operations (3rd ed.),* Naval Institute Press.

Keaney, T. A., and E.A. Cohen. 1993. *Gulf War Air Power Survey: Summary Report*. U.S., Government Printing Office.

Luttwak, E. N. 2001. Strategy: *The Logic of War and Peace,* Harvard University Press.

Mahnken, T. G., and B. D. Watts. 2014. *What the Evidence Shows: Defense Innovation and the Dynamics of Military Adaptation*, Center for Strategic and Budgetary Assessments.

McChrystal, S. 2019. *Team of Teams: New Rules of Engagement for a Complex World,* Penguin Business.

Nadimi, F. 2017. *Iran's Anti-access Strategy,* Washington Institute for Near East Policy.

NATO. 2022. *Allied Joint Doctrine for the Conduct of Operations (AJP-3),* NATO Standardization Office.

Perrow, C. 1999. *Normal Accidents: Living with High-risk Technologies* (Updated ed.), Princeton University Press.

Storr, J. 2022. *Something Rotten: Land Command in the 21st Century,* Howgate Publishing.

Toll, I. W. 2011. *Pacific Crucible: War at Sea in the Pacific, 1941–1942,* W. W. Norton & Company.

Toll, I. W. 2015. *The Conquering Tide: War in the Pacific, 1942–1944,* W. W. Norton & Company.

Toll, I. W. 2019. *Twilight of the Gods: War in the Western Pacific, 1944–1945*, W. W. Norton & Company.

Zaloga, S. J. 2007. *Soviet Air Defence Missiles: Design, Development, and Tactics,* Jane's Information Group.

Zysk, K. 2016. "Russia's Evolving Anti-access/Area-denial Capabilities". *The RUSI Journal,* 161(5), 8–15.

CHAPTER 2

Restoring NATO's Decision Dominance

Countering A2/AD's Evolution from Shields to Systems through Joint All Adaptation

Roy Hamans and Robert Bertrand

Abstract

Anti-Access/Area Denial (A2/AD) has evolved from static, localised defences into dynamic, multi-domain systems that enable adversaries to coerce NATO through hybrid warfare. This shift limits NATO's decision-making independence and operational flexibility, creating a systemic challenge to deterrence. This analysis contends that NATO must respond to A2/AD not by upgrading platforms but by developing joint, complex adaptive systems capable of learning and adapting faster than adversaries. Using evidence from recent U.S. live-fire exercises and Ukrainian battlefield adaptations, it advocates a bottom-up, interdiction-zone-based approach to regain decision-making dominance. The study outlines a method for integrating kinetic and non-kinetic functions across domains, emphasising realistic, iterative training over doctrinal overhauls to build resilience in contested environments.

Keywords: A2/AD; Joint all operations; Multidomain hybrid warfare; Command and control; Complex adaptive systems; Innovation, technology

Introduction

Publicly available information suggests that there have been 25 incidents against NATO military infrastructure in the first five months of 2025 alone.[1] Further data show a 46% increase in confirmed sabotage of critical infrastructure in NATO Member states from 2023 to 2024,[2] signalling a systemic shift in adversary strategy. Anti-Access/Area Denial (A2/AD) is no longer merely a tactical challenge, but an integral part of a strategy of pre-emptive coercion, dictating the terms of engagement and forcing NATO into reactive postures.

This evolution from static 'shields' to integrated, multi-domain 'systems' reflects Sun Tzu's ancient principle of imposing one's will on the enemy, but now with sophisticated hybrid warfare systems: missiles, cyber, electronic warfare, and sabotage. While NATO's 2022 Strategic Concept recognises this threat,[3] it is apparent that there is a need to learn how to operate synergistically to be able to counter it.[4] Therefore, this analysis addresses the question: How can NATO restore decision dominance amid A2/AD's shift from shields to systems? To explore this, a targeted literature review was conducted, focusing on recent academic publications, policy documents, and reports relevant to A2/AD's development and NATO counterstrategies. The findings suggest that the Alliance must move from platform-centric upgrades to a joint adaptation and learning system-of-systems, drawing on evidence-based lessons from US exercises and Ukrainian adaptations to prioritise bottom-up training and interdiction-zone-based operations.

A2/AD'S Evolution: From Shields to Systems

Understanding the evolution and use of A2/AD by adversaries is vital for developing effective counter-strategies. As previously mentioned, the principle of imposing one's will on the enemy is not new, but the methods by which this is achieved are. These methods have steadily developed over centuries, with significant shifts in response to advances in warfare. In recent decades, however, these developments have accelerated rapidly, as shown by the following table.

Table 2.1: Evolution of A2/AD

Era	Phase	Characteristics	Use	Targets	Examples
Pre-2000s	Static Shields	Local, physical, defensive structures	Defensive	Deployed military	Atlantikwall, Vauban forts
2000s – 2014s	Layered Systems	Missiles, sensors	Defensive	Military	A2/AD bubbles
2013s – present	Integrated Systems	Multi-domain systems	Offensive, strategic element in hybrid warfare	Military, dual-use, critical infrastructure	Increasing incidents; permanent A2/AD spheres

The terms A2 and AD were initially defined in a 2003 report by the US Centre for Strategic and Budgetary Assessments as follows:

- Anti-Access (A2) strategies aim to prevent US forces' entry into a theatre of operations,
- Area-denial (AD) operations aim to prevent their freedom of action in the confines of the area under an enemy's direct control.[5]

This report highlights the progress in missile systems and their possible implications for the US, such as the exposure of US overseas bases. Ten years later, the U.S. Department of Defence (DoD) officially adopted a definition for A2/AD in its Joint Operational Access Concept (2012). This definition differentiated between longer-range anti-access and shorter-range area-denial tactics, as well as between individual actions (like missiles) and broader capabilities[6] Subsequently, later definitions emphasised the growing threats across multiple domains, explicitly including the potential 'offensive' use of various ranges of missiles and other precision-guided munitions[7].

Since then, investments in A2/AD appear to have increased, and previously isolated opportunistic 'A2/AD bubbles' have become permanent positions, reinforced by electronic warfare and cyber tools. Starting around the geopolitical events of 2014, these positions have grown in significance to the point where they are no longer just tactical but central to adversaries' hybrid warfare strategies. During this period, technological advances have led to increased integration of sensors, electronic warfare, and cyber capabilities into A2/AD systems. Consequently, their range and effects have extended considerably, even reaching into space. Since 2022, A2/AD seems to have accelerated and is now deployed not only against military systems but also against targets of military importance. Many of these systems are dual-use and, when combined with sabotage and hybrid tactics, are aimed at critical infrastructure and private facilities.

The evolution of A2/AD shows how it has progressed from localised, defensive systems to expansive, multi-domain systems central to adversaries' hybrid warfare and coercive strategies. Developed in response to US military dominance, these systems increase the cost, complexity, and risk of intervention, thereby reducing the operational advantage of US and NATO forces. Therefore, NATO must adapt and overcome these challenges. It must develop joint all systems capable of matching adversary A2/AD architectures. These systems need to be all on the same scale, just as fast, and integrated to a similar standard. After examining the development of adversary A2/AD from tactical to strategic levels, the next section will focus on decision dominance as a countermeasure against such A2/AD.

Decision Dominance to Counter A2/AD

To counter A2/AD's coercive effects, NATO must regain decision-making dominance: the capacity to observe, decide, and act faster than adversaries, without being restricted or coerced[8]. This demands a shift not only in platforms or domains but also in how NATO conceptualises and operationalises joint operations.

From Joint to Joint All Systems

Conceptualising joint operations is one thing but implementing them is another. Integrating and orchestrating the capabilities required for this is considered as one of the most challenging military problems of our time and alliance members "must act collectively to solve" this.[9] Taking a systems approach to conventional military force is not new; this has already been done by the likes of John Boyd. Adversaries seem to have similarly employed systems thinking to develop integrated A2/AD architectures. Since this differs from traditional methods of coordinating forces, it is helpful to distinguish between traditional joint planning, multi-domain operations, and joint all approaches, as shown in the following table.

Table 2.2: Overview terms from joint planning to joint all

Concept	Domain	Non-Kinetic Non-Military	Primary Purpose	Counter to A2/AD	Implementation Mechanism
Traditional Joint planning	Army, Navy, Air Force	Possible	Plan, de-conflict, and concentrate effects across more than one domain for a single operation	Addresses localised threats	During planning phase of a single operation
Multi-Domain Operations MDO[10]	Land, maritime, air, space, cyber	Non-kinetic integrated; limited non-military	"The orchestration of military activities across all domains and environments, synchronised with non-military activities, to enable the Alliance to deliver converging effects at the speed of relevance".[11]	Designed to penetrate and dis-integrate more advanced A2/AD; Gain physical & psychological advantages, in the operational environment	Doctrine, concept and force design experiments; Combined exercises
'Joint All'-Domain Operations JADO	All available domains, military and non-military	All explicitly integrated and orchestrated	"Actions taken by the joint forces of two or more NATO nations, comprised of all available domains, integrated in planning and synchronized in execution, at a pace sufficient to effectively accomplish the mission."[12]	Here: to operate in contested or denied EMS environments; converge military and selected non-military capabilities to deter or defeat hybrid campaigns	Here: bottom-up experimentation at the tip of the spear, 'train as you fight' exercises and learning (e.g. US and Ukraine examples

Taken together, these concepts illustrate a trajectory from de-conflicted joint planning for a single operation, to ever more integrated systems orchestrated to converge effects across all available domains. This joint all approach is argued here to develop to a system-of-systems, focused on continuous adaptation[13] and learning, and designed to regain decision dominance in specific A2/AD interdiction zones. This approach aims specifically to converge effects across multiple systems and domains at designated times and locations, maintaining the necessary tempo. The targets of this convergence are interdiction zones: time- and space-bound segments of an adversary's overall A2/AD architecture. Effective counters to these interdiction zones can establish local decision dominance for a specific time and place, such as a corridor for manoeuvring.. This local decision-making dominance is defined as the ability to decide without being constrained or coerced.[14] Through local decision-making dominance, broader decision-making dominance may be achieved.

Regaining Decision Dominance

Decision dominance here fulfils three essential purposes:

- Descriptive: it shows how adversary A2/AD limits NATO's strategic and tactical choices, highlighting weaknesses in multi-domain conflicts.
- Normative: it seeks to enable NATO to deter and, if necessary, coerce adversaries with minimal risk, while remaining fully prepared for intervention if required.
- Evaluative: as a measure, it sets criteria to assess the effectiveness of counter-A2/AD strategies, ensuring these restore operational flexibility.

Executing operational flexibility happens at the 'tip of the spear', chosen here as the unit of analysis. This is where adversary A2/AD systems face allied countermeasures, and it's the point where decision dominance turns doctrine, material, manpower, and training into concrete results. Adversaries currently employ A2/AD to restrict NATO's options and decision-making. Following Wikingsson, these A2/AD systems can be seen as an 'umbrella' of multiple interdiction zones, each established against specific threats and with strategic goals.[15] Breaking these umbrella systems into time- and space-bounded zones clarifies why and how each is important, as well as which cross-domain effects need to be coordinated at the operational level and integrated tactically to enable coordinated countermeasures in time and space.[16] Effectively countering these interdiction zones is essential to regain decision dominance for a particular place and time. Having now defined decision dominance, the next step is to operationalise this.

Operationalising Decision Dominance

Forces achieve decision dominance by outpacing adversaries in what John Boyd called the 'OODA loop', the cycle of Observation, Orientation, Decision, and Action. In this loop, the quality of observation and orientation is as critical as speed. Maintaining coherence within this system-of-systems under attack is also vital. Modern A2/AD and hybrid warfare campaigns aim to disrupt this cycle by employing techniques such as spoofing, swarming, and multi-domain saturation, thereby increasing the cost and complexity of allied decision-making. Restoring decision dominance, therefore, requires a joint system-of-systems that can out-learn and out-adapt adversary A2/AD, and navigates its OODA loop more effectively and swiftly at the front where it confronts these challenges. While the OODA loop provides a useful framework for rapid adaptation and learning, a core challenge lies in integrating domains and capabilities into a synchronised whole.

From Platforms to Systems: Integration Challenges

Applying systems thinking to the military is not new, but so far has primarily focused on the tactical level, such as in close air-support paradigms. These paradigms assumed and relied on air superiority. With the proliferation of unmanned systems (UXVs), air dominance can no longer be taken for granted. Overcoming mid- to long-range, multi-domain integrated A2/AD requires integration beyond air and land, encompassing electronic warfare, cyber, and space capabilities to weaken and neutralise defensive positions, often in contested or denied environments.

While traditional joint planning focuses on coordinating platforms and services for a single mission, it does not combine them into a cohesive joint force or system of systems. Multi-domain operations (MDO) mark an advancement by adding and integrating domains, but progress is often slowed by 'component-specific problem solving', which maintains domain boundaries, and by emphasising 'coordinated joint actions' rather than developing into 'synchronised all-domain operations'.[17] Given the current landscape of integrated and coordinated multi-domain adversarial A2/AD, effectively countering it requires joint all approaches that go beyond service- and platform-specific boundaries to create a comprehensive all-sensor-to-all-effectors system of systems.[18] Only such a joint all system of systems offers NATO the integration, coordination, and responsiveness required to effectively counter modern A2/AD. Learning to synchronise across all domains, in order to match current A2/AD, and to prepare for a scenario where adversaries may compel NATO to evolve into a joint all force, is regarded as the 'most demanding military problem' facing allied nations and NATO,[19] and requires urgent attention.

Operationalising Joint All

As already established, regaining decision dominance requires NATO to outpace adversary A2/AD by targeting specific interdiction zones. This operational logic is already being tested in exercises in the US, elaborated upon below, demonstrating both the feasibility and urgency of this approach. Overcoming adversaries' sophisticated and evolving A2/AD systems is considered to require three interconnected shifts. First, experts and alliance discussions[20] highlight the need for a fundamental shift in perspective and thinking, emphasising integration and effective operations across all warfare domains. Second, joint all-systems-of-systems must integrate tactical forces and coordinate them at the operational level to produce effects in both time and space.[21] Third, this coordination must include non-kinetic capabilities, such as cyber and electronic warfare.[22] The multi-domain battle then centres on three main elements: calibrated force posture, resilient formations, and converging capabilities, which create windows of advantage enabling manoeuvre.[23]

To keep pace with the rapid development of modern A2/AD systems and their deployment in offensive hybrid campaigns, doctrinal reform alone will likely be insufficient. Instead, US and Ukrainian forces are already adapting through joint all and interdiction-zone approaches in practice and under contested A2/AD environments.

Evidence from Practice: The US and Ukraine

US live-fire joint exercises and Ukrainian battlefield adaptations demonstrate how forces can employ joint all, system-of-systems, interdiction-zone-based approaches to counter adversary A2/AD. In Ukraine, proliferating sensors have made the battlefield transparent[24] while A2/AD systems and precision fire complicate force massing and manoeuvre. Two case examples of combined-arms manoeuvres, in live-fire exercises and in actual battlefield conditions, are presented below, namely US Case: Live-Fire Joint Exercises and Ukraine Case: Battlefield Adaptation.

US Case: Live-fire Exercises: Structured, System-Level Adaptation

US forces employ live-fire exercises under the Joint Warfighting Concept (JWC) to integrate and develop the joint capability of land/air/cyber/electronic warfare (EW) and uncrewed systems (UXVs) against layered A2/AD defences. Recognising that air dominance can no longer be assumed, commanders train to coordinate multi-domain effects, creating temporary windows[25] of decision dominance with a "mission to gain the ability to converge effects across all warfighting domains: air, land, sea, cyber and space" and with this convergence of 'cross-cutting technologies' to yield the information advantage necessary for decision dominance and joint-force overmatch.[26]

Despite its experimental nature, the results mentioned are intriguing:

- OODA loop reduction: what "used to take us days and hours is just taking us minutes now" [...] "experimenting with some technologies that enable us to do that at machine speed".
- Joint all: "increased our ability to do Combined Joint All-Domain Command and Control and situational awareness synchronization with our multinational partners [...]".
- Resource efficiency and allocation: "prevent wasteful duplicate fires, ensuring efficient joint effects"[27].

This demonstrates that exercises can help adapt to joint all, integrate, and synchronise effects across domains and systems, and create decision dominance windows during simulated A2/AD, even when combined with allies.

Ukraine Case: Battlefield Adaptation: Bottom-up, Real-Time Learning

Ukrainian commanders have shifted from traditional close-deep-rear strategies to targeting contested interdiction zones.[28] This shift demands an increased number of arms to coordinate, greater coordination of arms at lower levels, a rebalancing of manoeuvres with fires, and the integration of capabilities within time-space-bounded actions against A2/AD.[29] These commanders employ a seven-phase approach: 1) survey, 2) isolate, 3) degrade, 4) fix, 5) suppress, 6) close, and 7) destroy and consolidate, enabling them to compress the entire operation into a 5–10 day cycle in A2/AD contested environments. According to two officers employing this method, this has drastically reduced casualties from around 50% to 5–10%.[30]

Some of the lessons from Ukrainian battlefield learning are:

- OODA loop: the seven-phase process condenses operations into 5–10 day cycles.
- Joint: enhanced coordination of arms at lower echelons (section/company level) through sharing and analysing live video feeds via UXVs.
- Resources at risk: lowering casualties from around 50% to 5–10%.

This illustrates that bottom-up learning driven by immediate battlefield demands can help adapt to greater coordination of arms and synchronisation under real-life A2/AD.

Evidence-based Practice

Although direct comparisons are limited by different operational contexts, US exercises being mainly system-level and organised while Ukrainian adaptations are driven by immediate battlefield needs, there are valuable insights to be gained from these. For NATO officers, these examples demonstrate the potential to blend structure with adaptability, fostering iterative learning and emphasising the importance of cross-domain orchestration at both tactical and operational levels.

Table 2.3: Side-by-side analysis of US and Ukraine examples[31]

	US	Ukraine	Insights for NATO
Environment	Planned experiments live fire exercises	Real-world battlefield adaptations	Train under realistic environments
Implementation	Bottom-up, system-level, structured	Bottom-up, driven by immediate battlefield needs	Prioritise bottom-up training to aid joint all integration and synchronisation
Adaptation & Learning	Iterative	Continuous	"Adaptability must become a system feature"[32]. Iteratively integrate domains and increase complexity of circumstances
Integration	Integration through technology	Continuously observing the battlefield, at all echelons	Prioritise technology and standards to integrate forces (beyond radio, manual relay)
Coordination	From edge to decider, whoever/ wherever C2 is located, to effector, at machine speed	At section- and company command level (coordinating EW, air defence and fires via live UAV feed)	Prioritise C2 to link sensors to shooters across organisations vs doctrine only; move to decentralised execution to be efficient[33]

	US	Ukraine	Insights for NATO
OODA Loop	From multiple sensors to multiple effectors, reducing decision cycles from days to minutes and seconds (machine speed) Common operational picture, a shared picture of the battlefield based on the same data for the joint force, including allies More efficient execution of operations (saving munitions)	Access to real-time info on enemy as key enabler of fires and crucial for planning manoeuvre Casualties reduced from approximately 50% to 5–10%. Ability to keep casualties low by halting operations if conditions are unfavourable. Seven-phase operation in 5–10 day cycles	Employ training, technology and decentralised execution to operate, adapt and learn at the required tempo (down to minutes, seconds, machine speed)

Combining key lessons from the US with the Ukraine evidence suggests that:

- a joint system-of-systems approach is both technically and organisationally feasible, and may enable NATO to attain decision dominance over interdiction zones and prevent adversaries from achieving joint all synchronised status before NATO does.
- Bottom-up, iterative development, through exercises and real-world battlefield adaptations, can enhance the integration and coordination of joint all forces, and more so than top-down doctrinal approach overhauls.
- Commanders can effectively coordinate effects within time- and space-sensitive interdiction zones, and can attain the required tempo and complexity through realistic, contested-environment exercises.

Ukraine's swift innovation, described as a "masterclass in accelerated adaptation," emphasises a key lesson for NATO: the side that learns fastest wins.[34] This study goes beyond theory by providing concrete examples of exercises and operationalised joint all, interdiction-zone approaches, offering a practical, actionable method for NATO implementation. This approach extends beyond merely adding and integrating new domains and capabilities. Effective operational-level integration of multi-domain systems requires innovative thinking, targeted training, and focused personnel development.

Policy Recommendations

Building on lessons from the US and Ukraine, NATO should systematically institutionalise joint all-integration, rapid adaptation, and continuous learning as core strategic capabilities. This requires:

- **Formalise adaptation**: establish a system-of-systems adaptation process within NATO's strategic command, driven by real-time insights from exercises and real-world operations.
- **Track progress**: implement relevant Key Performance Indicators (KPIs) such as decision cycle time, adaptation rate, and decision dominance to track progress and maintain progress at strategic, operational, and tactical levels.

For this to have the best chance of success as a deterrent and if necessary for interventions, alliance leaders must publicly confirm, politically resolve, and commit the resources necessary to develop these capabilities.

To start implementing immediately:

- **Education and training**: develop iterative exercises, increasingly integrating multi-domain exercises that simulate realistic, A2/AD environments and continually upgrade these exercises based on operations, allied and partner feedback and evolving threats.
- Mindset and skills: prioritise ongoing learning and adaptation, rather than a particular end-state, to continuously enhance joint all integration and coordination.
- **Interdisciplinary exercises**: forces should participate in interdisciplinary exercises as frequent as is practically possible, to operationalise and maintain these capabilities.

The speed and effectiveness with which NATO adapts to A2/AD will decide its ability to regain decision dominance and counter A2/AD's challenges.

Conclusion

The main issue here is not whether NATO *needs* to adapt, but how *quick* this should happen. Evidence from US live-fire exercises and Ukrainian battlefield adaptations shows that swift, evidence-based shifts to joint, system-of-systems approaches are achievable and can provide decision-making dominance over adversary A2/AD capabilities. Therefore, three main courses of action, based on these findings, are suggested to support this transition and help NATO regain operational superiority.

1) Reassess A2/AD as a strategic challenge instead of a tactical one, and craft a suitable response to counter it.
2) Adapt existing scheduled NATO exercises and include further exercises aimed at speeding up joint all integration, synchronisation, adaptation and learning, under realistic A2/AD conditions.
3) Measure progress: develop tactical, operational, and strategic Key Performance Indicators (KPIs) to assess progress towards decision dominance.

This study has aimed to provide specific, actionable implementation strategies rather than general advice; however, significantly more resources and effort are warranted to develop this approach further. Future research should explicitly address specific interoperability challenges faced among NATO allies, identify which technologies and command systems most effectively support rapid adaptation and learning, and create robust methodologies for integrating new insights gained from allied experimentation. Studies should also explore how training and simulation programmes can best facilitate real-time data integration across diverse sensors and actors to enhance decision-making, and effect generation and convergence on interdiction zones. It remains important for NATO to regularly assess progress in regaining decision-making dominance. Lastly, future research should explicitly evaluate the extent to which NATO forces can outperform adversaries in observation, orientation, decision-making, action, adaptation, and learning. The findings from this evaluation will be essential in guiding ongoing strategic priorities and operational adjustments.

Notes

1 Edwards and Seidenstein, *The Scale of Russian Sabotage Operations..*

2 Ibid.

3 NATO *Strategic Concept*, 3 "Authoritarian actors challenge our interests, values and democratic way of life. They are investing in sophisticated conventional, nuclear and missile capabilities, with little transparency or regard for international norms and commitments. Strategic competitors test our resilience and seek to exploit the openness, interconnectedness and digitalisation of our nations. They interfere in our democratic processes and institutions and target the security of our citizens through hybrid tactics, both directly and through proxies. They conduct malicious activities in cyberspace and space, promote disinformation campaigns, instrumentalise migration, manipulate energy supplies and employ economic coercion. These actors are also at the forefront of a deliberate effort to undermine multilateral norms and institutions and promote authoritarian models of governance".

4 Cochran and Willis, *Transitioning NATO..*

5 Krepinevich, Watts and Work, *Meeting the Anti-Access and Area-Denial Challenge*i.

6 Joint Air Power Competence Centre, *All Domain Operations in a Combined Environment.*

7 Williams, *The Russia-NATO A2/AD Environment.*

8 This is based on Lloyd's definition as "the capability to make better decisions, faster, enhanced by technology and convergence" where JADC2 provides information dominance and rapid convergence of effects to achieve "the critical advantage of speed" that underlies the future AI/ML competition. Lloyd, *Achieving Decision Dominance through Convergence: The U.S. Army and JADC2.*

9 https://www.japcc.org/flyers/all-domain-operations-in-a-combined-environment/

10 https://www.congress.gov/crs-product/IF11409.

11 https://c2coe.org/nato-c2coe-involvement-in-acts-mdo-definition/

12 Cochran and Willis, *Transitioning NATO,* 57.

13 Jensen, *Future Defense Systems and Strategic Flexibility.* "Adaptation is how forces survive and increase combat power in modern war".

14 Based on Lloyd and Rozman, *Achieving Decision Dominance.*

15 Wikingsson, "A2/AD in Combined Joint Multi-Domain Operations".

16 Ibid.

17 Cochran and Willis, *Transitioning NATO.*

18 NATO extends this concept to two or more nations' forces by using the term 'JADO'. The US has further extended this concept by adding 'Combined' to terms like JADC2, making it applicable to forces from two or more nations in CJAD2.

19 Ibid.

20 CZDEFENCE, "LANDEURO 2025".

21 Wikingsson "A2/AD in Combined Joint Multi-Domain Operations".

22 CZDEFENCE, "LANDEURO 2025"; Pettyjohn, *Competing in the Gray Zone.*

23 Pettyjohn, *Competing in the Gray Zone.*

24 Slusher, *Lessons from the Ukraine Conflict,* 28 "The Ukrainian experience of having a 'transparent battlefield' has shifted the advantage from 'hiders to finders,' creating a new dynamic in warfare."

25 Sdrakas, *An MDO Approach to NATO's Counter-IADS Strategy.*

26 Karako, "Project Convergence".

27 Pomerleau, "Army Evaluates Several Evolving Electronic Warfare Concepts", 5.

28 Watling, "Emergent Approaches to Combined Arms Manoeuvre in Ukraine".

29 Ibid.

30 Ibid.

31 Watling, "Emergent Approaches", Pomerleau, "Army Evaluates".

32 Jensen, *Future Defense.*

33 Cochran and Willis, *Transitioning NATO.*

34 Martin, "NATO Must Innovate Like Ukraine".

References

Abbe, Allison. 2023. *Critical Thinking in Military Training.* NATO Innovation Hub.

Cochran, Daniel, and Matthew Willis. 2021. *Transitioning NATO to an All-Domain Mindset.* Kalkar, Germany: Joint Air Power Competence Centre (JAPCC).

CZDEFENCE. 2025. "LANDEURO 2025: A New Doctrine for NATO's Eastern Flank." *CZDEFENCE,* July 19.

Edwards, Charlie, and Nate Seidenstein. 2025. *The Scale of Russian Sabotage Operations Against Europe's Critical Infrastructure*. London: International Institute for Strategic Studies, August.

Haggblom, Henrik. 2022. "A2/AD Bubbles and Hybrid Warfare: The Case of the Baltic Sea Region." *Chatham House Research Paper*, March.

Joint Air Power Competence Centre. 2021. *All Domain Operations in a Combined Environment*. Flyer, September.

Jones, Seth G. 2025. *Russia's Shadow War Against the West*. Washington, DC: Center for Strategic and International Studies (CSIS), March 18.

Jensen, Alex. 2025. *Future Defense Systems and Strategic Flexibility*. Washington, DC: Strategic Studies Press.

Karako, Thomas. 2022. "Project Convergence: An Experiment for Multidomain Operations." Washington, DC: Center for Strategic and International Studies (CSIS).

Krepinevich, Andrew F., Barry Watts, and Robert Work. 2003. *Meeting the Anti-Access and Area-Denial Challenge*. Washington, DC: Centre for Strategic and Budgetary Assessments.

Lloyd, Brittany, and Jeremiah Rozman. 2022. *Achieving Decision Dominance through Convergence: The U.S. Army and JADC2*. Spotlight 22-1. Arlington, VA: The Association of the United States Army, February 2.

Martin, Tim. 2025. "NATO Must Innovate Like Ukraine or Risk Losing a Fight with Russia: Swedish Official." *Breaking Defense*, November 4.

NATO. 2022. *Strategic Concept*. Brussels: NATO Public Diplomacy Division.

NATO Command and Control Centre of Excellence. 2022. "NATO C2COE Involvement in ACT's MDO Definition." NATO C2COE, August 10.

Osborn, Kris. 2021. "Army Makes Breakthrough in 'Sensor to Shooter' Time for Project Convergence 21." *The National Interest*, September 13.

Pettyjohn, Stacie L. 2019. *Competing in the Gray Zone: Russian Tactics and Western Responses*. Santa Monica, CA: RAND Corporation.

Pomerleau, Brent. 2024. "Army Project Convergence Speeds Up Data Sharing in Joint Exercises." *DefenseScoop*, March 8.

Pomerleau, Mark. 2025. "Army Evaluates Several Evolving Electronic Warfare Concepts at Project Convergence." *DefenseScoop*, April 15.

Praks, Henrik. 2025. *Russia's Hybrid Attacks in Europe: From Deterrence to Attribution to Response*. Policy Paper. International Centre for Defence and Security (ICDS), April.

Sdrakas, Athanasios. 2025. *An MDO Approach to NATO's Counter-IADS Strategy: Integrating Suppression of Enemy Air Defences and Cyber-Electromagnetic Activities*. Kalkar, Germany: Joint Air Power Competence Centre, December.

Smith, Aaron. 2025. "Warfare Experts Discuss Multi-Domain Operations at LANDEURO." *Army.mil*, July 17.

Slusher, Matthew. 2025. *Lessons from the Ukraine Conflict: Modern Warfare in the Age of Autonomy, Information, and Resilience*. Washington, DC: Center for Strategic and International Studies, May.

Sun Tzu. 2002. *The Art of War*. Translated by Lionel Giles. Mineola, NY: Dover Publications.

U.S. Department of Defense. 2012. *Joint Operational Access Concept (JOAC)*. Washington, DC: Joint Chiefs of Staff.

Watling, Jack. 2025. "Emergent Approaches to Combined Arms Manoeuvre in Ukraine." London: Royal United Services Institute (RUSI).

Williams, Ian. 2017. "The Russia-NATO A2/AD Environment." *Missile Threat* (CSIS).

Wikingsson, Jon. 2024. "A2/AD in Combined Joint Multi-Domain Operations." *Proceedings and Journal*, 1 (January/March): 77–81.

CHAPTER 3

Anti-Access/Area Denial

Prevailing Over the Threats and Challenges

Rémon ten Bhömer, Andreas Hörnedal, Simon Purton, Stian Betten, and Petteri Puusa

Abstract

Anti-Access and Area Denial (A2/AD) generally refers to the use of military capabilities aimed at preventing the deployment of opposing forces into a specific area of operations, or at limiting their freedom of manoeuvre once inside. Originally coined by the United States to describe a perceived Chinese threat, the concept gained traction in Europe, highlighting a Western European vulnerability versus Russia. The Western understanding of the A2/AD threat can be framed within the context of Russian strategic operations doctrine.

To better understand the impact on Europe and identify future capabilities essential for countering A2/AD, a NATO study was conducted. This study developed scenarios and vignettes, outlined several generalised kill chains, and proposed a diverse set of tentative counter-capabilities. To validate the analysis, assess the proposed capabilities, and gather further insights, a wargame involving NATO operational personnel was carried out. Several recurring themes emerged throughout the study: risk acceptance, the information space and situational awareness, and the integration of multi-domain and multi-effect operations. Key findings emphasised the importance of a timely and credible response to any A2/AD challenge, the need to concentrate efforts in both time and space, and the critical role of controlling the information environment. Ultimately, the A2/AD confrontation can be seen as a duel of wills and intellect, where intelligence and precision prevail over brute force. This chapter summarises and presents this work on an unclassified level with the goal to create awareness of A2/AD.

Keywords: NATO; Wargame; Anti-Access/Area Denial; A2/AD; Kill-chain

Introduction

The 2014 Russian invasion of Ukraine marked a turning point in European security and forced NATO to acknowledge that the possibility of a major war in or around Europe could no longer be dismissed. Prior to the invasion, the posture of NATO, its Allies, and European partners reflected a more peaceful era[1]. This period was characterised by limited forward-deployed forces and a reliance on reinforcement from the continental United States, together with in-theatre enablement for European-based forces during crises or conflict. This strategic assumption became increasingly untenable as Allies recognised that adversary Anti-Access and Area Denial (A2/AD) capabilities could critically threaten reinforcement and enablement operations. Such capabilities pose risks to the deployment of forces into contested areas and may also limit the freedom of action of forces already present.

A2/AD

Across the Alliance, a number of definitions for A2/AD coexist. In general, the concept refers to the employment of military capabilities intended to prevent the deployment of opposing forces into an area of operations or to reduce their freedom of manoeuvre once inside that area. Relevant descriptions can be found in sources such as Krepinevich, Watts and Work,[2] the USA Department of Defense,[3] and Simón.[4] Although such military capabilities have existed for decades, A2/AD initially emerged as terminology used to describe potential difficulties faced by the United States when projecting forces into various regions of the world, especially under constraints such as limited access to foreign bases or vulnerability of those bases. Early Allied thinking about A2/AD therefore focused primarily on possible crises in the Western Pacific involving the United States and China.[5]

Following Russia's invasion and occupation of parts of Ukraine, this conceptual lens was applied to the European theatre. Allies began recognising that Russia was increasingly adopting asymmetric capabilities designed to deny or impede the preferred Western approach, mostly in line with the already defined A2/AD concept. One long-standing strength of Western forces is the conduct of joint operations supported by overwhelming air power, illustrated clearly during the Gulf War through the devastation of the Iraqi army. For most NATO Allies and partners, detailed planning for conflict with a near-peer adversary had not been a priority since the end of the Cold War until Russia's actions in Ukraine brought the issue back to the forefront.

Today, adversaries possess highly sophisticated long-range A2/AD systems. At the beginning of SAS-147, few shared solutions existed to mitigate, suppress, or neutralise these threats. Nations therefore identified an urgent need to assess the dangers posed by current A2/AD systems and to evaluate potential technologies capable of countering them. NATO's 2022 Strategic Concept identifies Russia as the most significant and direct threat to Allied security in the Euro-Atlantic area. In a potential NATO-Russia conflict, Russian A2/AD strategies anchored in Integrated Air Defence Systems and long-range missile capabilities could impose high operational and strategic costs on Allied actions, challenge reinforcement and theatre enablement, and create local escalation dominance that undermines confidence among frontline states.

The System Analysis and Studies (SAS) Panel within the framework of the NATO Science & Technology Organization (STO), launched the Research Task Group SAS-147 Analysis of Anti-Access/Area Denial (A2/AD). This work resulted in the aptly named Technical Report "Analysis of Anti-Access Area Denial (A2/AD), Assessing the Consequences – Addressing the Challenge."[6]

The following chapter presents several key points from the SAS-147 work and discusses their relevance to the present analysis at the unclassified level. The intention of this is to make it accessible to a wider audience and to stimulate a wider debate on countering A2/AD.

Fitting A2/AD into Russian Strategy

Origins of the Debate

The contemporary debate on A2/AD emerged from advances in missile systems, sensors, and targeting capabilities, combined with lessons learned from Western air power in Iraq. These developments encouraged adversaries to adopt asymmetric strategies designed to deny access and restrict manoeuvre in key areas of operation. Although the mainstream discussion initially centred on potential crises between the United States and China in the Western Pacific, Russia's actions in Ukraine shifted attention toward the European theatre. A successful Russian A2/AD posture could significantly complicate NATO's capacity to reinforce its eastern flank.

Russian Strategic Thinking

Russia lacks the economic strength required to sustain a military capable of full-spectrum dominance. Consequently, it must be selective in choosing conflicts and must exploit advantages related to geography and political freedom of action. Russian leadership can act rapidly because it is not constrained by democratic processes, although economic and human resources remain limited.

Russia does not explicitly use the term A2/AD, and Russian strategy is not purely defensive in nature. Instead of waiting for a NATO air campaign to degrade Russian forces, Russian strategy seeks to disrupt early, create uncertainty, impose costs, and weaken Allied cohesion. Russian strategists view the opening phase of a crisis as decisive. Through pre-emptive action, Russia aims to prevent NATO from executing its preferred approach to warfare. Russian planners anticipate an intense air offensive by NATO and understand that static defence would quickly fail. As a result, they intend to strike strategically important targets, impose heavy costs, and threaten prolonged conflict to drive de-escalation on terms favourable to Russia. Russian thinking therefore treats war as coercion directed at political will, where defence is undertaken only to enable offensive action.

Russian Operations

Russian doctrine organises force employment into strategic operations. These include strategic operations to repel aerospace attack and strategic operations for the destruction of critically important targets, commonly referred to as *SODCIT*.[7] Execution is multi-domain, involving air, land, maritime, cyber, and electromagnetic effects. These operations target all levers of state power.

Russia has invested heavily in long-range missile systems launched from land, sea, and air. Such systems extend operational reach, complicate NATO targeting, and often have dual conventional nuclear capability, which introduces an added layer of deterrence. Their long range allows launch platforms to operate deep within Russian territory, which reduces their vulnerability to Allied strikes.[8]

While A2/AD constructs in Eastern Europe might not withstand a full NATO air campaign in the context of total war, they remain important due to their psychological and political implications. They could give Russia local escalation dominance in areas such as the Baltics, the Black Sea, or the Kola Peninsula. This perception may influence NATO decision-making by increasing the expected political and military costs of intervention. Even the threat of A2/AD alters strategic calculations by challenging assumptions about automatic escalation to a larger NATO-Russia conflict.

Methodology

The study encompassed strategic, tactical, and technical dimensions, which required a multi-domain perspective across geographic scales ranging from local to regional and temporal scales ranging from slow developments to rapid responses. As a result, the methodological framework adopted for the study was non-trivial and multi-layered. Drawing on the expertise of analysts and scientists, the group adopted a pragmatic approach that combined simplifications, such as the use of generic target types and streamlined game cards, with detailed analysis in areas where greater depth was required.

A comprehensive understanding of the adversary formed a central element of the approach. This understanding extended from strategic concepts to technical capabilities. Given the group's composition, emphasis was placed on emerging technologies. Scenario-based discussions served as the primary vehicle for structuring the group's work. Although simulation tools were considered, the group selected a wargame format as the most effective way to use available expertise.

Incorporating external experts through a wargame provided a crucial validation mechanism, and added to the body assessments. Their participation helped ensure completeness and reduced the likelihood of analytic oversights. The study group developed a master question list that identified key areas of analysis and supplementary questions. Each question was mapped to a set of primary, secondary, and tertiary methods of analysis.

Primary methods included preparatory work, such as scenario development and centre-of-gravity (CoG) analysis (Fig. 3.1). These were presented during the wargame for participant validation, which represented a secondary method. Tertiary methods included examination of additional sources, follow-on workshops, and comparisons with prior classified studies. Capability cards prepared before the wargame served as a primary method for assessing operational and tactical challenges. Their use during the wargame produced secondary insights, while post-game analysis across vignettes and kill chains provided tertiary insights, in support of the primary method. This structured, multi-tiered methodology ensured robustness and completeness in evaluating A2/AD challenges.

Scenarios, Vignettes, and Centres of Gravity

To analyse the A2/AD problem, the study group developed three large-scale scenarios. These scenarios and vignettes examined NATO's core tasks of deterrence and defence across the spectrum from pre-conflict to collective defence. The regions and situations reflected the strategic concerns of countries participating in the study. The scenarios also aligned with elements of NATO's 2022 Strategic Concept.

Due to the expertise and preferences of participating nations, scenarios focused primarily on the European theatre. This focus became even more relevant toward the end of the study as Russia launched its full-scale invasion of Ukraine. Although the study did not address the ongoing war directly, it highlighted the relevance of the analytical focus.

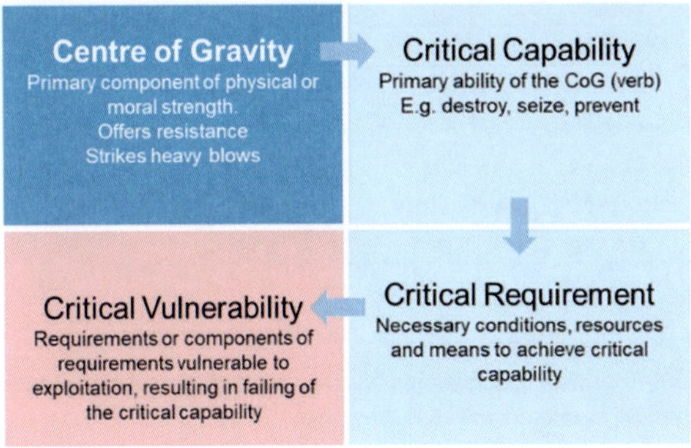

Figure 3.1: Centre of gravity analysis

The scenarios were as follows:

- The first scenario focused on the High North region and Russia's strategic submarine bases associated with the Bastion defence.
- The second scenario examined the Baltic Sea region, emphasising challenges related to Sea Lines of Communication.
- The third scenario explored an adversary invasion of European territory aimed at establishing a land corridor to the Baltic Sea, supported by a large missile attack consistent with Russian SODCIT concepts.

As preparatory work for the wargame, the group conducted CoG analyses for each vignette. These analyses identified adversary goals, critical capabilities, critical requirements, and critical vulnerabilities, specifically in relation to A2/AD. The wargame participants reviewed these analyses to validate them.

Kill Chain Analysis

To degrade or destroy an adversary's systems, a series of actions must be performed successfully. This sequence is known as a kill chain or effects chain. Descriptions of kill chains were used to assess Allied capabilities and to identify potential capability gaps. To streamline analysis and reduce reliance on classified information, each kill chain was divided into four sub-tasks, all of which needed to be performed reliably:

- Reaching the target with sensor and effector platforms.
- Detecting the target, which corresponds to the Find, Fix, Target, Track portion of the F2T2E2A model.[9]
- Engaging the target with kinetic or non-kinetic effects to degrade or eliminate it.
- Assessing the resulting effect with sufficient confidence to update situational awareness.

					GENERIC, INITIAL EVALUATION			
#	Domain	Mobility	Emission	Example	Reach	Detect	Effect	Assess
1	Land	Mobile	Emitting	SAM	Yes	No	Yes	Maybe
2	Land	Mobile	Non-emitting	SSM	Maybe	No	Yes	No
3	Land	Static	Emitting	Radar	Yes	Yes	No	Yes
4	Land	Static	Non-emitting	C2 node	Yes	Yes	Maybe	Yes
5	Air	Mobile	Emitting	AEW	Yes	Yes	Maybe	Yes
6	Air	Mobile	Non-emitting	Backfire/Bear	No	Maybe	Maybe	Yes
7	Maritime	Mobile	Emitting	Air Defence	Yes	Yes	Maybe	Yes
8	Maritime	Mobile	Non-emitting	Corvette	No	Maybe	Yes	Yes
9	Submarine	Mobile	Non-emitting	SSGN	No	No	Maybe	Maybe

Figure 3.2: Generic target categories with examples and an assessment of current capabilities across the kill chain against those as initially made by the group.[10]

Generic Target Types

To encompass the wide variety of A2/AD relevant targets without requiring detailed analysis of specific adversary systems, the group defined a set of generic target categories (Fig. 3.2) based on the following properties:

- Operational domain: land, air, maritime, or sub-surface.
- Emissions: emitting or non-emitting.
- Mobility: mobile or static.

Nine generic target categories were created, including categories such as land emitting mobile, which includes ground-based air defence systems. These categories covered relevant A2/AD targets in sufficient detail and are found in Figure 3.2. Initial assessments by the study group of kill chains for each target category provided a baseline for further analysis.

Capability Cards

Capability cards were created to represent potential solutions to specific challenges in kill chains and target types. Most capabilities described plausible future systems based on advanced technologies. A total of forty cards were developed and organised into six categories:

- Intelligence, surveillance, and reconnaissance.
- Kinetic effects.
- Non-kinetic effects.
- Protective measures.
- Command and control.
- Enabling technologies.

Wargame

In June 2022, the study group conducted a three-day wargame at the NATO Joint Warfighting Centre in Stavanger, Norway. The event included military personnel from across the Alliance, as well as the then non-aligned Finland and Sweden. These participants represented a diverse set of operational domains and expert areas, such as air, sea and ground warfare, together with special operations, logistics and space. A notable omission was the lack of cyber specialists.

The wargame was set up as a Disruptive Technology Assessment Game (DTAG) and included several gaming sessions using the above-mentioned scenarios, vignettes, and capability cards. The setup was non-exhaustive, in the sense that not all combinations of vignettes and capability cards were used. However, each vignette and game card respectively was used in at least one session. The wargame was supported with TARTAN, TNOs *Tool to Assess Revolutionary Technologies and Assets for NATO.*[11]

The wargame validated previous findings, evaluated the utility of different capability cards, and elicited expert assessments on A2/AD challenges and solutions. Several initial conclusions were revised as a result, and new findings emerged. The wargame is best described as a one-sided scenario-based discussion in which the players represented NATO while the study group acted as the adversary when relevant.

While sacrificing some ability to quantitatively estimate effects and impacts, this format allowed for more open-ended discussion and elicitation of judgment from the assembled experts. In turn, it enabled subsequent integration of material across sessions and straightforward identification of overarching themes and insights. The wargame included scenario and vignette briefings, as well as after-action reviews to allow the players to reflect upon their actions and what motivated them. Data collection at the wargame was performed by analysts taking notes as free text, together with a bespoke tool for systematic data collections and analysis.

Wargame Limitation

An identified limitation during the wargame was that both operational experts as well as researchers in the field of A2/AD were very oriented towards kinetic effects-based capabilities, seeing the problem mainly as neutralising threatening kill chains.

For this reason, SAS-147 also started with quite a kinetic approach, however later trying to introduce some playing cards providing non-kinetic effects into the wargame, e.g. cognitive effects. Unfortunately, as expertise in the field of cognitive warfare was somewhat lacking among players, these capability cards were generally seen as not very helpful in the foreseen schemes to handle A2/AD.

Findings

After the conclusion of the wargame, the combined results from the pre-wargame analysis and the wargame were combined into an integrated synthesis. This resulted in a number of primary findings and important insights.

The challenge posed by A2/AD systems is significant and may create severe operational and strategic difficulties for the Alliance. However, the challenge is not insurmountable. The study identified promising technologies and concept adaptations that can mitigate these threats. Four principal findings are highlighted.

Timeliness

Rapid response is essential when confronting an A2/AD challenge. Delay would allow the adversary to manoeuvre freely and establish a favourable end state. Protracted force build-up would cede initiative to the opponent, allow them to influence the strategic narrative, and give them the ability to dictate the terms of escalation. Demonstrating both the capability and the willingness to act early is therefore critical as a deterrent.

Credibility

During escalation, potential responses must be perceived as credible. This credibility requires a combination of effects that degrade the adversary's systemic ability to function. Relevant capabilities must penetrate integrated air defences, blend multiple technologies, and strike critical nodes with high precision, including targets located deep within enemy territory.

Concentration of Effects

The third finding emphasises the importance of concentrating effects across time and space. Operations must employ a coordinated multi-domain approach that combines different types of effects and aligns them within a coherent operational framework.

Information Superiority

The information domain is decisive. Operational success depends on access to actionable and shareable intelligence regarding the adversary's force disposition, system topology, interconnections, and vulnerabilities. When this information is available, even a force that is not overwhelming can avoid broad engagement and instead direct kill chains precisely against critical nodes within the adversary's architecture. Ultimately, the contest may become a duel of cognition and will, in which intelligence and precision overcome brute force.

General Considerations

Throughout the study, several themes emerged that extended beyond the central analytic framework. These themes emerged from the aggregation and synthesis of the above-mentioned wargame after-action reviews and discussions. Three themes were considered in more depth.

Risk Acceptance

If an aggressor attacks an Alliance member and cannot be repelled immediately, the Alliance may experience severe difficulty in maintaining cohesion. A natural reaction might be to withdraw, consolidate forces in a secure area, and wait until favourable conditions permit a counter-offensive supported by significant overmatch and broad domain dominance. Although this approach minimises immediate losses, it may have severe long-term consequences for the affected population. Early defeat can result in humanitarian and material suffering and may lead to a frozen conflict in which the adversary consolidates gains. The presence of A2/AD capabilities may deter the Alliance from action and force a diplomatic settlement due to the threat posed by SODCIT-capable systems.

An alternative would be to maintain forces at high readiness and adopt an aggressive posture, even without overwhelming advantage. A better long-term outcome may require higher initial risks. In this case, adversary A2/AD capabilities must be degraded sufficiently, even if only within limited areas and for short periods, to enable necessary counter-operations. The fight against A2/AD would then become a continuous effort for the full duration of hostilities.

Information Space and Situational Awareness

Rapid and decisive reaction requires accurate and timely situational awareness. This depends on a robust set of command, control, communications, intelligence, surveillance, and reconnaissance capabilities. A highly effective adversary A2/AD posture may seek to introduce ambiguity and confusion in order to delay timely counteraction, especially during early escalation. Adversaries may combine cognitive effects such as feints, decoys, electronic warfare, and cyber operations to create uncertainty. Cognitive effects can influence military, political, and social environments and can be used by both attackers and defenders. Strict control over media within authoritarian states limits the Alliance's ability to influence adversary populations. At the same time, maintaining credibility within Allied information ecosystems is vital. Online platforms and data-driven algorithms that support decision-making can also be targeted for cognitive manipulation.

Similar to how Endsley[12] defines a higher level of situation awareness as the ability to project future situations, the Alliance must progress from reactive detection to continuous understanding of adversary dispositions and probable intentions, and stay ahead of the adversary. This may require tracking threats and targets even in peacetime. Achieving high levels of situational awareness requires the collection, fusion, and analysis of large volumes of data using advanced analytical techniques such as pattern-of-life analysis and big data methods.

Multi-domain, Multi-effect Operations

In addition to improved situational awareness, effective responses to A2/AD systems require coordinated application of multiple effects across all domains. Multi-Domain Operations (MDO) must achieve sufficient threat degradation in specific areas and at specific times. Implementing such operations poses major challenges for traditional planning and battle management methods, which will require significant adaptation.

Practical, Policy, and Doctrinal Implications

The following implications to NATO and members are neutral and do not reflect on current NATO capabilities or approaches for A2/AD, as this would need to introduce classified material.

To counter the challenges posed by A2/AD systems, the Alliance must adopt a comprehensive and adaptive approach. NATO must be capable of fielding a flexible set of capabilities that can scale as a crisis develops. Flexibility ensures that the Alliance can position forces effectively and grow in strength to meet emerging threats before they become unmanageable.

Balance across capabilities is equally important. Forces require a diverse portfolio of systems optimised to suppress, degrade, and disrupt A2/AD networks. Integration of air, land, maritime, cyber, and electronic warfare capabilities is essential to deny adversaries the freedom of action they seek.

Precision strike capability remains a cornerstone of this approach. NATO must maintain the ability to engage critical components of A2/AD networks using long-range, high-precision stand-off weapons. Such weapons allow the Alliance to neutralise key nodes while minimising risk to its own forces.

Information superiority is also decisive. Persistent intelligence, surveillance, and reconnaissance capabilities must be enhanced across the Alliance. Multi-source and multi-platform collection combined with rapid analysis and dissemination will provide actionable insights. These insights must be shared effectively among Allies. Timely and unambiguous information exchange will support a common operational picture and enable anticipation of adversary actions.

Reliable and secure communications systems are essential for achieving these goals. The Alliance must invest in technologies that support resilient command and control in contested environments. Without such systems, even advanced capabilities may not achieve their intended effects.

Beyond the physical domain, the Alliance should explore capabilities that degrade an adversary's cohesion and ability to perceive, comprehend, and predict

NATO actions. This includes non-kinetic effectors and cognitive warfare tools intended to disrupt decision-making.

NATO forces must also be prepared to act rapidly and operate effectively even without complete domain dominance. Early stages of a crisis will likely be characterised by uncertainty and distributed operational conditions, and adaptability will be essential.

Geography adds further complexity. A2/AD challenges in the Arctic and Baltic regions are more manageable if the Alliance has full integration and access to relevant NATO and partner territories. Flexible use of these regions will be important for rapid reinforcement and manoeuvre.

Finally, NATO must develop clear policy, guidelines, and plans for countering A2/AD threats across all phases of conflict. This framework will ensure coherence among operational concepts, force development, and political decision-making and will strengthen the Alliance's ability to meet the challenge effectively.

Reflections on the Study in 2026

SAS-147 published its report in 2021. In 2026 the study appears to remain as relevant as at the time of publication. Countering A2/AD remains a critical operational challenge at all conflict levels, including full-scale war. If left unchecked, this threat would clearly generate enduring operational constraints and long-term strategic disadvantages.

Already at the outset of the SAS-147 study, there was a general notion that the A2/AD threat requires a comprehensive multidomain approach, including both operational and strategic perspectives and timelines. Early on the study group acknowledged this, but decided to limit the scope of this study to a somewhat more operations-centred perspective. Due to the composition of the study group, where participants were drawn mainly from the Air Force and Navy, there was significant emphasis on countering various missile systems together with sensors and launching platforms. The study methodology as such became somewhat centred on 'day one' operational problems, with mainly technical solutions. This bias was perhaps compounded by the composition of the game participants, as detailed above.

If the Alliance intends to advance toward Multi-Domain Operations in its most comprehensive sense, the conceptualisation of A2/AD must evolve as well. Therefore, any future studies and wargaming efforts would be wise to scope their A2/AD problem clearly and invite an even broader range of subject matter experts accordingly. The operation should also be complemented by a separate study with a strategic scope. This would probably require an entirely different composition of competences, compared to the SAS-147 study. However, such diversity would provide synergies to both strategic and operational thinking and contribute to a holistic approach to the A2/AD problem.

Notes

1 Caveat: this chapter reflects the personal views of the authors and not the official views of NATO.
2 Krepinevich, Watts and Work, *Meeting the Anti-Access and Area-Denial Challenge.*
3 USA Department of Defense, *Joint Operational Access Concept.*
4 Simón, *Demystifying the A2/AD Buzz.*
5 Krepinevich, Watts and Work, *Meeting the Anti-Access and Area-Denial Challenge.*
6 SAS-147, *Analysis of Anti-Access Area Denial (A2/AD), Assessing the Consequences – Addressing the Challenge.*
7 Kofman et al., *Russian Military Strategy: Core Tenets and Operational Concept.*
8 Dalsjö et al., *Bursting the Bubble.*
9 NATO, *NATO Standard AJP-3.9.* U.S. Air Force, *Air Force Doctrine Publication 3–60.*
10 SAS-147, *Analysis of Anti-Access Area Denial (A2/AD), Assessing the Consequences – Addressing the Challenge.*
11 NATO, *Assessment of Technologies with a Disruptive Effect on Defence and Security.*
12 Endsley, "Towards a Theory of Situation Awareness".

References

Dalsjö, R., C, Berglund, and M. Jonsson. 2019. *Bursting the Bubble.* FOI-R—4651—SE, Stockholm, Sweden: FOI.

Department of Defense, USA. 2012. *Joint Operational Access Concept (JOAC).* Department of Defense, USA.

Endsley, M.R. 1995. "Towards a Theory of Situation Awareness in Dynamic Systems." *The Journal of the Human Factors and Ergonomics Society* 37(1) 32–64.

Kofman, M., A. Fink, D. Gorenburg, M. Chesnut, J. Edmonds, and J. Waller. 2021. *Russian Military Strategy: Core Tenets and Operational Concepts.* Stuttgart, Germany: CNA.

Krepinevich, A., B. Watts, and R. Work. 2003. *Meeting the Anti-Access and Area-Denial Challenge.* Washington DC, USA: CSBA Center for Strategic and Budgetary Assessments.

NATO. 2016. *NATO Standard AJP-3.9 Allied Joint Doctrine for Joint Targeting Edition A Version 1.* NATO Standardization Office (NSO).

NATO. 2022. *NATO 2022 Strategic Concept.* NATO.

NATO RTO. 2009. *Assessment of Technologies with a Disruptive Effect on Defence and Security, NATO RTO Technical Report RTO-TR-SAS-062.* Neuilly-sur-Seine, France: NATO Research & Technology Organization.

SAS-147. 2023. *Analysis of Anti-Access Area Denial (A2/AD), Assessing the Consequences – Addressing the Challenge.* NATO STO.

Simón, L. 2017. "Demystifying the A2/AD Buzz." *War on the Rocks.* 4 January. Geopend 2025. https://warontherocks.com/2017/01/demystifying-the-a2ad-buzz/.

U.S. Air Force. 2021. *Air Force Doctrine Publication 3–60 Targeting.* U.S. Air Force.

CHAPTER 4

The Threat of Hypersonic Glide Vehicles in A2/AD Scenarios

Ralph Savelsberg, Mitchell Kampert, and Arthur Vermeulen

Abstract

Hypersonic glide vehicles travel at speeds exceeding Mach 5 while gliding unpowered through the atmosphere. Unlike conventional ballistic missiles they can alter their trajectory in flight, making them unpredictable and extremely challenging to intercept. This makes them particularly suitable for Anti-Access/Area Denial (A2/AD) strategies. For instance, the Chinese DF-17 could be deployed for such purposes in the South China Sea. In this study, we employ a simplified two-dimensional model to explore the area that can be reached by a North Korean hypersonic glide vehicle tested in January 2022. Large areas can be reached within short timeframes, and the same target may be approached at different times via multiple flight paths. This uncertainty largely negates existing outer missile defence layers.

Keywords: North Korea; Hypersonic glide vehicle; Ballistic missiles; Missile defence; Hypersonic flight; Flight performance

Introduction

In recent years, the Russian Federation, the People's Republic of China and the Democratic People's Republic of Korea (the DPRK or North Korea) all claim to have introduced hypersonic weapons. A commonly used definition requires such a weapon to fly a significant part of its trajectory through the atmosphere at more than five times the speed of sound, i.e. the Mach number (the ratio of its velocity over the speed of sound) exceeding five. Hypersonic cruise missiles are powered throughout much of their flight, while hypersonic glide vehicles are only powered during the first part of their flight, i.e. their boost phase. Their subsequent flight through the atmosphere is unpowered.

Two examples are shown in Figure 4.1. The Russian Kinzhal (Dagger), NATO codename RS-AS-24 Killjoy, shown in Figure 4.1(a) suspended from a MiG-31K aircraft, is derived from the surface-launched Iskander-M (NATO codename RS-SS-26 Stone). The missile has aerodynamic control fins that, in theory, enable it to

(a) (b)

Figure 4.1: Examples of hypersonic weapons. (a) A MiG-31K aircraft carrying a Kinzhal missile on its centreline, and (b) Transporter Erector Launchers carrying DF-17 missiles during a parade in Beijing in 2019. (Images: Militaryleak.com, People's Liberation Army)

manoeuvre in flight and to fly range-extension manoeuvres inside the atmosphere. Since the aircraft that launches the missile effectively serves as its first stage, the Kinzhal can fly at hypersonic speeds for part of its trajectory through the atmosphere. Based on open sources, the Kinzhal's maximum range is 900–1400 km.[1] Russia uses it in its war against Ukraine, but it could also play a role in a future conflict with NATO, for instance in the Baltic Sea[2] or in Northern Norway.[3] The Chinese DF-17 (NATO codename CH-SS-22), shown in Figure 4.1(b) has a solid-propellant rocket booster and a wedge-shaped glide vehicle. The booster launches it on an initial ballistic trajectory. After booster separation, during the descent, the glider can start to generate aerodynamic lift, subsequently fly a prolonged flight through the atmosphere and perform manoeuvres. It can attack high-value assets and potentially US aircraft carrier battlegroups in a conflict with Taiwan and, as such, would be part of a Chinese A2/AD strategy.[4]

The DPRK has also presented several types of (potentially) hypersonic missiles and has performed several flight tests, twice launching a new missile type in January 2022. Its Korean designation is unknown, but it may be called the Hwasong-12Ga, with 'Ga' being the first letter of the Korean alphabet.[5] It has a liquid-propellant booster and an axisymmetric glide vehicle with cruciform control fins. In a conflict on the Korean peninsula, the DPRK could potentially use such a weapon to strike ports and airfields in the Republic of Korea and/or Japan, to prevent US forces from entering the conflict, as part of an anti-access strategy.

Layered Defences

Traditional threats to ports or airfields and to carrier battlegroups include combat aircraft, cruise missiles and (anti-ship) ballistic missiles. The defence against these threats is layered. Against combat aircraft and cruise missiles, the outer defensive layer consists of fighter aircraft supported by airborne early-warning aircraft. The relatively slow speed of the threats allows sufficient time to position the fighters to intercept them far from their intended targets. Surface-to-air missiles and shorter-range air defences, such as cannons, form the inner layers, intended to intercept threats that may have penetrated the outer layer. However, fighter aircraft cannot perform area defence against hypersonic missiles. The missiles' high speeds drastically reduce the reaction time. Furthermore, their flight altitude exceeds the maximum engagement altitude of typical air-to-air missiles.

The outer defensive layer against ballistic missiles consists of long-range or area missile defence systems, such as the US Aegis ballistic missile defence system with SM-3 interceptor missiles or Israel's Arrow-3, that intercept missiles in space. When entering a contested area, a US carrier battle group would likely rely on Aegis ships and SM-3 missiles to defend against anti-ship ballistic missiles, such as the Chinese DF-21D.[6] A limited number of terminal defences such as Patriot (from land) or SM-6 (from ships), for intercepting threats in the final descent, form the inner layer. Long-range defence against ballistic missiles is possible, since a ballistic missile, by definition, flies a ballistic trajectory. Consequently, if it is tracked early in its flight, for instance using a long-range radar, its future position is highly predictable. A single long-range ballistic missile defence system can therefore launch interceptor missiles over substantial distances, in the knowledge that they will get close to their intended targets in the terminal phase of the intercept. To compensate for errors in the initial track, the interceptors can steer towards their targets using thrusters, guided using an infrared-sensor that is intended to identify and track the missile's re-entry vehicle. Consequently, a small number of such exo-atmospheric missile defence systems can defend a large area. The so-called European Phased Adaptive Approach, for instance, is intended to protect Europe against ballistic missile attacks from the Middle East using just two land-based Aegis sites and a handful of Aegis ships.[7] Similarly, the Japanese Maritime Self-Defence Force operates eight Aegis ships to defend Japan from ballistic missiles, with two more planned.[8]

Like ballistic missiles, hypersonic weapons can cover a large area in a relatively short period of time. However, unlike ballistic missiles, hypersonic missiles may never leave the atmosphere. This makes them impervious to exo-atmospheric interceptors. Their lower flight altitude, compared to ballistic missiles that fly similar ranges, also makes them more difficult to detect using air-defence radars because they can fly below the radar horizon for a longer part of their trajectory.[9] Furthermore, since hypersonic missiles can manoeuvre inside the atmosphere, their trajectory is difficult to predict. This too challenges the defence.

Defence against Hypersonic Weapons

Aerodynamic drag will slow down a hypersonic glide vehicle during its flight. This makes defence in the terminal phase the flight possible, because the vehicle will have lost much of its velocity and its trajectory becomes easier to predict. However, terminal defences, such as the Patriot missile defence system have a relatively short range. Additionally, during the final phase of the gilder's flight, the higher atmospheric density at lower altitudes may allow it to outmanoeuvre terminal interceptors. Recent reports from Ukraine indicate a reduction in the interception rates of Kinzhal missiles (and the Iskander-M), using Patriot. This may, in part, be due to manoeuvres.[10]

To cover a large area with a limited number of missile defence systems, a future long-range interceptor intended against hypersonic missiles, such as the American Glide Phase Interceptor or the European HYDEF and HYDIS interceptor concepts, will have to be able to travel long distances. However, during the flight, its target may manoeuvre. So, instead of being able to fly towards a predictable intercept position and only having to compensate for measurement errors, such an interceptor may have to continuously adjust its course. While the target loses velocity due to manoeuvres, so does the interceptor. Even if the likely future position of a hypersonic glider were to be known, for instance because it is assessed to be heading towards a particular high-value target, the time at which it arrives near that target can vary, depending on the exact route it flies to get there.

This raises the question whether area defence, i.e. an outer layer of a layered defence system, is possible against hypersonic gliders. This depends on what area a hypersonic glider can reach, as well as the range of possible arrival times at any given point in that area. Predicting this would help determine where a hypothetical long-range interceptor would need to go. In open sources, little quantitative information is available about real-world hypersonic weapons. However, using claims by the DPRK for the second flight of their January 2022 hypersonic missile, reported observations of the missile in flight from Japan and South Korea, an analysis of an image of the missile, and engine data, the hypersonic missile's performance could be evaluated in a flight simulation.[11] This simulation provides the initial conditions for a simplified two-dimensional model of the glider in level flight.

The next section explains the initial performance analysis, as well as the two-dimensional model for the glider's flight. This is followed by results of computer simulations of the area the missile can cover, depending on how it steers, as well as the different arrival times that result from different trajectory choices. Subsequently, we will look at implications of the results for missile defence systems aimed at defending large areas by intercepting the glider in its glide phase, particularly for A2/AD scenarios.

Modelling the DPRK's Hypersonic Missile

To assess the performance of a real-world hypersonic missile, we use a two-dimensional model to simulate the glide phase of the hypersonic missile. The missile consists of two parts: the booster (with a liquid-propellant rocket engine) and the glide vehicle (top section); these parts can clearly be seen in Figure 4.2(a). The trajectory of the missile consists of two main phases as illustrated in Figure 4.2(b).

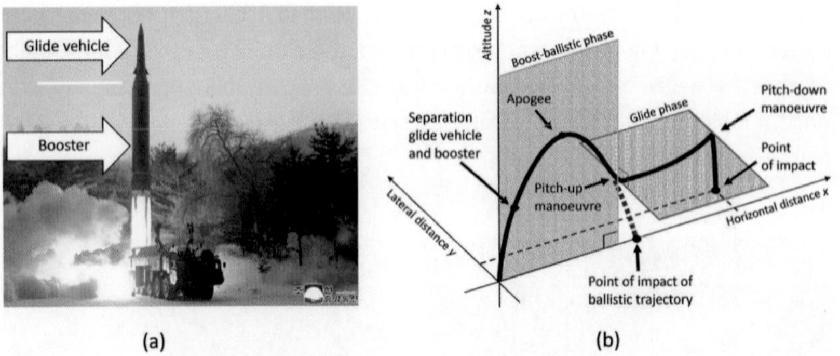

Figure 4.2: (a) The 5 January 2022 launch of a missile that North Korea claims is hypersonic, from a mobile launcher. A typical hypersonic missile consists of two parts: a booster and a glide vehicle. (b) Decomposition of the three-dimensional trajectory of the hypersonic weapon in the boost-ballistic phase (vertical plane) and, after the pitch-up manoeuvre, the glide phase (horizontal plane). Trajectories in this horizontal plane are the subject of study of the present work. At the end of the glide phase, the glide vehicle performs a pitch-down manoeuvre towards the target. (Original image: Korean Central News Agency)

The first phase is called the boost-ballistic phase. In this phase the booster accelerates the missile to its burnout velocity, after which the glide vehicle separates from the booster. It subsequently follows a ballistic trajectory (as a ballistic missile). This ballistic trajectory is in the vertical plane. Figure 4.2(b) shows the point of impact if the glide vehicle does not manoeuvre. However, the glide vehicle will typically not follow this trajectory because during the descent, at a certain altitude, it uses its aerodynamic controls to generate aerodynamic lift and perform a pitch-up manoeuvre. This starts the second phase of the trajectory: the glide phase. In this phase, the glide vehicle maintains (approximately) level flight and only manoeuvres in the horizontal (two-dimensional) plane. The glide phase finishes when the glide vehicle can no longer maintain level flight due to its velocity decreasing, because of aerodynamic drag, or if the glide vehicle starts a pitch-down manoeuvre towards its target. One of the challenges of trying to assess the performance and planning the design of defences against hypersonic missiles is the large number of different variables involved and possible scenarios.[12] We limit them by choosing trajectories with a constant flight altitude during the glide. This corresponds to the trajectory claimed by the DPRK.

Finding the Initial Velocity and Altitude

After the missile's second flight, on 11 January 2022, the DPRK claimed that they had tested a hypersonic missile and that the missile flew 600 km before transitioning into a gliding trajectory. During the glide, it performed a 'lateral manoeuvre' over a distance of 240 km before landing in the Sea of Japan.[13] This description closely matches a visualisation of the missile's trajectory shown in a photograph of Kim Jong Un looking at a bank of monitor screens, published by the (official) Korean Central News Agency (see Figure 4.3(a)).

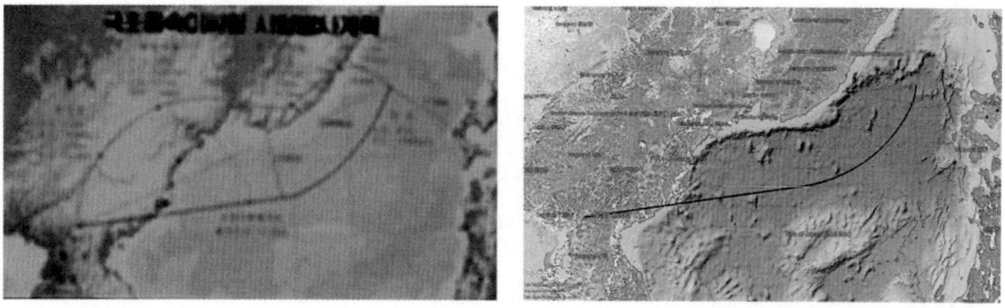

(a) (b)

Figure 4.3: Visualisation of the second flight of the Hypersonic Missile in January of 2022 as shown on a monitor screen, in an official photograph released through the Korean Central News Agency, (a) deliberately blurred by the DPRK, and (b) the ground projection of the trajectory from a computer simulation. (Images: Korean Central News Agency and Savelsberg)[14]

The booster's ballistic trajectory has a straight-line projection on the Earth's surface, after which the re-entry vehicle follows a gliding flight, turning north over the Sea of Japan, prior to pitching down. This may be propaganda, but the ballistic phase was observed. The Japanese Ministry of Defense reported that the missile flew a ballistic trajectory over a range of 700 km, reaching a maximum altitude of 50 km, while South-Korean media reported a maximum height of 60 km, as well as a maximum speed equivalent to Mach 10.

The missile's booster uses an engine cluster previously used for the DPRK's Hwasong-12 missile. Scaling an image of the missile with the nozzle diameter and measuring the shape of the re-entry vehicle from images allowed for estimating its aerodynamic properties and to perform a flight simulation that matches the reported observations of the flight and the visualisation. The simulations were done with a three-degrees-of-freedom modified point-mass model. The coefficients required to model the aerodynamic forces on the glide vehicle (i.e. the lift and drag coefficients) were calculated using an aero-prediction model for the glider trimmed in the pitch direction.[15]

Figure 4.3(b) shows the projection of the simulated trajectory of the missile's flight. For this simulation, the initial ballistic trajectory was aimed at 700 km, i.e. the range reported by the Japanese Ministry of Defense. The simulation program iteratively calculated the missile's pitch program during the boost phase, required for this range, and this resulted in an apogee at 55 km, i.e. between the reported values for the test flight. The burnout velocity is 3.4 km/s, equivalent to Mach 10 as reported in South Korean media. In the simulation, at about 4 minutes into the flight, the missile performs a pitch-up manoeuvre, achieving level flight at 600 km from the launch site, in agreement with the North Korean claim. During this glide, the simulated re-entry vehicle generates a lateral aerodynamic force to also turn north. At the start of the glide, the velocity is 3 km/s at an altitude of about 28 km. Just prior to the final pitch-down, it is still 1.7 km/s, or Mach 5. Thus, if the missile did fly this trajectory, it would have been hypersonic.

The conditions at the start of the glide phase serve as the initial conditions for the two-dimensional simulations of the area that the missile can reach during the glide phase.

Modelling the Glide Phase in Two Dimensions

In the glide phase, the glide vehicle performs a level flight, and its trajectory is entirely in a two-dimensional plane, i.e. the horizontal plane or *x*-*y* plane. Therefore, simulating the glide phase does not require the full three-dimensional model used for the analysis of the 11 January flight test. The free body diagram of the glide vehicle is shown in Figure 4.4.

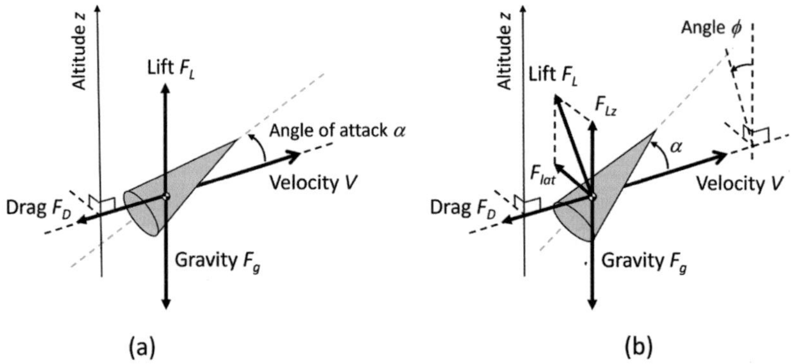

Figure 4.4: Free body diagram of the glide vehicle for level flight in the glide phase (a) without turning, or (b) with a turn. In both cases the velocity vector V and the drag F_D are orientated in the horizontal plane, and the vertical component of the lift F_L compensates the gravitational force F_g. The lift and drag depend on the angle α between the velocity vector and the body axis; this angle is in (a) in the vertical plane, or in (b) under the angle ϕ with this plane.

The velocity V is horizontally orientated (level flight) and, consequently, the drag F_D which, by definition, is exactly in the opposite direction, is also horizontally orientated. The lift F_L is perpendicular to the velocity vector and points upwards, provided that the angle α between the length axis and the velocity vector is positive. This angle is called the angle of attack.

The aerodynamic lift follows from

$$F_L = \frac{1}{2} \rho V^2 S C_L, \tag{1}$$

with ρ the atmospheric density, S a reference area (i.e. the cross-sectional area at the glide vehicle's base), and C_L the glide vehicle's lift coefficient. Similarly, the aerodynamic drag follows from

$$F_D = \frac{1}{2} \rho V^2 S C_D, \tag{2}$$

with C_D the glide vehicle's drag coefficient. The lift and drag coefficients are functions of the vehicle's shape, the Mach number, the flight altitude (because the viscosity of the air depends on the altitude) and the angle of attack α.

The gravitational force is modelled as

$$F_g = \frac{GMm}{(R + z)^2}, \tag{3}$$

with G the gravitational constant, M the mass of the Earth, m the mass of the vehicle, R the radius of the Earth and z the (constant) flight altitude. This equation accounts for gravity decreasing at higher altitudes but does not include effects of the curvature of the Earth.

The glide vehicle maintains level flight if the lift F_L compensates the gravitational force F_g, as shown in Figure 4.4(a). In this figure there is no lateral force on the vehicle. This situation corresponds to a glide vehicle that flies in a straight line (no turn).

To turn in the horizontal plane, the vehicle needs to generate a lateral aerodynamic force. The real vehicle angles its control surfaces such that, in addition to the angle between the length axis and the velocity vector in the vertical plane, an angle is generated in the horizontal plane (the so-called sideslip angle). In the three-dimensional model described in the previous section and in the two dimensional model, this effect is approximated by rotating the lift vector over an angle ϕ with the vertical, see Figure 4.4(b). For an axisymmetric vehicle with relatively small fins, this is a reasonable approximation. To maintain level flight in the turn, the vertical component of the lift F_{Lz} should still counteract gravity, while the horizontal component of the lift F_{lat} causes a lateral force on the glide vehicle.

The lift coefficient is modelled as

$$C_L = C_{L\alpha}\,\alpha, \tag{4}$$

i.e. it is proportional to the angle of attack (the vehicle is symmetric, so it does not generate lift if α= 0°). In this, $C_{L\alpha}$ is a constant that solely depends on the vehicle's geometry, since for high Mach numbers, the aerodynamic coefficients do not depend strongly on the Mach number, and the flight altitude is constant in the simulations.

Since the vertical component of the lift F_L should compensate for the gravitational force, the combination of α and the roll angle ϕ determines the trajectory, under the condition that

$$F_{Lz} = F_L \cos\phi = F_g\,. \tag{5}$$

For a value of α that meets this condition the drag coefficient then follows from

$$C_D = C_{Do} + C_{D\alpha}\,\alpha^2\,. \tag{6}$$

In this C_{Do} is the drag coefficient for α= 0° and the second term on the right-hand side accounts for the drag increasing with the angle of attack. Like $C_{L\alpha}$, both C_{Do} and $C_{D\alpha}$ are constants that solely depend on the vehicle's geometry.

How far a hypersonic vehicle can glide largely depends on the velocity and altitude at the start of the glide phase and on the lift-to-drag ratio C_L/C_D of the vehicle. For very efficient gliders, this is close to 4.[16] For our vehicle model, we find that the ratio (calculated from equations 4 and 6) has a maximum for $\alpha = \sqrt{C_{Do}/C_{D\alpha}}$; this value corresponds to what we will call the optimal lift-to-drag ratio in the remainder of the chapter.

Under the condition that the vehicle is in level flight, the equations of motion become two-dimensional. They are non-linear, however, so they are solved numerically, using a time-marching integration scheme (in MATLAB/ Simulink®). The angle of attack is limited to the maximum angle of attack at which the vehicle can be trimmed. The simulation ends when the vehicle can no longer maintain sufficient lift to remain in level flight.

These simplifications result in a far quicker computer model than the three-dimensional model. This facilitates running a large number of simulations required to assess where the vehicle can go during the glide phase.

Parameters of the Glide Vehicle

The aerodynamic parameters were derived from the parameters used in the three-dimensional model, with a linear fit to the (tabulated) lift coefficients from the aeroprediction code, and a quadratic fit to the drag coefficients, respectively, averaged over a range of Mach numbers and interpolated for the correct flight altitude.

The various model parameters and the initial conditions for the simulations are listed in Table 4.1. These values have been calculated by Kampert[17] and are slightly different from Savelsberg[18] because Kampert's model is less detailed. Using the aerodynamic parameters in Table 4.1, we find that the DPRK vehicle has an optimal lift-to-drag ratio of 2.44, for an angle of attack of 9.31°. This is considerably less than for an optimised glider, but reasonably close to the estimated lift-to-drag ratio for a US hypersonic test vehicle (HTV-2), of 2.6, and of 2.7 for a Kinzhal-like missile.[19]

Table 4.1: Parameters used in the simulation of the glide phase

Earth parameters	
Gravitational constant G [-]	$6.67430 \cdot 10^{-11}$
Mass Earth M [kg]	$5.9722 \cdot 10^{24}$
Radius Earth R [km]	6371
RV mass and aerodynamic specifications at glide altitude (27.5 km)	
Mass RV m [kg]	900
C_{Do} [-]	0.104

C_{Da} [-]	0.0012
C_{La} [-]	0.0546
Reference area S [m²]	0.5411
Maximum angle of attack [°]	19.5
Initial conditions for simulation of the glide phase	
Horizontal distance travelled in boost/ballistic phase [km]	700
Simulation start time [s]	266
Initial velocity [km/s]	2.861 (approximately Mach 8)
Initial angle of attack [°]	10.4

Results

Simulations are performed to calculate trajectories in the glide phase and to construct the area that the glide vehicle can cover with two successive turns of different durations, and a varying duration of straight-line flight between the two turns (i.e. turn-straight-turn-straight). Furthermore, two conditions are considered for the turns:

1. Turns in which α for the optimal lift-to-drag ratio is used to calculate the angle ϕ. The advantage of using the optimal ratio is that the glide vehicle loses less possible energy (due to drag) to have the largest range. As explained in the previous section, the optimal lift-to-drag ratio is obtained for $\alpha = \sqrt{C_{Do}/C_{Da}}$, and the angle ϕ can be calculated from the corresponding lift F_L (equation 1) and the condition of level flight (equation 5).
2. Turns with the maximum possible lateral force. In this situation the glide vehicle uses its maximum angle of attack (see Table 4.1) for each turn, which gives the largest (possible) lift F_L. This results in the sharpest possible turn. The angle ϕ is again calculated from equations 1 and 5.

The first condition (optimal lift-over-drag ratio) is considered, in combination with a single turn that starts at the beginning of the glide phase. The duration of this turn varies from 0 to the maximum time, at regular time steps, and can be both to the right or to the left. Simulation results for 75 trajectories are shown in Figure 4.5(a). Three trajectories are highlighted in the figure as reference trajectories. Reference 1 is a trajectory without a turn; the vehicle flies in a straight line until the angle of attack reaches its maximum value and the velocity is too low to generate enough lift to compensate for gravity. This is the situation at all the endpoints of the trajectories. Reference 1 gives the maximum range of the glide vehicle, of 1350 km.

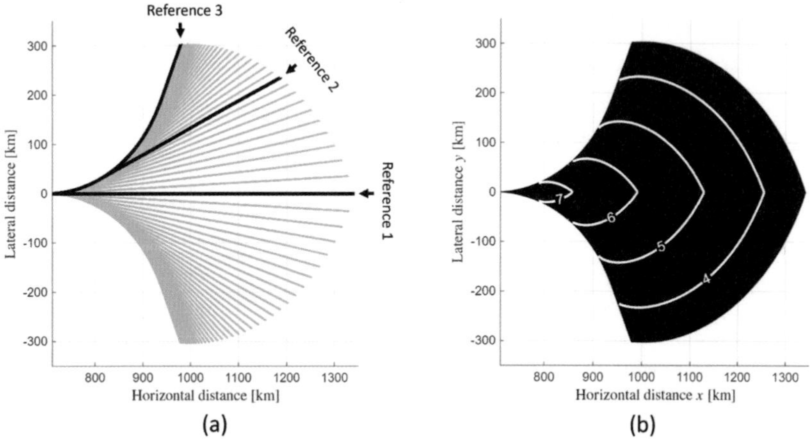

Figure 4.5: (a) Trajectories of the glide vehicle (in the horizontal plane) for a single turn of different durations. The turn always starts at the end of the pitch-up manoeuvre (700,0) and the lateral force corresponds to the optimal lift-over-drag ratio. Three reference trajectories are shown: reference 1 corresponds to no turn (turn duration is zero), reference 2 corresponds to a turn duration of 46 seconds, and along the reference trajectory 3, the glide vehicle is turning all the time. All trajectories end when the glide vehicle can no longer maintain level flight. (b) The area in the horizontal plane covers all possible trajectories and is large.

Along the reference trajectory 3, the glide vehicle turns continuously during its level flight. This gives the maximum time duration of the manoeuvre. No locations on the left side of the reference 3 line can be reached with manoeuvres whilst maintaining the optimal lift-to-drag ratio. Reference trajectory 2 has been chosen as an example of a turn with a limited duration (specifically 46 seconds).

If the number of time steps is increased, the trajectories will blend to an area, which is shown in Figure 4.5(b). It covers all possible trajectories for a turn with the optimal lift-to-drag ratio. The contour lines in the figure indicate the velocity of the glide vehicle (in Mach) and show the decrease of the velocity due to drag along the trajectories. Evidently, at a little above Mach 3, the vehicle cannot generate sufficient lift to maintain level flight at its maximum angle of attack.

Two examples of the trajectories obtained after two turns are shown in Figure 4,6. Figure 4.6(a) corresponds to a manoeuvre in which the vehicle initially flies reference trajectory 1 of Figure 4.5(a), i.e. no turn, but after several seconds of straight-line flight (at 25 second intervals) performs the second turn (again with different durations). The area that can be reached after the second turn is shown as a polygon, and a greyscale is used to distinguish the different time steps. The figure shows that the size of the polygon decreases when the second turn is performed later. This was to be expected, because the vehicle continues to lose velocity when flying in a straight line. Nonetheless, the area that it can reach remains relatively

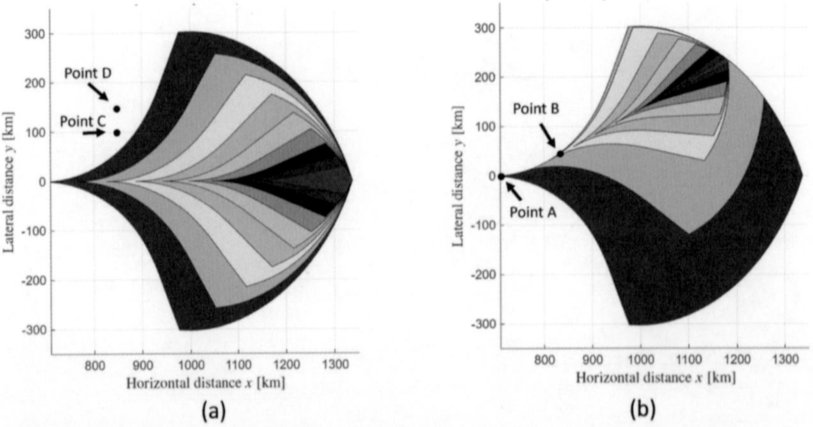

Figure 4.6: Examples of trajectories of the glide vehicle for a second turn with variable duration, if the glide vehicle follows (a) reference trajectory 1 of Figure 4.5, or (b) reference trajectory 2. The lateral force is calculated with the optimal lift-over-drag ratio. This figure shows that even if the glide vehicle can be tracked until the second turn, a large area can still potentially be reached.

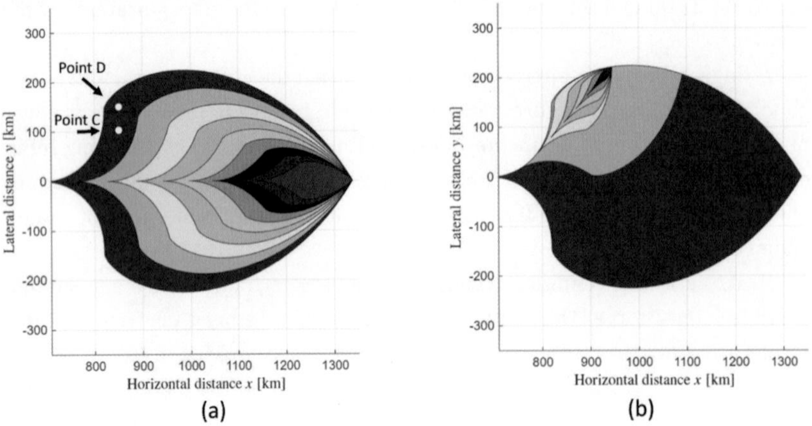

Figure 4.7: As Figure 4.6 but here the lateral force corresponds to the maximum possible lateral force. This figure shows, for example, that the size of the area the vehicle can reach during the glide phase depends on the lateral force used for the turn.

large. Furthermore, it can be noticed that the trajectory of the second turn never leaves the area covered by the first turn, as shown in Figure 4.5(b). Figure 4.6(b) shows areas that the glide vehicle covers if the first turn was performed at the start of the glide phase and lasts 46 seconds; this corresponds to reference trajectory 2 of Figure 4.5(a). A similar pattern (around the reference trajectory) can be observed

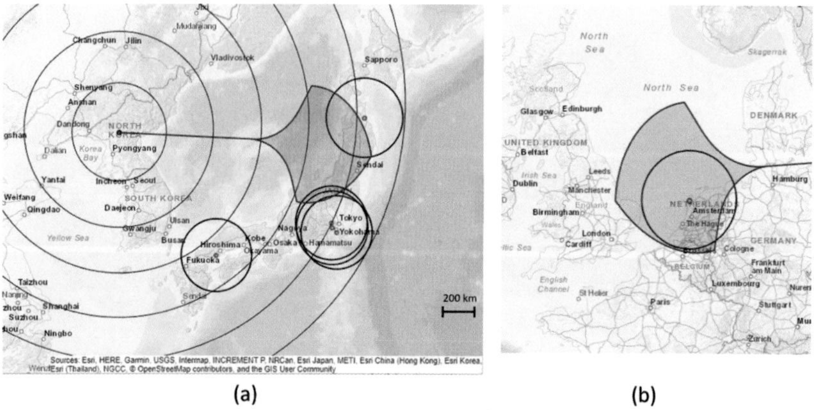

Figure 4.8: Projection of the area calculated in Figure 4.5(b) on (a) the Sea of Japan, or the North Sea. Launch sites are (a) North-Korea and (b) Kaliningrad. The centres of the thick circles correspond to (a) known American military bases in Japan, or (b) the Den Helder Naval base in the Netherlands. Each thick circle has a radius of 200 km, indicating the potential interception range of the THAAD system. Figure generated with the Mapping toolbox of MATLAB®.

in Figures 4.6(a) and (b), and this will also be the case for other durations of the (first) turn.

A single mid-course interceptor launched when the glider is at the start of its glide phase, point A in Figure 4.6(b), will have to be able to manoeuvre to cover much of the black area. Launching 50 seconds later, when the glider flying reference trajectory 2 has turned and reached point B, reduces the area the interceptor needs to cover (part of the light grey area), but this is still thousands of square kilometres.

Next, the second condition is considered, where each turn uses the maximum possible lateral force. The same set of simulations has been performed. Some results are shown in Figure 4.7. This figure should be compared to Figure 4.6. The major differences observed are related to the shape of the area: on the one hand, the glide vehicle can make a sharper turn if the maximum possible lateral force is used, for example the point C (850,100) or even D (850,150) can now be reached. On the other hand, the lateral distance covered is smaller for the turn with the maximum possible lateral force, because its angle of attack is larger during the turn, and consequently so is the aerodynamic drag coefficient (equation 6). Each turn has its own advantages, and a large area can be attained in the glide phase with both manoeuvres.

The area covered by the glide vehicle is large. This is illustrated in Figure 4.8, where the area has been projected onto the Sea of Japan, as well as on the North Sea. The launch sites chosen were the site of the launch of January 2022[20] for

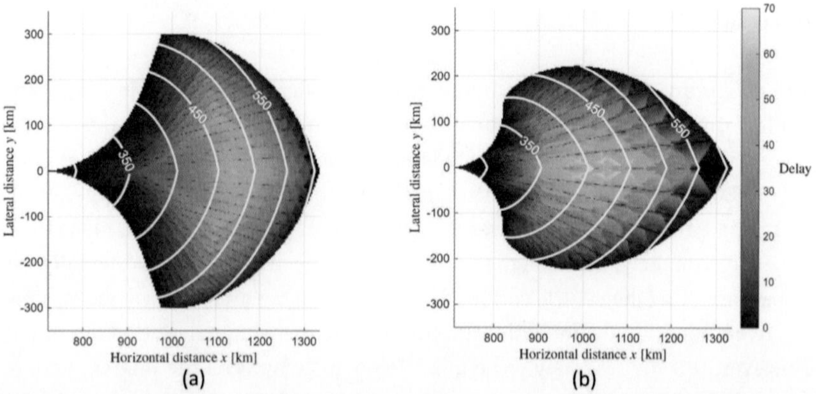

Figure 4.9: Uncertainty in arrival time at different locations in the impact area if the lateral force corresponds to the optimal lift-over-drag ratio (a) or the maximum possible lateral force. The contour lines indicate the minimum arrival time [s] and the greyscale corresponds with the possible maximum delay[s]. The relatively large possible delays make it difficult to counter glide vehicles with an interceptor.

Figure 4.8(a), and Kaliningrad for (b). Figure 4.8(a) shows that the glide vehicle can reach a large part of the sea (area denial) and threaten several ports and American facilities in Japan (anti-access). The maximum interception range of the THAAD (Terminal High Altitude Air Defence) missile defence system (200 km)[21] is indicated by the thick circles in Figure 4.8. It can cover only a part of the area under threat.

Finally, the uncertainty in the arrival time at each location in the area has been considered. Most locations can be reached with more than one trajectory, resulting in a range of possible arrival times, which makes interception more difficult. Figure 4.9 shows the minimum arrival time (contour lines) and the maximum time delay (greyscale-coded) observed in the simulations, for two turns with the optimal lift-over-drag ratio, or for two turns with the maximum possible lateral force. The time delay is relatively large, about 60 seconds, at a horizontal distance of 1100–1200 km from the launch site for the optimal lift-to-drag ratio, and even between 900 and 1200 km for the maximum lateral force. This large delay extends to a lateral distance of about 100 km and results in an arrival time between seven and eight minutes, with a minimum time of seven minutes for example (i.e. the uncertainty is 1/8 = 12.5%). Depending on the assumption about the glider's manoeuvres, an interceptor launched towards the centre of the area may arrive up to one minute too soon or too late.

Discussion

The maximum range in the two-dimensional model for this vehicle is 1350 km, while using the three-dimensional model it is about 1500 km.[22] However, the three-dimensional simulations include Earth rotation and the missile flies from the DPRK towards Japan, such that the Earth rotation adds to its range. While including Earth rotation increases the accuracy of the results, it does make them dependent on the location on the Earth and the launch direction. Additionally, the impact area that the vehicle can reach is larger than the areas shown above, since the two-dimensional simulations do not include the terminal dive to the target. However, this dive is likely to be quite steep, to complicate the interception, so not much distance will be covered during the dive. Furthermore, we are mostly concerned with the area where the missile can be intercepted by a (hypothetical) glide-phase interceptor.

There are uncertainties in the model parameters. The original model could reproduce the North Korean trajectory visualisation, so that parameter set is plausible. The mass estimate is based on the approximate size of the glider and the mass of other gliders of similar sizes. A larger mass would have two effects: the heavier vehicle would have a higher ballistic coefficient ($\beta = m/S\,C_{D0}$), which means that it slows down less due to drag. This could increase the range. However, the pitch-up manoeuvre prior to the glide phase would take longer and the vehicle would level off at a lower altitude and at a lower velocity at the start of the glide phase. Consequently, the area the missile can reach during the glide would be smaller. Similarly, there is uncertainty in the aerodynamic parameters. However, the aerodynamics of the same vehicle were also studied using a more advanced aerodynamic model that includes thermal effects associated with hypersonic flow. This resulted in lift and drag coefficients that differ from the values found using the aero-prediction model by only a few percent.[23] As stated earlier, in the two-dimensional model the Mach-dependence in the aerodynamic coefficients can be neglected if the Mach number remains high. Since the simulations end at Mach values above three, this condition is met.

The current model does not include limitations that may arise due to heating of the vehicle. For certain combinations of altitudes and Mach numbers, the surface temperature could become too high for the materials available to the DPRK. For the test flight, this is likely not the case.[24] A more detailed study is hampered by the lack of information on the vehicle's internal structure.

Rather than flying a glide trajectory at a constant altitude, the vehicle could fly at the optimal lift-to-drag ratio and use the aerodynamic lift to fly manoeuvres in the vertical plane. This results in a so-called phugoid or skipping trajectory, in which the vehicle repeatedly gains altitude and drops down again due to the decreasing atmospheric density at higher altitudes. The DPRK may have tested some of its missiles on such trajectories.[25] They can cover a longer maximum range than gliding trajectories. However, at high altitudes the vehicle's manoeuvrability is limited by the low atmospheric density. Consequently, there, the trajectory is somewhat more predictable. Furthermore, one of the advantages of flying a glide at a near-constant lower altitude is that, depending on the geography, the vehicle may remain below the horizon of surface-based air-defence radars, while manoeuvres in the vertical plane could result in it flying above the horizon. Conversely, performing a vertical manoeuvre may allow the vehicle to overfly the range/ altitude envelope of a terminal defence system along the route, to reach targets that are behind it.

The overall performance and the optimal lift-to-drag ratio of the DPRK glide vehicle is comparable to the performance of the Kinzhal. Thus, the simulated areas can also represent the Kinzhal. However, since Kinzhal is air-launched, its launch point does not have to be on land. With a launch off the Norwegian coast, for instance, it could cover much of the North Sea. The more complicated wedge shape of the DF-17 likely increases its lift-to-drag ratio to a value closer to four. Modelling such more complicated shapes requires a more advanced aerodynamic model than the aero-prediction code, because it is limited to vehicles with circular or elliptical cross-sections. The likely consequence of a larger lift-to-drag ratio is that the vehicle can fly longer ranges during its glide phase, although this obviously also depends on its initial velocity.[26]

Conclusions

Even a relatively simple hypersonic glider with a moderate optimal lift-to-drag ratio can cover hundreds of kilometres in minutes. Steering at the maximum lateral force is used to cover the most lateral distance, at short ranges up to 900 km. Steering using the optimal lift-to-drag ratio can be used to travel the largest lateral distance, but it is only useful at medium ranges, from 900 to 1250 km. How much the vehicle can manoeuvre later in flight depends on previous manoeuvres and on the steering method used. Using the maximum lateral force drastically reduces the range but can be used for targets at close range and can cause large speed and time differences, complicating an intercept.

The results illustrate the inherent challenges in developing an interceptor that can protect large areas from hypersonic missiles. To protect a large area, an interceptor should be launched soon after the hypersonic missile is detected. However, because the glider can manoeuvre, its future position can shift by hundreds of kilometres in a few minutes. Launching a glide-phase interceptor later in the glider's flight reduces the area that it should be able to manoeuvre to, but launching later also reduces the area that it can defend. Area defence against this type of threat requires an interceptor that has long range and a large divert capacity or using multiple (simpler) interceptors. To compensate for the possible different arrival times, it should preferably also have variable thrust. Such an interceptor may be prohibitively expensive and may not even be practically feasible.

Implications

Defending high-value assets inside the area that an enemy hypersonic glider can cover, such as ports and bases in Japan from a DPRK anti-access attack using hypersonic missiles, may require each to have its own terminal-phase interceptors. Nonetheless, the threat may be able to outmanoeuvre the interceptors. Similarly, a NATO maritime task force operating in the Baltic could face a Russian area-denial attack using hypersonic missiles and a US carrier battle group operating near Taiwan a Chinese attack using DF-17s. These weapons largely negate the outer layers of the defences, greatly increasing the risk of operating in those areas.

Notes

1 See Halswijk, "Potential and Practical Performance of Hypersonic Rockets" and Dekker, "Intercepting the 'Unstoppable' Kinzhal Missile".
2 Newdick, "Russian MiG-31s".
3 Lavikainen, "Strengthening Russia's Nuclear Forces in the Arctic".
4 Nouwens, "Long-range Strike Capabilities in the Asia-Pacific".
5 Kristensen, "North Korean Nuclear Weapons 2024".
6 Savelsberg, "Performance Estimates for the DF-21D".
7 Sankaran, "The United States' European Phased Adaptive Approach".
8 Japanese Ministry of Defense, "Defense of Japan 2024".
9 Elshove, "Detecting Hypersonic Glidevehicles".
10 Cranny-Evans, "An Improved Russian Missile".
11 Savelsberg, "Performance Assessment of the DPRK's" and Savelsberg "The DPRK's January 2022 'Hypersonic' Missile".
12 Paukner, "Challenges in Expanding Defended Area Analyses".
13 Korean Central News Agency, "Success in Another Hypersonic Missile Test-fire".
14 Savelsberg, "Performance Assessment of the DPRK's".
15 An aero-prediction model is a semi-empirical model to estimate aerodynamic properties of a vehicle based on its geometry. The model used was Missile DATCOM, see Blake, "Missile DATCOM User's Manual".
16 Acton, "Hypersonic Boost-Glide Weapons".
17 Kampert, "Modeling the Potential Impact Area".

18 Savelsberg, "Performance Assessment of the DPRK's".
19 Acton, "Hypersonic Boost-Glide Weapons" and Halswijk, "Potential and Practical Performance".
20 Latitude 40.0292° N and Longitude 126.1926° E, based on the DPRK's visualisation; see Savelsberg, "Performance assessment of the DPRK's".
21 This currently seems to be the only available system to counter hypersonic glide vehicles in (the final part of) their glide phase, see Missile Defense Project, "Terminal High Altitude Area Defense (THAAD)".
22 Savelsberg, "Performance Assessment of the DPRK's".
23 van Niersen, "Re-entry Analysis of a North-Korean Hypersonic Glider".
24 Ibid.
25 Savelsberg, "Performance Assessment of the DPRK's" and Savelsberg, "Assessing the Maneuverability of North Korea's".
26 Acton, "Hypersonic Boost-Glide Weapons".

References

Acton, J.M. 2015. "Hypersonic Boost-Glide Weapons", *Journal of Science & Global Security*, 23 (3), 191–219.

Blake, W.M. 1998. "Missile DATCOM User's Manual", Air vehicles directorate, Air Force Research Laboratory, February.

Cranny-Evans, S. and S. KaushalIskander. 2025. "An Improved Russian Missile Tests Ukraine's Air Defence", Royal United Services Institute, 18 November. https://www.rusi.org/explore-our-research/publications/commentary/iskander-improved-russian-missile-tests-ukraines-air-defence.

Dekker, E. 2024. "Intercepting the 'Unstoppable' Kinzhal Missile: An Assessment of Flight- & Intercept Challenges for This Maneuverable, Hypersonic Missile Threat". MSc., Netherlands Defence Academy.

Elshove, N. 2024. "Detecting Hypersonic Glidevehicles Using Phased Array Radar: An Open-Source Analysis of the Plausibility of Detection of North Korean Hypersonic Missile Tests above the Sea of Japan Using an AN/TPY-2". BSc. Netherlands Defence Academy, April.

Halswijk, W. 2024. "Potential and Practical Performance of Hypersonic Rockets". In *Proceedings of the AVT-SET-396 Research Symposium on Hypersonic Weapons*, Koblenz, Germany, 15 October.

Japanese Ministry of Defense. 2024. "Defense of Japan 2024", August, https://s3.documentcloud.org/documents/25088868/doj2024_en_full-1.pdf.

Kampert, M. 2024. "Modeling the Potential Impact Area of Hypersonic Boost-Glide Weapons". BSc. Netherlands Defence Academy, May.

Korean Central News Agency. 2022. "Success in Another Hypersonic Missile Test-fire". KCNA Watch, 11 January. https://kcnawatch.org/newstream/1642042882-312269642/success-in-another-hypersonic-missile-test-fire/.

Kristensen, H.M., M. Korda, E. Johns, and M. Knight. 2024. "North Korean Nuclear Weapons 2024", *Bulletin of the Atomic Scientists*, 14 July. https://www.tandfonline.com/doi/full/10.1080/00963402.2024.2365013.

Lavikainen, J. 2021. "Strengthening Russia's Nuclear Forces in the Arctic, The Case of the Kinzhal Missile". Center for Strategic and International Studies, 14 September. https://www.csis.org/analysis/strengthening-russias-nuclear-forces-arctic-case-kinzhal-missile.

Missile Defense Project. 2018. "Terminal High Altitude Area Defense (THAAD)," Missile Threat, Center for Strategic and International Studies, June 14. https://missilethreat.csis.org/system/thaad/.

Newdick, T. 2022. "Russian MiG-31s Armed with Air-Launched Ballistic Missiles Have Arrived in Kaliningrad". *The Warzone*, 8 February. https://www.twz.com/44205/russian-mig-31s-armed-with-air-launched-ballistic-missiles-have-arrived-in-kaliningrad.

van Niersen, S. 2025. "Re-entry Analysis of a North-Korean Hypersonic Glider: an Aerothermal Approach". BSc. Netherlands Defence Academy, April.

Nouwens, V., T. Wright, E. Graham, and B. Hertzinger. 2024. "Long-range Strike Capabilities in the Asia-Pacific: Implications for Regional Stability". IISS, January.

Paukner, D. 2023. "Challenges in Expanding Defended Area Analyses from Ballistic Missiles to Hypersonic Threats", Paper Presented at 3AF IAMD Conference, Porto, Portugal, 13–15 June.

Sankaran, J. 2015. "The United States' European Phased Adaptive Approach Missile Defense System, Defending Against Iranian Threats Without Diluting the Russian Deterrent". RAND Corporation, 13 February. https://www.rand.org/pubs/research_reports/RR957.html.

Savelsberg, R. 2015. "Performance estimates for the DF-21D Anti-Ship Ballistic Missile". Paper Presented at 3AF Missile Defense Conference, Barcelona, Spain, 9–12 June.

Savelsberg, R. 2023. "The DPRK's January 2022 "Hypersonic" Missile Tests: An Open-source Analysis", Paper Presented at 3AF IAMD Conference, Porto, Portugal, 13–15 June.

Savelsberg, R. 2024. "Performance Assessment of the DPRK's January 2022 Hypersonic Missiles". In *Proceedings of the AVT-SET-396 Research Symposium on Hypersonic Weapons*, Koblenz, Germany, 15 October.

Savelsberg, R. 2026. "Assessing the Maneuverability of North Korea's KN-23/Hwasong-11A SRBM". 38 North, 3 March 2026, https://www.38north.org/2026/03/assessing-the-maneuverability-of-north-koreas-kn-23-hwasong-11a-srbm/.

CHAPTER 5

Multi-band Neutralisation of Drones with High-Power Microwave RADAR Technology

A Multi-Domain Approach for Anti-Access and Area Denial

Joost van Oostrum

Abstract

This chapter presents a novel approach to neutralising drones using High-Power Microwave (HPM) technology, combining multiple radar frequency bands to enhance effectiveness. The main contribution is the development of a radar system that utilises HPM technology, leveraging multiple radar bands to increase the energy, power and efficiency. The key argument is that using multiple radar frequency bands can significantly increase the effectiveness of HPM technology, allowing for more precise and powerful drone neutralisation. This chapter discusses the theoretical foundations of HPM technology and its application in drone detection and neutralisation, as well as the benefits and challenges of combining multiple radar frequency bands. Modelling results demonstrate the increased effectiveness of multi-band radar-based HPM technology in neutralising drones. The implications for future Anti-Access and Area Denial (A2/AD) operations are discussed, including enhanced protection of High Valuable Units and increased operational effectiveness in contested environments. By outlining the potential of multi-band radar-based HPM technology, this chapter aims to contribute to the development of more effective A2/AD strategies and enhance the security of modern military operations. The results of this research have significant implications for the development of future drone detection and neutralisation systems.

Keywords*:* HPM; RADAR; DEW; Drone; Counter-UAS

Introduction

The military environment is changing rapidly. In response to Russian aggression, the development of drones has accelerated. Because drones are inexpensive, they can be produced in large numbers and employed in swarm attacks. The primary steps in countering these drones are detection and neutralisation. Drones vary from small (NATO class I) to large sizes (NATO class III). Traditional weapons such as missiles and grenades are extremely costly, however, for large drones (NATO class II to III) these are the best way to counter them.

A high-power microwave (HPM) system which is a directed energy weapon (DEW) only incurs the cost of the energy it uses, meaning the operating expense is minimal. Another advantage of these systems is the ability to neutralise drones quickly and in large numbers, i.e. the engagement is at the speed of light with an electromagnetic beam with non-negligible spatial footprint. For small drones (NATO class I) directed energy weapons are the best way to neutralise them. HPM systems are developed for land-based operations like the U.S. Army's Indirect Fire Protection Capability (IFPC)-HPM.[1]

The principal constraint on a vessel is the scarcity of usable volume and weight. Each cubic metre of deck or hull space is a limited resource. Consequently, naval architects must continually balance multi-functionality, redundancy, and occupied space against stringent weight-and-balance limits. In many cases, the platform cannot accommodate additional hardware without compromising stability or exceeding the design-gross-weight allowance. Under these circumstances, system reuse becomes the only pragmatic means of introducing an extra, potentially mission-critical capability.

A possibility for navies is the use of existing volume search and fire control radar systems as HPM weapons, since these radars already generate high-power electromagnetic signals and are already installed aboard its ships. The primary purpose of HPM systems is to disable hostile drones by inducing a large current in their electronic components such as the antennae, engine or electric circuits. This can be done with front-door coupling (at the operating frequencies of the drone) or back-door coupling (through the body of the drone).[2] The weapon is deemed successful when it impedes the drone's task execution, regardless of the specific effect it induces (e.g. crash, propulsion shutdown, communications loss, or sensor failure).

The drone can be shielded against HPM effects; consequently, more advanced, larger and expensive military Unmanned Aerial Vehicles (UAVs) may not be readily neutralised by such HPM weapons. The main research question is: Can multiple radars in different frequency bands work together as a combined HPM system? The method to answer the research question is by deriving equations and modelling. The purpose of this study is to synthesise a systematic analysis of the technological, organisational and strategic developments for military planners, decision-makers and policymakers for HPM weapons; in addition, examine the implications for future doctrinal amendments and capability development programmes.

This chapter begins with a global system overview, describing the current radar and HPM systems, followed by a section that explains high-power microwave fundamentals and the technologies that generate them. Next, we examine the correlation between HPM and radar characteristics, identifying similarities and differences. Building on this foundation, a novel HPM-combined radar contribution is presented, detailing the integrated architecture, simulation, and performance trade-offs. To demonstrate the real-world viability of the proposed HPM-combined radar architecture, we conduct a detailed case study. After these technical considerations the last sections focus on strategic considerations, starting with the frequency-spectrum implications, analysing allocation, interference, and coexistence issues. Finally, this study explores the practical, policy, and doctrinal implications, addressing operational use-cases, legal constraints, and doctrinal integration before concluding with a summary of contributions and suggestions for future work.

Global System Overview

This section presents further theoretical background on the research topic, starting with Anti-Access and Area Denial (A2/AD) and the functional chains of the sensor and HPM weapon.

Anti-Access and Area Denial (A2/AD) is a layered defence system. For the maritime environment a swarm attack of low cost commercial off-the-shelf UAS systems can occur in an Above Water Warfare (AWW) scenario to protect a High Valuable Unit (HVU) or own platform. A swarm attack can occur in the last defensive layer, closest to the warship, and typically consists of low-elevation targets that are detectable only at short distances. A confidential parameter is the keep-out distance. The weapon system's range must meet or exceed the drone's keep-out distance. This will not be explained further in order to keep this study unclassified.

The functional chains of the UAS and counter UAS system are explained in this section. In Figure 5.1 the functional chain of the radar system is presented. The UAS can also be detected with another sensor like an electronic support measure, and electro-optic or infrared system. The propagation media of these sensors will depend on the operating frequency in the electromagnetic spectrum. After detection, the object's parameters – such as its position, velocity, heading, and other relevant attributes – are tracked and transmitted to the Combat Management System (CMS), which then displays this plotted track to the operator.

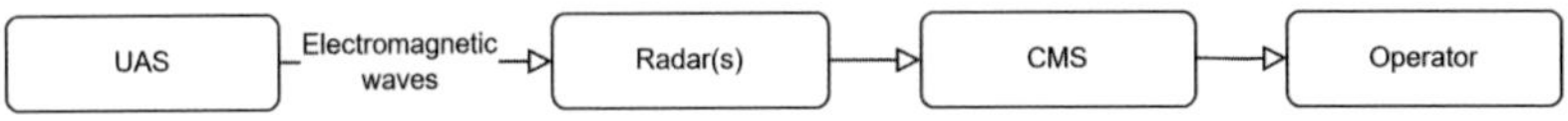

Figure 5.1: Functional chain radar tracking

Once the UAS is identified as hostile, the weapon chain will be used. The functional chain of the HPM weapon system is presented in Figure 5.2. This is the opposite of the functional chain of the sensor system. It starts with the operator who will trigger a weapon in the CMS. In this case, DEW or HPM weapons can be used but also Electronic Counter Measures (ECM) like jamming or spoofing and guns. Different propagation media are used depending on the used weapon. The preferred order of weapon selection is as follows: first, ECM; second, DEW such as HPM and as a final option conventional firearms. After the use of a weapon system, a kill assessment should be executed in order to see if the engagement was successful.

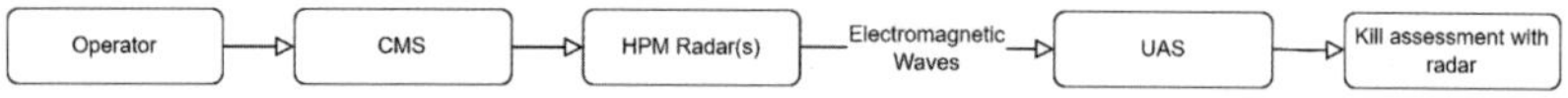

Figure 5.2: Functional chain HPM engagement

In this study, we focus for the detection chain on one or multiple radar systems. For the weapon, we use an HPM weapon, which is also the radar system. Therefore, kill-assessment shall be performed with the ship's radar system, although at short ranges an Electro-Optical/Infrared sensor can also be used to verify target destruction.

With the functional chains presented in Figure 5.1 and Figure 5.2 an overview of the HPM and radar functions are given. In the following section, an HPM weapon is explained.

HPM Explained

High-power microwaves are electromagnetic waves occupying the microwave portion of the spectrum like conventional communication or radar transmitters. The peak power densities are orders of magnitude larger than the conventional transmitters that enables the generation of an intense electromagnetic field that can interact strongly with electronic systems and other matter.

The objective of high-power microwave weapons is to neutralise a drone. This can be accomplished in several ways,[3] for example by:

- destroying the drone's electronic circuitry,
- disabling its propulsion system (engine or motor),
- damaging the battery, or
- damaging the communication and position-navigation-timing (PNT) circuits through the antennas.

The operating principle of an HPM source is analogous to that of a microwave oven: in both cases, electromagnetic energy is produced with an antenna. The critical distinction lies in the environment and intent of the radiation. A microwave oven is enclosed within a metallic, wave-guide-type shield that confines the energy to the cooking cavity, whereas an HPM system deliberately radiates its high-power bursts into free space. Consequently, HPM devices are engineered to couple efficiently to free-space propagation, often-employing high-gain antennas or phased-array apertures to direct the energy toward a target. The next section outlines the characteristics of HPM weapons and radar systems.

HPM and Radar Characteristics

In this section, we compare the electromagnetic equations that describe high-power microwave sources and conventional radar systems. The comparison identifies the parameters that govern the effective power or energy incident on the drone.

Electromagnetic energy is produced by a current on an antenna. It can be transmitted as a communication measure or for radar purposes. The main equation for transmission of electromagnetic energy is the Friis equation. With this equation, a one-way propagation between a transmitter and receiver at distance R in [m] is described.

The received power P_r in [W] is:

$$P_r = P_t \cdot G_r \cdot G_t \left(\frac{\lambda}{4\pi \cdot R} \right)^2 \qquad (1)$$

This equation[4] depends on the antenna gain of the receiver G_r and the transmitter G_t, which are both dimensionless. It is also depending on the wavelength λ in [m] (which is related to the frequency f) of the electromagnetic wave. The received power is proportional to $(1/R)^2$. This equation is the principal equation of an HPM weapon where the transmitter is the HPM antenna and the receiver is the target drone with a target gain.

It must be noted that at the drone the energy couples in on different frequencies with different antenna gains. The antenna gain depends on the angle-of-arrival θ at the drone and on the polarisation of the antenna. If there is a polarisation mismatch, the receiver's gain drops dramatically, making the coupled energy negligible (and, in the worst case, essentially zero). This will result in a target gain $G_{r,f,\theta}$ that depended on a frequency and angle-of-arrival.

The total energy E_r in [J] at the drone is related to the received power and the exposure time t in [s]:

$$E_r = P_r \cdot t \qquad (2)$$

A radar receives the reflected energy of an object-of-interest. Therefore, the received power back at the radar will be reduced with a ratio $(1/R)^4$. Due to this relation a radar needs a high transmit power to operate and to achieve adequate performance. The radar equation is given in different books.[5] For the primary radar, equation 3 is the most well-known one. Where the signal-to-noise ratio *SNR* (dimensionless often-calculated in dB) is calculated as follows:

$$SNR = \frac{P_t \cdot G_t \cdot G_r \cdot \lambda^2 \cdot \sigma}{(4\pi)^3 \cdot R^4 \cdot k \cdot T_0 \cdot F \cdot B} \qquad (3)$$

Where in this expression: P_t is the transmitted power in [W], G_t is the gain of the transmit antenna, G_r is the gain of the receive antenna, λ is wavelength of the carrier in [m], σ is the radar cross section from the object-of-interest towards the radar in [m^2], R is the range from the radar to the object-of-interest in [m], k is Boltzmann's constant (1.38×10^{-23} W/K/Hz), T_0 is the standard temperature (290 K), F is the noise figure of the receiver (dimensionless) and B is the bandwidth of the radar signal in [Hz].

The power at the receiver of a radar system must be large enough to determine if there is an object-of-interest. The difference between radar and a regular HPM weapon is the receive path of a radar. This path has delicate receivers, which operate at a dynamic range of 40 to 80 dB.[6] This receive path must not be harmed by the own HPM waveforms if the radar is used as an HPM weapon.

With both equations, we can see that the effective power or energy level at the drone depends on:

- Transmit power;
- Wavelength (frequency);
- Antenna gain of the transmitter;
- Antenna gain at the drone (or target gain);
- Polarisation;
- Distance of the drone;
- Exposure time.

Most of these parameters are confidential and will determine the confidential performance and range for an HPM weapon. Another important radar parameter is the transmitted waveform. In the present work the signal frequency varies with time according to a conventional linear-frequency-modulation (LFM) scheme, which is the standard waveform employed by most surveillance radars.

With this background information, we can proceed to the novel HPM combined radar contribution in the next section.

Novel HPM Combined Radar Contribution

This section contains a novel way to combine electromagnetic power in order to upgrade the defence capability of multiple radar systems. The main contribution is that we can add up the energy of different radars for HPM purposes under the right circumstances. Therefore, several conditions must be satisfied.

In order to create an unclassified study a few parameters are not given. In this study, the confidential frequencies will not be used. Instead, fabricated frequencies will be used. In addition, the power levels of the different radar systems will not be used. Instead, an amplitude without dimension will be used.

The first condition is that the signals must be in the far-field. In the far-field region, the electromagnetic waves are locally planar. Other measures such as phase-difference techniques must be employed to focus the energy in the intermediate-field or near-field region.

We start with the mathematics and within the electromagnetic pulse of a radar signal s_n in the far-field as:

$$s_n(t) = a_n \cdot e^{j(2\pi \cdot f_n \cdot t + \varphi_n)} \tag{4}$$

Where f_n is the carrier frequency of the radar system n, a_n the amplitude of radar n, t is time in [s] and φ_n the phase shift in [rad].

To maximise the energy at the target, the signals must arrive at the target in phase. Consequently, the exact distance from each radar system to the target must be determined. With this information, the appropriate phase offset for each radar can be set relative to the others, thereby achieving the desired gain advantage.

The second condition is that the signals must be harmonics of one another. This means that the different frequencies that are combined share the same periodicity. For example, $2 \cdot f_1 = f_2$ and $f_3 = 4 \cdot f_1$. Where the higher frequencies are always an integer multiplication of the lowest frequency.

In the following examples three radar systems with different carrier frequencies $f_1 = 1$ GHz, $f_2 = 3$ GHz and $f_3 = 9$ GHz are used. The sum of three different signals $s_\Sigma(t)$ with equal amplitudes (1.0 [dimensionless]) is presented as:

$$s_\Sigma(t) = e^{j(2\pi \cdot 1 \cdot t + \varphi_1)} + e^{j(2\pi \cdot 3 \cdot t + \varphi_2)} + e^{j(2\pi \cdot 9 \cdot t + \varphi_3)} \tag{5}$$

In this equation, time t is in [ns]. In this study, propagation losses between the transmitter and the receiver are ignored, so the amplitudes of the three signals are assumed equal.

The final requirement is that the signals be phase-coherent, a condition illustrated by the figures that follow.

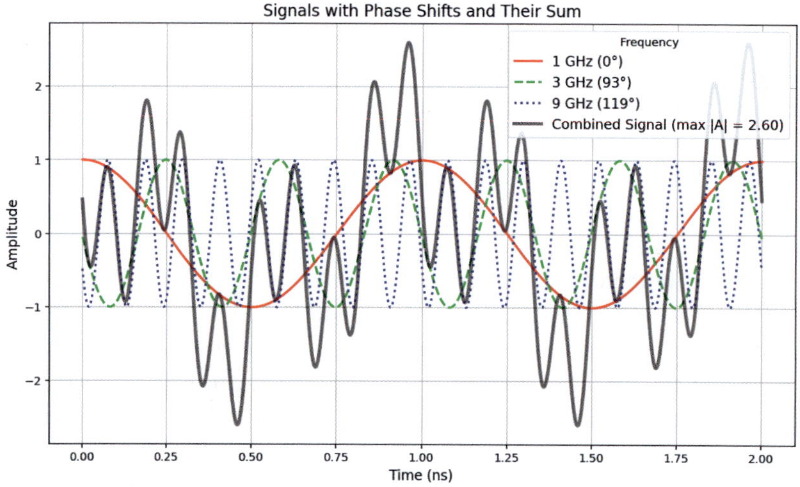

Figure 5.3: Sum of harmonic signals not in phase (with peaks at 2.60 and -2.60)

Figure 5.3 depicts the signal when phases are not aligned in-phase. The 1 GHz signal is the signal with the lowest frequency and has zero phase offset, while the 3 GHz and 9 GHz components are phase-shifted by 93° and 119°, respectively. In this particular configuration, the maximum absolute amplitude of the summed signal is approximately 2.60, which is close to, but below, the theoretical maximum of 3.0 that would occur if all three signals were fully phase-aligned. In a simulation comprising 10000 independent trials with the three harmonic frequencies, the peak amplitude of the response was characterised as follows. The average (mean) peak amplitude was 2.68, whilst the observed extremes spanned a minimum of 2.24 and a maximum of 3.00.

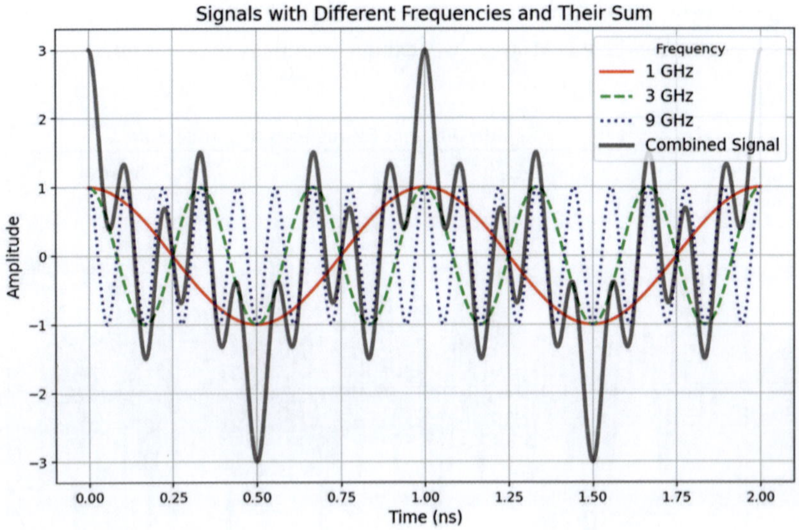

Figure 5.4: Sum of harmonic signals in phase

In Figure 5.4 the signals are phase-aligned in cosine functions, resulting in a higher observable amplitude. The summed waveform reaches its peak absolute value of 3.0 (the sum of the three unit-amplitude components) at 0.0 ns, 0.5 ns, 1.0 ns, etc., which coincide with the peaks of the 1 GHz carrier. In this example, the gain for the lowest frequency is a factor three which corresponds to 4.77 dB.

In addition to the principal peak of 3.0 at 3 GHz, an additional gain of 1.5 (i.e. 1.0+1.0–0.5) is observed at the instants 0.17 ns, 0.33 ns, 0.66 ns, 0.83 ns, and so forth. However, if a further gain of 2.0 is desired for this particular frequency, the lowest frequency should be omitted. Only 3 GHz and 9 GHz should be used. In that case, the gain is a factor two which corresponds to 3.0 dB.

In summary, the transition from the phase-misaligned condition of Figure 5.3 to the phase-coherent arrangement of Figure 5.4 demonstrates how deliberate phase alignment can substantially amplify the signal amplitude and the corresponding gain at the individual carrier frequencies. The analysis also highlights the trade-off between multiband gain and frequency-specific optimisation when certain frequencies are omitted.

With the three conditions set out above, the maximum amplitude gain a_{Σ} for multiple radar systems on the lowest frequency is the sum of the amplitudes of the signals as presented in Figure 5.4.

In Figure 5.5 we evaluate the influence of three representative linear-frequency-modulation waveforms on the target drone. Because the interaction mechanism depends principally on the sweep characteristics of the chirp, any radar that employs a comparable LFM sweep will exhibit essentially the same effect.

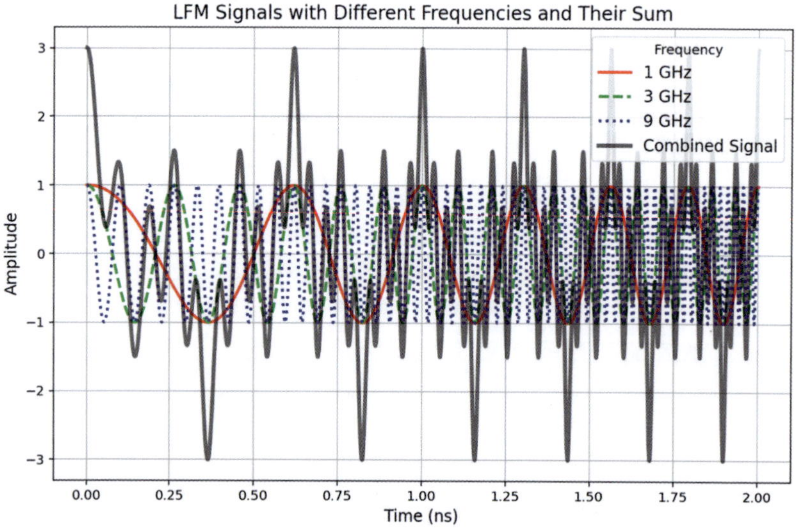

Figure 5.5 Sum of harmonic LFM signals in-phase

All three signals have an identical sweep duration of 2.0 ns. The fundamental chirp spans 1 GHz to 3 GHz, the second harmonic signal is a three-times-scaled replica (3 GHz to 9 GHz), and the third harmonic signal is a nine-times-scaled replica (9 GHz to 27 GHz). Because each higher-frequency chirp is obtained by multiplying the instantaneous frequency of the fundamental by an integer factor, the instantaneous phase of every chirp remains a perfect integer multiple of the fundamental phase throughout the sweep, ensuring full phase coherence across the three frequency components.

Concluding this section, without exposing confidential parameters, a clear gain can be calculated. With the following conditions:

- The target is in the far-field;
- The radar frequencies must be harmonics of each other;
- The signals must arrive in phase at the target;
- LFM sweeps for the different frequencies which results in in-phase signals.

The amplitude of the combined signal at the lowest frequency can be calculated as the sum of the different amplitudes. The main idea is that the energy spike at the lowest frequency (1 GHz) will likely disable the drone through one or more of the mechanisms described earlier: destroying its electronic circuitry, disabling the propulsion system, compromising the battery, or impairing the communication and PNT circuits via the antennas. However, confidential experiments should conform to these assumptions. At this point, we have a useful method to create a more effective radar-based HPM weapon; we now turn to examine how the electromagnetic wave propagates within copper.

Inside the UAV

When the electromagnetic waves reach the drone, they couple onto copper wires or printed-circuit boards. These materials are "good conductors". The first limiting factor is the skin-depth, which depends on frequency and material of the properties. The skin-depth introduces additional attenuation at higher frequencies. Consequently, higher frequency components do not penetrate as deeply into the material as lower frequency components.

A second limitation arises from the phase velocity within the conductor, which affects the phase of the propagating wave. This behaviour is illustrated in Figure 5.6. All the employed frequencies start at the interface of free space propagation and the copper with equal amplitudes and in-phase. After propagating to a depth of 0.7 μm, the signals are no longer in phase; nevertheless, distinct peaks with a value of 1.6 remain observable at 0.0 ns, 1.0 ns, 2.0 ns, etc. which is almost exactly at 1 GHz frequency. Even when the harmonic signals are out of phase, a noticeable energy peak still shows up at the lowest frequency; however, it's just not as high as it is when the signals are perfectly in-phase.

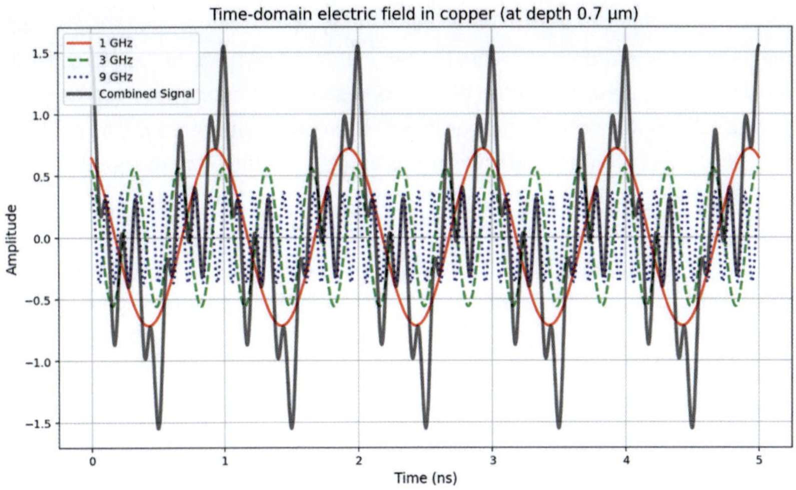

Figure 5.6: Electric field inside copper

The phase shift and the skin-depth in a good conductor both scale with the square-root of the frequency. Consequently, harmonic components that start with the same phase at the interface do not remain perfectly coherent as they propagate; each frequency acquires a different phase delay. With this knowledge, the set 1 GHz, 3 GHz and 9 GHz will never be exactly in phase. This is presented in Figure 5.7. The penetration depth shown in this figure is for the higher frequencies practically unattainable, as signal attenuation caused by skin-depth damping prevents the waves from reaching that depth.

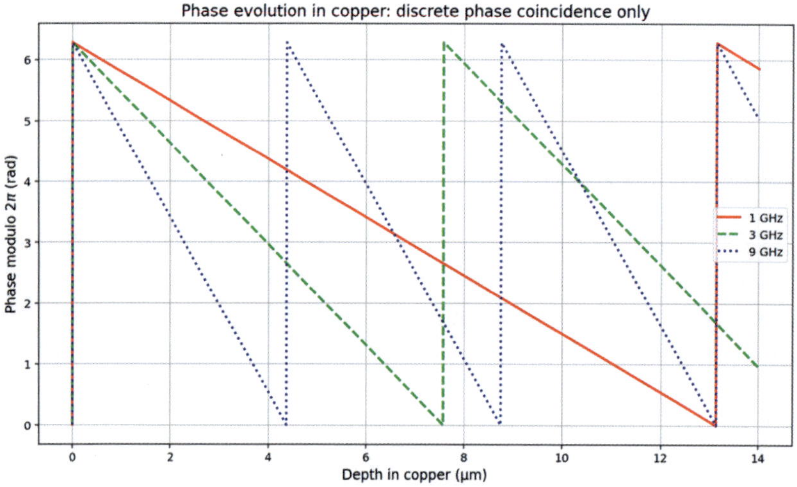

Figure 5.7: Phase in copper of 1 GHz, 3 GHz and 9 GHz signals

Exact phase coherence can be restored only if the constituent frequencies obey a specific relationship. For the particular set of frequencies 1 GHz, 3 GHz and 9 GHz the rate at which the phase propagates through copper can be tuned with an LFM signal. When higher-order effects are taken into account, a non-linear frequency modulated (NLFM) sweep can bring all of the electromagnetic waves travelling in copper into phase.

To conclude: inside the copper structures of the drone, the incident radar waves experience a frequency dependent penetration (skin-depth) and a concomitant phase shift. Despite the attenuation, measurable field enhancements are still observed when the coherent waves that arrive from free space enter the copper body of the vehicle.

Because the phase advance depends on frequency, achieving a prescribed field distribution at a particular depth within the copper requires careful selection of the operating frequencies. Moreover, the desired effect can only be realised with LFM or NLFM signals that satisfy the phase-coherence condition.

Further confidential experimental and numerical investigations are therefore required to identify the optimal frequency sets and waveforms for which constructive interference can be exploited in practical UAV radar interaction scenarios.

Case Study

In this section, a case study is presented, illustrating the combined signal generated from multiple aspect angles toward the drone and the drone's velocity relative to the radar's characteristics.

HPM Radar Layout

In the situation described above, three radar systems are installed on a vessel. For the purposes of this example, we assume they are positioned along the centreline of the ship, from bow to stern: first the 3 GHz system, then the 1 GHz system, and finally the 9 GHz unit at the aft. The radars are equipped with Active Electronically-Steered Antenna (AESA) arrays, which allow them to steer their radar beams. The spacing between the 3 GHz and 1 GHz systems is taken as 25.1 m, and the distance from the 1 GHz to the 9 GHz system as 37.3 m. A drone is situated exactly 1009.0 m perpendicular to the 1 GHz antenna, mid-ship. A schematic overview of the layout is shown in Figure 5.8.

One might wonder whether the drone is travelling faster and whether phase-alignment is feasible when the various antennae are not situated at exactly the same point. In this example, a pulse-repetition frequency (PRF) of 10 kHz is employed. The lowest-frequency radar typically operates at the highest PRF, and a conventional radar pulse length is about 1.0 μs – a value that can be used across all three radars. However, the 1 GHz system would normally employ a longer pulse, and, of course, it can transmit a longer-duration pulse if required. With these parameters, we can calculate the field at the specified distance.

Combined HPM Signal

Figure 5.9 (a)–(c) displays the individual field patterns generated by each radar system, with the drone positioned on the centreline at a distance of 1009.0 m from the ship. Figure 5.9 (d) shows the resultant field obtained by summing the three individual contributions. In this panel, the red/blue colour map has been expanded to accentuate the extrema; the normalised Electric-field (E-field) amplitude reaches +3.0 at the positive peak and -3.0 at the negative peak of the 1 GHz component.

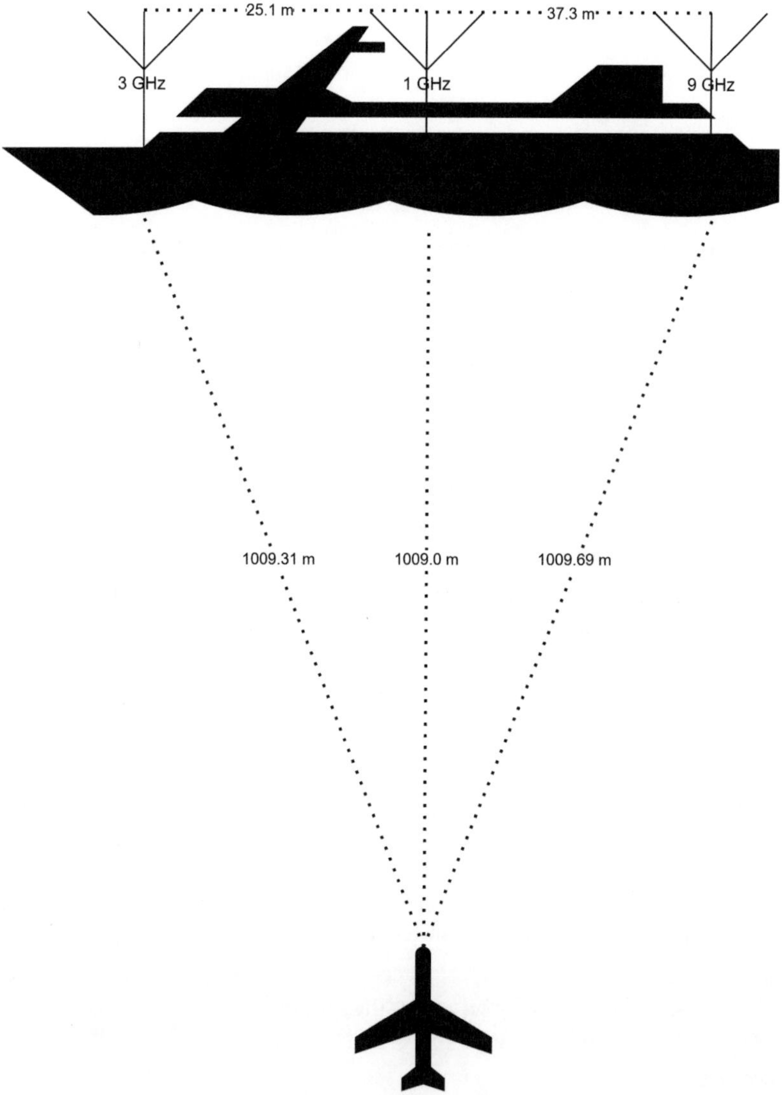

Figure 5.8: Schematic (not to scale) of a ship equipped with three radar systems and a drone

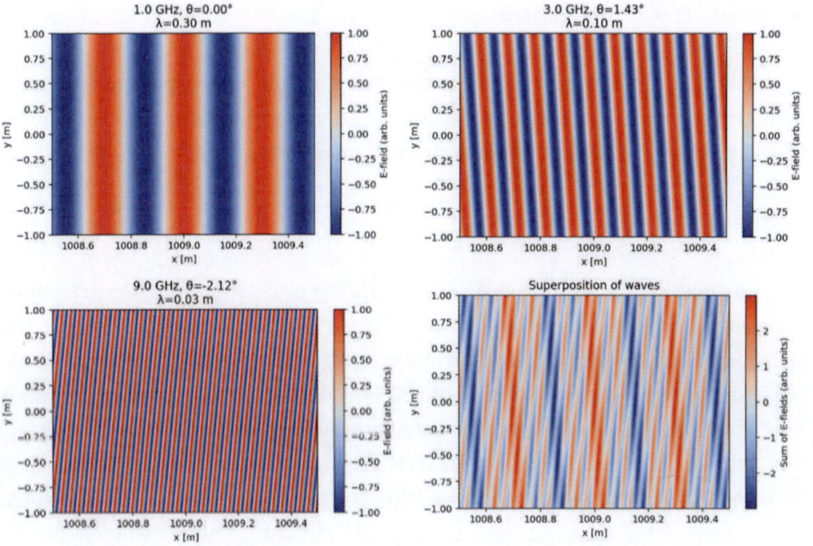

Figure 5.9 Three radar signals (a-c) and their combined signal at 1009 m distance (d)

Drone Velocity

In this part of the case study, we explore how an HPM burst hits a drone, ensuring the radar-guided beam never loses its target despite the drone's rapid motion at a distance of 1009.0 m. With electromagnetic waves propagating at the speed of light and a pulse length that corresponds to roughly 300 m (≈ 1.0 μs), the HPM burst occupies a spatial extent far larger than the distance the drone travels during a single pulse. Even at a ground velocity of 136 km/h (≈ 38 m/s), the drone moves only about 38 μm during the pulse duration, so it remains entirely within the HPM wave area for the whole transmission. If the PRF of 10 kHz is used the drone moves 3.8 mm between consecutive pulses. With a radar track-update time of 1.0 s, the drone will have moved 38 m. This means that only when the drone is moving laterally could the electromagnetic waves miss the drone, depending on the beam-width of the radars that are used as HPMs. The radar tracking filters should be able to guide the radar beam to keep the drone in track and guide the HPM waves in a similar way.

In the following section the frequency spectrum and proportionality/legality will be discussed when we want to use an HPM weapon.

Discussion Frequency Spectrum and Legality

In this section the electromagnetic-interference implications, spectrum-management constraints and proportionality-and-legality considerations arising from the use of radar as a high-power microwave system are examined.

Electromagnetic Interference

With the purpose of harming electronic components of the drone, risks to friendly systems and civilian infrastructure are present. If the radar is used as HPM it is possible that it is also used for detection and ranging if the receiver inside can handle large amounts of power. Another possibility is electromagnetic protection when the radar is in HPM mode. For other friendly systems present on board a warship, mitigation methods can be taken. Also friendly drones or other friendly units can be protected if the location is known. With the purpose of destroying electronic components, it will also influence civilian infrastructure. Safety distances can be calculated where the HPM should not be used.

Spectrum Management

Every electromagnetic transmitter is bounded by its spectrum. For military radar systems, a wide frequency spectrum can be used; however, due to regulations of the National Inspection Service for Digital Infrastructure (Rijksinspectie Digitale Infrastructuur (RDI)) only certain frequencies with a limited amount of power are assigned. For military systems the National Radio Frequency Agency (NARFA) will determine together with the RDI which frequencies can be used. For HPM systems, some of the most effective frequencies that will affect the drones are frequencies that are used for civilian applications like the Wi-Fi frequency of 2.4 GHz. To be able to counter different threats, the frequency range of the HPM system must be designed in a wide range. The military perspective for HPM is to use the widest range of frequencies while the civilian perspective is to prevent the HPM from harming civilian systems.

Proportionality/Legality Considerations

Decisions to use the radar as an HPM system shall be made by a human. If only hostile drones are affected, there is no issue. If civilians and/or combatants are present in the direction of HPM, electromagnetic wave proportionality considerations have to be made by the operator. Electromagnetic waves constitute a radiation hazard for personnel. Their principal effect is thermal heating of the body, for which exposure limits have been established. At a warship, the commanding officer has the responsibility to decide if a weapon is allowed to be used. In this case, the radar in HPM mode will be treated as a weapon.

Besides the frequency discussion, also the practical, policy, and doctrinal implications will be discussed. This will be addressed in the following section.

Practical, Policy, and Doctrinal Implications

This section is aimed at military planners, decision-makers, and policymakers to reflect on implications for future doctrine, capability development, or operational planning.

Capability Development

In future ship designs, high-power microwaves can be incorporated within the existing radar suites. Accurate detection and ranging of the drone are essential to achieve the in-phase combination of multiple radar arrays. A further prerequisite is that the radars operate in harmonic relationship with one another; for example, carrier frequencies of 1 GHz, 3 GHz and 9 GHz could serve as design drivers. To realise the requisite power boost, the radars should be tightly integrated into the combat management system. Once the range to the target has been determined, each contributing radar must know the exact phase offset at which its energy should be transmitted.

A radar HPM system necessarily consumes electrical energy, and therefore power budgeting must be an integral part of the design of both current and future radar platforms. Even when the individual frequency components are coherently combined to increase the resultant amplitude, the total radiated power may still be insufficient to induce the desired effect on the target drone. Consequently, the design principle is straightforward: the greater the available transmitted power, the more favourable the performance. In practice, higher transmitted power translates directly into a larger effective range of the HPM weapon system. This relationship should therefore be reflected explicitly in any system-level trade-off analysis.

Safety

One of the advantages is that the existing ship design can be retained together with the current safety precautions for radar systems. In normal operations, radiation hazards to fuel, ammunition and personnel can be managed. For example, when the navy's own helicopter is landing on the flight deck, regulations and mitigation measures are applied to lower the radar energy directed toward the aircraft, thereby preventing injuries or accidents. Similarly, the HPM effects can be managed in the current ship designs.

Future Doctrine

The possibility of employing HPM as a weapon should be considered at the operational level on each ship. The energy budget allocated for counter-UAS (C-UAS) missions cannot simultaneously be used for conventional radar tasks. Consequently, when HPM is integrated into the radar suite, decisions regarding weapon employment must be made. Operators should have access to a range of options, including electronic warfare measures, soft-kill tactics, and hard-kill solutions (both HPM and kinetic ammunition).

With the practical, policy, and doctrinal implications described, a wide view on the radar-based HPM topic is presented and will be concluded in the next section.

Conclusion

The rapid escalation of drone capabilities driven by recent conflicts and the low cost of swarm-type platforms has fundamentally reshaped the modern battlespace. High-power microwave systems that exploit existing ship-borne radars therefore represent a cost-effective, rapidly deployable counter-measure against small (NATO Class I) unmanned aerial threats.

In this chapter, we investigated whether radars operating on disparate frequency bands can be combined into a single HPM weapon. Using analytical derivations and numerical modelling, we demonstrated that the effective HPM power (or energy) incident on a drone is a function of transmit power, operating wavelength (or frequency), transmitter-antenna gain, drone-antenna gain, polarisation, and exposure time. It is crucial that, when the radar frequencies are harmonic, the emitted waveforms arrive in phase at a far-field target. The analysis is carried out on the lowest-frequency band. Under these conditions, the attainable gain can be expressed as the sum of the individual signal amplitudes. This results in a theoretical gain of approximately 4.77 dB for a configuration with three radars used as an HPM weapon and equal power at incidence. The foregoing calculation is also applicable to a linear frequency modulated radar waveform. Similar effects are observed in copper; a tailored LFM or non-LFM sweep can generate a harmonic, phase-aligned signal within the copper, producing high-energy peaks that are sufficient to disable the drones.

These technical results carry several operational implications. Deploying radars in HPM mode introduces notable electromagnetic-interference effects, imposes stringent spectrum-management constraints, and raises proportionality-and-legality considerations that must be resolved before fielding. Consequently, future naval doctrine and ship design should explicitly accommodate HPM integration within existing radar suites, ensuring precise phase-coordinated operation, adherence to established safety regimes, and allocation of a dedicated energy budget that balances electronic warfare and kinetic neutralisation options for effective C-UAS capability.

The present study considered radars co-located on a single platform. A natural extension is to examine synergistic operation of radars on multiple platforms, or of land-based HPM systems in a distributed architecture. Further research should also explore the influence of alternative waveforms and pulse-modulation schemes on target vulnerability. Another promising line of research concerns the use of radar systems as ECM platforms. Modern military radars already incorporate a range of Electronic Counter Counter Measure (ECCM) techniques, such as frequency hopping and Low Probability of Intercept (LPI) waveforms. The potential advantages of integrating high-power ECM capabilities (for spoofing and jamming) directly into a high-power radar architecture as counter drone capability should be researched.

Acknowledgements

The author used the assistance of artificial-intelligence tools, which were employed to improve the chapter's language owing to the author's dyslexia. Special thanks to many colleagues for their valuable discussions, peer reviews and the information they provided. This work is unclassified; consequently, the specific technical parameters have been omitted.

Notes

1 Sandall, "US Army, Philippine Air Force Test".

2 Li et al.,"Evaluation of Front-Door and Back-Door".

3 Weinberg, "Prediction of UAV Swarm Defeat".

4 Balanis, *Antenna Theory: Analysis and Design.*

5 Skolnik, *Introduction to Radar Systems;* Edde, *Radar: Principles, Technology, Applications*; Richards, *Principles of Modern Radar.*

6 Richards, *Principles of Modern Radar.*

References

Balanis, Constantine A. 2016. *Antenna Theory: Analysis and Design*. Hoboken, NJ: Wiley.

Edde, B. 1993. *Radar: Principles, Technology, Applications*. Englewood Cliffs, NJ: Prentice Hall.

Li, J., D. Gómez-García, F. Rodríguez-Morales, and C. Leuschen. 2023. "Evaluation of Front-Door and Back-Door RFI Impacts on Small UAS Operation Safety." In Proceedings of the IEEE Symposium on Electromagnetic Compatibility & Signal/Power Integrity (EMC+SIPI), 712–717. Grand Rapids, MI: IEEE, July 29–August 4, 2023. https://doi.org/10.1109/EMCSIPI50001.2023.10241597.

Richards, M., W. Holm, and J. Scheer. 2010. *Principles of Modern Radar: Basic Principles*. Scottsdale, AZ: SciTech Publishing.

Sandall, I. (Maj.). 2025. "US Army, Philippine Air Force Test Counter-Drone Systems at Balikatan 2025." U.S. Army, May 1. Accessed November 24, 2025. https://www.army.mil/article/285095/us_army_philippine_air_force_test_counter_drone_systems_at_balikatan_2025.

Skolnik, M. 2001. *Introduction to Radar Systems*. 3rd ed. New York: McGraw-Hill.

Weinberg, G. V. 2022. "Prediction of UAV Swarm Defeat with High-Power Radio-Frequency Fields." *IEEE Transactions on Electromagnetic Compatibility* 64. 2157–2162. https://ieeexplore.ieee.org/document/9852009.

CHAPTER 6

Electromagnetic Dominance

Invisible Cornerstone of Modern A2/AD

Susan Varenbrink and Alexander Schmets

Abstract

Since the late 19th century, wireless electromagnetic signals have fundamentally reshaped the practice of military operations. Innovations in the use of the electromagnetic spectrum for communication, navigation, and sensing created a new tactical and strategic necessity in the practice of warfare: to secure spectrum access for friendly forces while denying it to adversaries. This requirement underlies what is now termed electromagnetic warfare (EW), a domain that has become indispensable to modern warfare and underpins the effectiveness of almost all contemporary military operations. Unlike purely geographic domains, the electromagnetic environment is inherently multidimensional. It spans vast ranges of frequency, power, and timing parameters. As a result, the contested area extends across many orders of magnitude beyond its physical boundaries. This work takes initial steps toward providing a unified framework for future quantitative analysis to support electromagnetic situational awareness and countermeasure planning across all phases of military operations.

Keywords: Electromagnetic warfare (EW); Electromagnetic dominance; Contested electromagnetic area; Contested electromagnetic environment; Anti-access/area denial (A2/AD)

Introduction

As Richard Feynman observed in the 1960s, “From a long view of the history of mankind, there can be little doubt that the most significant event of the 19th century will be judged as Maxwell’s discovery of the laws of electrodynamics”.[1] Two decades later, in the 1880s, Heinrich Hertz experimentally confirmed the existence of electromagnetic waves by producing and detecting them for the first time. The military soon recognised the potential of the electromagnetic domain. Initially, it reshaped the battlefield through wireless communication and sensor-assisted navigation. Over time, it evolved into a battlespace in its own right.[2]

Because both military and civilian systems increasingly depend on the use of the electromagnetic (EM) spectrum, the EM domain has become a contested environment, giving rise to what is now termed electromagnetic warfare (EW). Because control of the spectrum directly affects freedom of movement and access to operational space, the EM domain is closely related to A2/AD strategies. Rather than functioning as independent military operations, electromagnetic warfare serves as an essential supporting capability embedded within broader campaigns. Securing control of the electromagnetic domain is therefore essential for achieving and sustaining operational advantage in the theatre of conflict. Under conditions of electromagnetic dominance, friendly forces can deny or constrain the opponent's operations, degrade their EM capabilities, and limit their use of EM-dependent systems. Such electromagnetic operations align closely with anti-access/area-denial (A2/AD) strategies, which aim to deny, degrade, or exploit access to contested areas. This dynamic is clearly visible in contemporary conflicts. Current conflicts illustrate how access to parts of the electromagnetic spectrum is actively contested, for example, through GPS jamming in specific areas[3] or the disruption of wireless communication links used for drone control.[4] In addition, news reports suggest that some actors prepare in advance to dominate the electromagnetic spectrum, as indicated by reports on China's strategic planning.[5]

However, electromagnetic warfare remains largely the domain of specialists, and its effectiveness is often opaque and difficult to quantify. The EM spectrum lies beyond unaided human perception and depends entirely on instrumentation and analysis. Because most electromagnetic signals are imperceptible to the senses, EM operations remain mediated, abstract, and difficult to observe or attribute. When embedded within broader operations, quantifying electromagnetic activity and its effects is essential for generating situational awareness and guiding countermeasures.

This chapter begins with a historical overview of how the electromagnetic domain has shaped past operations and the lessons derived from them. It then defines the electromagnetic battlespace, incorporating not only position but also frequency, signal strength, and time as additional, operational dimensions. Next, it discusses and briefly analyses examples of electromagnetic anti-access and area-denial (A2/AD) activity in current conflicts. Finally, this chapter examines how to integrate all electromagnetic warfare disciplines to achieve area dominance, emphasising the need for broader knowledge and interdisciplinary education.

Historical Context

Understanding the role of the electromagnetic (EM) domain in anti-access and area-denial (A2/AD) strategies requires a historical perspective. In its early military applications, electromagnetic warfare operated in a sparsely used and largely uncontested spectrum. Since the advent of wireless communications,[6] the EM spectrum has been used in warfare, but its strategic significance expanded dramatically during the Second World War.

During this period, the use of the EM spectrum rapidly evolved beyond communications to include navigation, detection, and deception. Military operations became increasingly dependent on systems that relied on electromagnetic signals, and both sides began to exploit the spectrum not only to enhance their own capabilities but also to deny or disrupt those of the adversary. This marked the emergence of the EM spectrum as a contested battlespace, a precursor to what is now recognised as electromagnetic warfare.

A comprehensive account of this evolution is provided in Most Secret War,[7] written by Reginald Jones, one of the British scientists pioneering EM operations during the Second World War. His work documents not only the technological innovations of the time but also the operational secrecy required to stay ahead of the opponent. The Second World War thus offers a unique and well-documented case study of early electromagnetic warfare, as many records have since been declassified.

One of the most striking examples from Jones' account is the so-called "Battle of the Beams." German engineers developed a system of electromagnetic beams to guide bombers to precise targets in the United Kingdom, enabling accurate night-time bombing. In response, British forces transmitted deceptive beams to mislead the bombers, causing them to deviate from their intended flight paths. Once the Germans realised their beams were being jammed, believing the British had somehow 'bent their beams', they adapted by shifting frequencies and deploying new beam strategies. The British countered in turn, triggering a dynamic cycle of measure and countermeasure that exemplified the tactical interplay of electromagnetic warfare.

Another example of EM deception in the Second World War occurred during the German naval operation known as the Channel Dash. To conceal the withdrawal of their battlecruisers *Scharnhorst*, *Gneisenau*, and *Prinz Eugen* through the English Channel, the Germans employed a jamming strategy. In the days leading to the operation, they gradually increased jamming power to blend with routine background noise, thereby masking their activity.[8] The British failed to detect the gradual rise in background noise and did not register the cruisers' passage electromagnetically. Although other factors contributed to the operation's success, and the precise impact of the jamming remains subject to debate,[9] this case illustrates how electromagnetic actions were deliberately integrated into operational planning.

A contrasting form of electromagnetic deception, aimed at fabricating rather than denying information, was employed during Operation Overlord. During this operation, the Allies executed two deception operations, Glimmer and Taxable, that created the illusion of large naval armadas approaching the French coast at different locations.[10] These fleets did not exist physically and were not optically visible. Nevertheless, they appeared on German radar as credible threats. The resulting false alarms disrupted decision-making on the German side and diverted attention from the actual landing zones.

While the Second World War marked the birth of electromagnetic warfare as an indispensable tool, later conflicts continued to demonstrate its evolving application. After the Second World War, EM contestation continued to evolve during the Cold War. The Soviet 'Woodpecker' radar system reportedly projected an electromagnetic interference grid over the United States, with the full knowledge of the CIA.[11] This system illustrates how long-range EM interference became a strategic tool. After 9/11, EM operations increasingly targeted asymmetric threats. In conflicts, such as those in Afghanistan and Iraq, electromagnetic signals were used to disable adversaries' improvised explosive devices (IEDs), underscoring the continued relevance of EM operations in modern countermeasures.[12]

Together, these historical cases illustrate the transformation of electromagnetic warfare from a reactive, experimental practice into a deliberate and integrated component of military strategy, laying the foundation for its central role in contemporary A2/AD operations.

The Nature and Scope of Electromagnetic Warfare

Understanding electromagnetic warfare requires clarifying its conceptual boundaries, particularly in relation to the definition of the electromagnetic area as a contested battlespace. In this context, *area* does not refer to geographic space, as in most traditional forms of area denial, but to a multidimensional construct defined by parameters that govern how access, presence, and control are established and denied within the electromagnetic domain. Denial within this area constitutes a distinct form of anti-access and area denial (A2/AD). This connection arises because control of the EM area directly shapes an actor's ability to restrict or enable access to this operational space. At the same time, electromagnetic operations can also support other A2/AD strategies, for example by diverting projectiles from their intended targets or by disrupting navigation and targeting systems. The term 'electromagnetic area' is used here to emphasise this structured, contested character, where control over these parameters is central to military effectiveness. The conceptualisation of this electromagnetically contested area will be developed further in what follows.

To understand the operational impact of the EM domain, it helps to start with its basic characteristics. Historical examples reveal several fundamental characteristics of the electromagnetic domain. First, interventions within the EM area are essentially non-kinetic, manipulating sensors and information, yet they can enable or trigger kinetic outcomes under certain conditions and are often operationally decisive. For example, navigation interference can divert a bomber towards a friendly target, producing a kinetic effect. Similar misdirection is a current concern in Ukraine, where Russian electromagnetic operations are feared to affect targeting.[13]

Secondly, the EM area is perceived only through instruments that augment human senses, as most electromagnetic activity is instrument-mediated rather than directly observable. This explains how the Germans concealed the Channel passage of their cruisers by gradually blending a jamming signal into background noise, and how radar deception operations created virtual armadas to mislead the Germans during Overlord.

In military usage, electromagnetic warfare refers to actions that contest an opponent's use of the electromagnetic area. These include efforts to control, jam, spoof, seize, or destroy electromagnetic functions. From an operational perspective, such actions are typically categorised by purpose: (a) surveillance, (b) attack, and (c) defence.

Electromagnetic warfare is never a military goal in itself, but always part of a broader mission or strategy, as demonstrated by the examples above, where electromagnetic activity served as a means to bomb, escape, or invade. Since general definitions of warfare also apply to electromagnetic warfare, it is the contested nature of the area that makes it distinct. If electromagnetic warfare is defined by battles fought in the electromagnetic area, then other forms of warfare must also be included, such as (satellite) navigation warfare, communication warfare, and optronic warfare, all of which rely on wireless access to the area. Less formally categorised applications should also be considered, such as counter-IED operations that aim to prevent improvised explosive devices (IEDs) from being triggered via electromagnetic signals.

Although it is precisely the concept of a contested area that makes electromagnetic warfare a distinct form of conflict, it is not commonly analysed from an anti-access and area denial perspective. Existing research tends to focus on operational and technical solutions for specific scenarios.[14] While such solutions expand the tactical toolbox for acting on the electromagnetic battlefield, they do not alter the nature of the area itself. A more foundational analysis of the electromagnetic area is therefore essential to guide the effective deployment of countermeasures and to support strategies aimed at dominating or denying access to this space.

The electromagnetic battlefield being both multidimensional and used by multiple actors, calls for a general conceptual model. Such a model should support future developments in operational preparation and provide a basis for assessing the impact and effectiveness of electromagnetic actions across different operational contexts.

The Present State of the Electromagnetic Domain

Despite being imperceptible to human senses, the electromagnetic area is extensively used for both military and civilian applications. These applications depend on the assumption that access to the electromagnetic area is both available and unimpeded. But what constitutes 'unimpeded' use? And what problems arise when diverse users, military and civilian alike, operate simultaneously in the same electromagnetic environment? Before turning to a more abstract and conceptual definition of the electromagnetic area – one that allows electromagnetic warfare to be situated within an anti-access and area denial context – it is essential to examine the contemporary challenges that define its operational character.

These challenges can be summarised by the so-called 5Cs: the area is (a) connected, (b) congested, (c) contested, (d) constrained, and (e) cluttered as introduced by the UK Ministry of Defence Strategic Trends Programme.[15]

First, the electromagnetic area is increasingly connected. A growing number of systems, both military and civilian, rely on wireless connections for communication, navigation, detection, and collision avoidance. This dense web of interdependent systems makes the electromagnetic area a critical enabler of modern life and operations.

Second, this connectivity contributes directly to congestion. The proliferation of electromagnetic signals, especially from high-bandwidth applications such as modern communication systems, places growing demands on available spectrum. The result is a crowded environment in which access becomes increasingly difficult to coordinate and guarantee.

Third, this congestion leads to contestation. Multiple actors seek unhindered use of the electromagnetic area, and their interests often conflict. In military contexts, this contest takes the form of deliberate interference such as jamming, spoofing, or seizing control of adversary systems. In the civilian domain, contestation emerges from the expanding demand for spectrum, as seen in the rollout of 5G and future wireless technologies. The contested nature of the electromagnetic area is therefore not only a matter of military confrontation but also of competing civilian priorities. This contested character is closely tied to the logic of anti-access and area denial operations.

Fourth, the electromagnetic area is constrained. In response to growing demand and potential interference, the International Telecommunication Union (ITU) has established a global spectrum allocation framework.[16] This regulatory structure assigns specific frequency bands to defined applications, limiting how and where certain systems may operate.

Finally, the electromagnetic area is cluttered. Even within regulated and coordinated environments, uncontrolled activity persists. This clutter arises from both technical and natural sources, including electromagnetic interference from electronic devices, spurious emissions, intermodulation products, signal reflections, and environmental phenomena such as atmospheric,[17] ionospheric, or solar disturbances.[18] These factors introduce unpredictability and noise into an already complex environment.

Together, the 5Cs illustrate the operational conditions that shape the electromagnetic area today. They also underscore the need for a more precise conceptual understanding of this area, not only as a medium of transmission but as a contested and dynamic battlespace. Overall, this shows that the electromagnetic area must be understood not only technically but also conceptually.

Having been characterised as dynamic, multidimensional, constrained, congested, and heavily contested by various actors, conceptualising the electromagnetic area through a generalised model enables shared operational use. Such a structured approach supports the integration of the electromagnetic area into mission planning, execution, and adjustment at both tactical and strategic levels.

In the context of anti-access and area denial (A2/AD), such a model must also support the ability to assess electromagnetic dominance. This requires an operationally derived form of situational awareness. It must integrate one's own mission requirements with the observed behaviour and intent of the opponent, by quantifying four categories: (1) the parts of the electromagnetic area necessary for the unhindered use of one's own systems, (2) the parts used by the opponent's systems, (3) the parts cluttered by unintentional or uncontrolled activity and (4) the parts of civil importance.

Once this electromagnetic situation is established, i.e., when situational awareness has developed through the synthesis of operational needs and adversary activity, it becomes possible to formulate measures that block or deny the opponent's access to specific parts of the area. Tracking and assessing these measures requires them to be quantifiable and integrated into the operational situational picture.

To support this, we propose a conceptual model of the electromagnetic area defined by two structuring elements: its dimensions and its actors. As illustrated in Figure 6.1, the electromagnetic area can be shaped along four key dimensions: position, time, signal strength, and frequency. It is shaped by the presence, activity, and aims of four types of actors: friendly forces (ally), adversaries (opponent), civil users (civil), and the natural environment (nature).

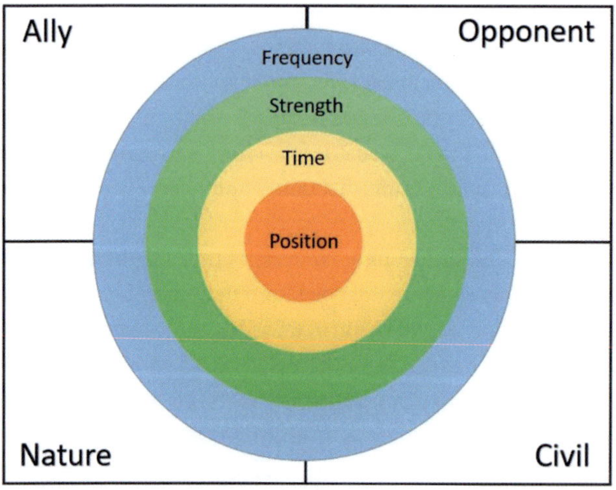

Figure 6.1: Conceptual model of the electromagnetic area (EMA) as created by the author of this article

The circular core of Figure 6.1 represents the four-dimensional electromagnetic area as it is used, contested, or influenced by the surrounding actor types. This area spans the space of all relevant values of position ($\mathscr{P}$), time ($\mathscr{T}$), frequency ($\mathscr{F}$) and signal strength ($\mathscr{S}$), and is shaped by the presence, activity, and aims of the actors positioned around it. The total electromagnetic area (EMA) comprises the full set of all possible combinations of these four dimensions. This is expressed symbolically as:

$$EMA = \mathscr{P} \times \mathscr{T} \times \mathscr{F} \times \mathscr{S} \quad (1)$$

This formulation defines the electromagnetic area as the product space of all possible values across the four structuring dimensions. It represents the theoretical totality of the electromagnetic environment: every frequency at every signal strength, at every moment in time, and at every position in space. Consequently, the EMA is vast and inherently unbounded.

From an operational perspective, however, it is neither feasible nor meaningful to dominate the entire EMA. Instead, mission planning requires identifying those specific subsets within the EMA that are relevant to a given objective, such as the subset required for the uninterrupted operation of own systems, or the subset on which the opponent depends.

To operationalise this theoretical space, the EMA can be decomposed into discrete elements, each corresponding to a specific combination of the four dimensions: The actual EMA refers to any constrained portion of this space, such as a frequency band, a time window, a geographic sector, or combinations thereof, representing the subset of conditions under which an effect is generated or observed.

Beyond this formal definition, the EMA can also be described in operational terms. The electromagnetic area can then be conceptualised as a four-dimensional design space defined by position, time, frequency, and signal strength (P, T, F, S). The actual EMA is the subset of this space that is occupied at a given moment by, for instance, real emissions, receivers, and propagation conditions. In practice, the EMA is continuously shaped by transmitters (intentional and unintentional), sensed by receivers, and constrained by geometry and environmental factors.

Operationally, the EMA is best treated as a shared and contested environment structured by four interacting actor perspectives (Figure 6.1):

- Ally and Opponent: competing military uses and countermeasures,
- Nature: propagation, attenuation, clutter, ionospheric and weather effects, noise,
- Civil: shared spectrum use, infrastructure dependencies, regulatory constraints, and related forms of civil electromagnetic activity.

These perspectives determine which portions of the EMA are usable, vulnerable, or denied at a given position, over time, across frequency and as a function of signal strength. Thereby it connects the abstract (P, T, F, S) space to a fused operational situational picture.

This formulation provides a conceptual framework for assessing electromagnetic dominance in contested environments. By explicitly linking the physical dimensions of the electromagnetic area to the intentions and activities of relevant actors, it enables planners to identify critical dependencies, anticipate interference, and prioritise protective or disruptive measures. The model thus supports both situational awareness and decision-making in mission planning and execution.

Structuring the Electromagnetic Area

To operationalise the previously introduced concept of the electromagnetic area (EMA), it is necessary to specify both its physical dimensions and the actors that shape, exploit, or contest it.

Physical Parameters: Position, Time, Frequency and Signal Strength

The EMA, as defined in Figure 6.1 and formalised by (1), is structured by four physical parameters: position, time, frequency and signal strength. Among these, position is particularly salient from an anti-access/area denial (A2/AD) perspective, the central focus of this volume. Here, position denotes the physical location or spatial domain in which electromagnetic signals manifest. It is usually specified using a three-dimensional geodetic coordinate system, defined by latitude, longitude, and altitude.

To further clarify the nature of electromagnetic signals, it is instructive to consider the concept of the electromagnetic spectrum, which classifies signals by their frequency or, equivalently, their wavelength. While the electromagnetic area is defined as a four-dimensional construct, the spectrum is characterised solely by one of these parameters. Frequency and wavelength of electromagnetic signals are inversely related through the speed of light.

As illustrated in Figure 6.2, the wavelength (or equivalently, the frequency) of an electromagnetic signal determines the functional domain of that spectral segment. It defines not only the physical characteristics of the signal but also the type of application or system that operates within that range. Signals of similar wavelength are susceptible to mutual interference, making spectral proximity a critical factor in electromagnetic conflict. Frequency also plays a central role in delineating the operational domains of electromagnetic warfare. For example, navigation warfare targets the frequency bands used by satellite-based positioning systems. Other subfields such as communication warfare and optronic warfare are similarly defined by their dependence on distinct portions of the spectrum.

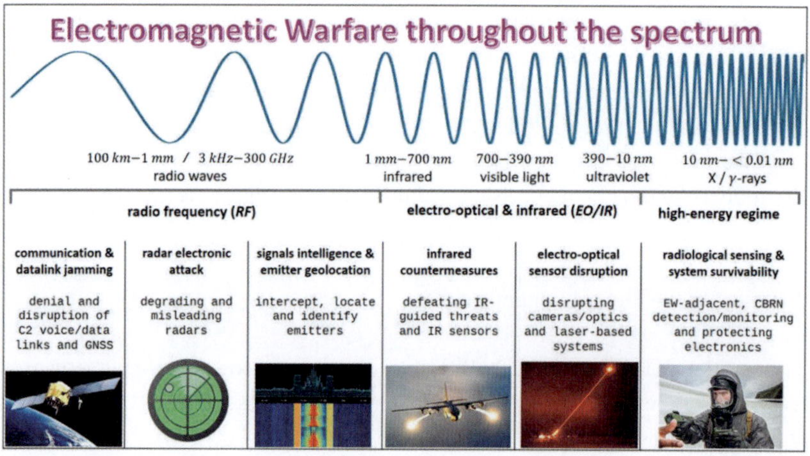

Figure 6.2: The electromagnetic spectrum with its applications as created by the author of this article with images obtained from Wikimedia Commons[19]

Another defining parameter of the electromagnetic area is signal strength, or more precisely, power density. This refers to the power carried by a signal at a specific geographic position. It depends on factors such as transmitted power (P_t), distance from the source (R), atmospheric attenuation (L_{atm}), and effective receiving area (A_e). The signal strength at a given position (S_p) is described by the Friis transmission equation,[20] expressed as:

$$S_P = \frac{P_t A_e}{4\pi R^2 L_{atm}} \tag{2}$$

Time introduces a uniquely strategic dimension to the electromagnetic area. Unlike the other parameters, it governs not only when signals are present but also how patterns evolve, escalate, or remain concealed. Signals are transmitted at discrete moments and propagate through space, making temporal control essential for synchronisation, coordination, and deception. Time is referenced using Coordinated Universal Time (UTC), which provides a standardised framework for aligning electromagnetic activity across platforms and domains. As previously discussed, time can also be exploited to shape perception – for example by the gradual modulation of background noise to mask operational movements.

Actor Types and their Operational Roles

The following overview categorises key actor types by their typical use of the electromagnetic area, their operational stakes, and the reliability of their behaviour within it. Actors give operational meaning to the electromagnetic area by shaping how it is used, contested, or protected. Each engages with it to pursue specific objectives – whether to communicate, sense, disrupt, deceive, or deny. These include state and non-state entities, military and civilian users, and both friendly and adversarial forces.

Actors can be distinguished by their intent, capability, and mode of engagement. Some aim to exploit the electromagnetic area for operational gain – through communication, navigation, or sensing – while others seek to disrupt or deny access, using techniques such as jamming, spoofing, or physical destruction. The same actor may shift roles depending on context, mission phase, or technological capability.

From a planning perspective, mapping the actor landscape is essential. It helps anticipate contested zones, identify critical dependencies, and assess risks of interference or escalation. Combined with the physical parameters of the EMA, actor analysis supports electromagnetic situational awareness and mission design and informs the prioritisation of protective or disruptive measures. Table 6.1 summarises this actor landscape across four representative categories – allies, opponents, civilian users, and natural phenomena – highlighting their typical use of the electromagnetic area, their operational stakes, and their expected reliability.

Table 6.1 Representative actor categories by use, stake, and reliability

	Use	Stake	Reliability
Ally	Intentional for: • Communication • Data sharing • Situational picture • Navigation	High, winning the battle and disrupting the opponent	Very high, as it is assumed that the allied parties work together for optimal use of the electromagnetic area.
Opponent	Intentional for: • Communication • Data sharing • Situational picture • Navigation	High, winning the battle and disrupting the allies	Very low, as it is assumed that the enemy parties aim for disrupting use of the electromagnetic area by the allied parties.
Civil	Intentional for: • Communication • Data sharing • Situational picture • Navigation • Payments • Door locks • Park assists • Additional civilian applications	Very High, disruption of society of great impact	Moderate, as civilian use generally adheres to regulatory constraints.
Nature	Unintentional natural processes	Low	Unreliable, its effects are unintentional. Nevertheless, forecasting algorithms can give for example (space) weather predictions.

Operational Control of the Electromagnetic Area

Identifying the actors in the electromagnetic area is only a first step. To shape, contest, or dominate that space, actors must deploy technical systems that enable sensing, manipulation, and coordination across the EMA's defining parameters. What follows outlines the operational requirements for such control, from sensor and effector technical capabilities to their strategic and tactical, adaptive deployment.

Sensing, Effecting, and Command in the Electromagnetic Area

NATO's APP-34 standard on area access control emphasises the need for integrated sensors, effectors, and command systems to manage physical space. By analogy, it is reasonable to assume that these same principles apply to the electromagnetic area: operational control depends on the ability to observe and influence signals across time, space, frequency, and strength.

Effective control begins with sensing. Systems such as Electromagnetic Support Measures (ESM) and Electromagnetic Intelligence (ELINT) are essential for detecting and characterising signals from actors in the area. As a minimum, these systems must be able to measure frequency and signal strength across time and location. No single system can monitor the entire electromagnetic spectrum at all power levels; trade-offs are therefore unavoidable and should be guided by operation-specific priorities.

Effectors are similarly constrained and must be matched to the spectral characteristics of their targets. Jammers, for example, may be optimised for radar disruption, while lasers target electro-optical systems. Both operate within the electromagnetic domain, but in distinct regions of the spectrum. Therefore, effective area denial depends on aligning effector capabilities with the relevant spectral segment and the operational context.

Command and Control (C2) systems are critical for orchestrating sensing and effecting functions in real time. Changes in frequency or power occur near-instantaneously, exceeding human reaction times. Therefore, automated or semi-automated C2 architectures are essential for timely counteraction, adaptive coordination, and sustained control. The model proposed in this contribution provides a structured framework to assess which parts of the electromagnetic area can be sensed or manipulated by specific systems, clarifying both operational possibilities and the inherent limitations of A2/AD.

Strategic and Tactical Application of A2/AD in the Electromagnetic Area

The PAIM loop (Predict–Act–Indicate–Model), developed by Mulder,[21] provides a structured framework for strategic and tactical decision-making in the electromagnetic area. Figure 6.3 is adopted from Mulder and visualises the structure of the loop and the interconnections between its states.

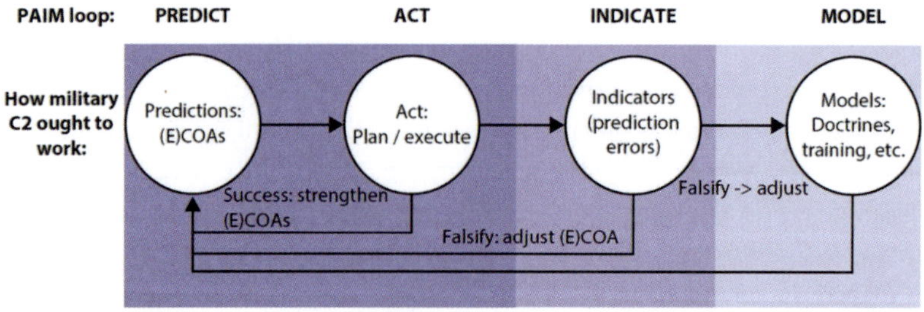

Figure 6.3: The PAIM loop, a framework for electromagnetic A2/AD operations from Mulder[22]

The loop begins with prediction, PAIM's most distinctive feature. Operational advantage depends on a priori knowledge of the electromagnetic environment, not to achieve dominance over the entire spectrum, but to control the portion relevant to a specific operation.

Based on this situational awareness, actors can take targeted actions, such as jamming selected frequencies with sufficient power to deny the opponent access. These actions must align with technical capabilities and prior intelligence.

Electromagnetic conflict is increasingly adaptive;[23] opponents may shift to other spectral regions in response. Detecting such shifts, represented in the model as prediction errors, requires continuous sensing, remodelling of key parameters, and robust decision support. This feedback loop enables new or adjusted denial measures and helps sustain control over the electromagnetic area of interest. The model presented in this article serves as a structured foundation for such management by identifying the parameters that govern access and control.

Visualising the electromagnetic area, for example through a dashboard, supports real-time awareness and operational planning. Mapping parameter values and operational constraints enables faster, better-informed decisions. A standardised model also facilitates interoperability and combined operations by enabling partners to share information and coordinate actions across domains.

Electromagnetic Dominance and Operational Application

Operational control enables actors to sense, manipulate, and coordinate activity within the electromagnetic area; the broader objective is to achieve electromagnetic dominance, the ability to shape the environment to support operational objectives while denying adversaries equivalent access. This section explores how the proposed EMA model supports dominance across three operational domains: manoeuvre warfare, navigation warfare, and information warfare.

Manoeuvre Warfare

Incorporating the electromagnetic area into manoeuvre warfare, particularly in the context of A2/AD operations, requires particular attention to an actor's physical position within the theatre of battle. Synchronising electromagnetic effects with manoeuvre elements constrains the model's *position* parameters to the physical area of battle. At the same time, the parameters frequency, signal strength and time are adjusted to improve electromagnetic dominance.

Consider an incoming missile equipped with an active radar seeker. Transmitting a stronger signal at the same frequency can jam the seeker, effectively denying the missile access to that portion of the EMA and preventing target acquisition. However, the missile may adapt by switching frequency bands, reactivating its radar in another part of the spectrum. This requires rapid detection and reconfiguration of countermeasures. The principal A2/AD challenge is therefore not only situational awareness and reaction speed but also the availability of technical systems capable of detecting and countering adaptive threats in real time.

But dominance in the EMA can also limit an opponent in more traditional manoeuvre warfare. By disabling the opponent's use of wireless communications, coordinated manoeuvres on the battlefield area are much more complicated and possibly completely denied.

Navigation Warfare

Navigation warfare involving Global Navigation Satellite Systems (GNSS) is a core component of electromagnetic warfare. Beyond its importance for radio navigation, GNSS also acts as the primary source for precision timing. Precision timing is indispensable in battle as it is necessary for interoperability between systems, platforms, and partners. GNSS signals operate within fixed, internationally allocated frequency bands, making them predictable and therefore vulnerable in A2/AD scenarios. Usability of the GNSS signal is determined by its presence (or absence) at a given power level, time, and location. Denying GNSS access therefore involves manipulating these parameters to disrupt or degrade signal reception. Lubbers examines the effects and effectiveness of manipulating power density to jam or spoof GNSS signals; his findings are presented in Chapter 7 of this volume.

Information Warfare and CEMA Integration

In today's interconnected battlespace, information is decisive. Electromagnetic warfare increasingly integrates cyber operations under the umbrella of Cyber and Electromagnetic Activities (CEMA), addressing both the transmission of information via electromagnetic means and the integrity and authenticity of that information.

Electromagnetic dominance is typically defined as the ability to access and use the EMA. In A2/AD contexts this often translates into denial measures. When the objective is information control, dominance may also involve spoofing, causing the opponent to receive signals that appear normal but contain false or misleading content.

Detecting spoofing differs fundamentally from detecting jamming. While jamming can be detected by measuring anomalous electromagnetic activity, the detection of spoofing activities requires analysis of signal content. Although spoofing detection is not fully captured by the current EMA model because it depends on information-level attributes beyond the model's physical parameters, future integration of spoofing and jamming detection capabilities is strongly recommended in order to enhance electromagnetic situational awareness and operational resilience.

Conclusion

After initial framing and providing some historical context of electromagnetic warfare, this chapter states that for A2/AD operations, dominance in the electromagnetic area (EMA) is crucial. However, full control of the EMA is neither necessary nor realistic. Instead, operations concentrate on mission-relevant subsets of the EMA. The EMA is modelled as a four-dimensional space parameterised by position, time, frequency, and signal strength. This enables systematic mapping and fusion of observations and effects. The EMA is used by various actors: by friendly and opponent forces, civil users and environmental conditions. The latter is here considered as an unintentional actor, whose effect on the EMA includes propagation conditions such as terrain, clutter, weather, and noise.

Operationally, EMA understanding is sustained through the operational cycle. This loop includes information harvesting, situational observations, decision making and the creation of effects. This framework helps to identify the relevant EMA subset. This is briefly illustrated in the context of manoeuvre, navigation, and information warfare. In this view, electromagnetic dominance is achieved when control of the relevant EMA conditions is realised, enabling friendly operations while denying opponents access.

Practical, Policy and Doctrinal Implications

The most prominent practical implication of this work is to make the electromagnetic area visible and measurable in routine planning and execution. This allows the mission-relevant EMA subset, and the degree of electromagnetic dominance within it, to be assessed and managed alongside other operational conditions. In line with this chapter's title, the intent is to make electromagnetic effects observable for commanders and staff, rather than implicit and easily overlooked.

This requires translating the EMA model into a shared graphical representation, for example a dashboard integrated into the common operational picture. The display should combine friendly, adversary, civil, and environmental factors, and update as conditions change. Before implementation, further work is needed to make the model practically usable. This includes refining parameter definitions and quantifying dominance for both planning and execution, supported by validation in exercises.

EMA conditions may vary rapidly in time, space, and frequency, while detectability is governed by signal strength; therefore, the functionality should be embedded in existing situational awareness systems. Standardised representations and terminology will also improve interoperability in joint and combined operations by enabling unambiguous information sharing.

Finally, electromagnetic dominance should be embedded in routine practice. This requires inclusion in doctrinal and tactical publications, and sustained education and training across ranks and disciplines within the armed forces. Military educational institutions have a key responsibility for building and sustaining this knowledge base.

Notes

1 Feynman, Leighton, and Sands. *The Feynman Lectures on Physics, Vol. II.*
2 Price, *The History of US Electronic Warfare.*
3 *GPSJAM.*
4 TWZ.
5 Mizzima.
6 Hezlet, *The Electron and Sea Power.*
7 Jones, *Most Secret War.*
8 Ford, *Run The Gauntlet.*
9 Giessler, "The Breakthrough of the 'Scharnhorst'"; Watson-Watt, "Battle Scars of Military Electronics".
10 Price, *The History of US Electronic Warfare.*
11 Freeland, *Under an Ionized Sky.*
12 Ambroziak, "Analysis of the Usage Possibility of the Electromagnetic Curtain".

[13] Withington, "Jamming JDAM: The Threat to US Munitions from Russian Electronic Warfare".

[14] Kong, "Analysis of Electromagnetic Denial Technology and Characteristics."; Hofstetter and Wojciechowski, "Electromagnetic Spectrum Survivability"; Clifford, "Maneuver in the Electromagnetic Domain"; Clark, *Winning in the Gray Zone.*

[15] Defence, UK Ministry of, *Strategic Trends Programme*.

[16] *International Telecommunication Union.*

[17] Norin, "Anomalous Propagation and the Sinking of the Russian Warship Moskva".

[18] Galvan-Tejada, "Space Weather Observations".

[19] NASA, *GPS Satellite (artist's impression)*; Vignoni, *Radar Icon Green*; f4exb, *SDRpp FM Subcarriers*; The National Guard, *107th Lights up the Sky*; UK Ministry of Defence, *Dragonfire Laser System Test Firing*; AEA Imagebank, *Radiation Check.*

[20] Friis, "A Note on a Simple Transmission Formula".

[21] Mulder, "How to Think and Act Faster Than the Enemy."

[22] Ibid.

[23] Haigh and Andrusenko, *Cognitive Electronic Warfare.*

References

Ambroziak, R., S. J. Ambroziak, and M. Milczanowski. 2017. "Analysis of the Usage Possibility of the Electromagnetic Curtain." *Scientific Journal of Polish Naval Academy,* 210 (3), 5–12.

Clark, Bryan, Mark Gunzinger, and Jesse Sloman. 2017. *Winning in the Gray Zone; Using Electromagnetic Warfare to Regain Escalation Dominance.* Center for Strategic Budgetary Assessments.

Clifford, John. 2025. "Maneuver in the Electromagnetic Domain." *Journal of Electromagnetic Dominance (JED).*

f4exb. *SDRpp FM subcarriers.* Wikimedia Commons.

Feynman, Richard P., Robert B. Leighton, and Matthew Sands. 1964. *The Feynman Lectures on Physics, Vol. II: Mainly Electromagnetism and Matter.* Reading, Massachusetts: Addison Wesley.

Ford, K. 2012. *Run The Gauntlet; The Channel Dash 1942.* Osprey Publishing.

Freeland, E, and C Carnicom. 2018. *Under an Ionized Sky From Chemtrails to Space Fence Lockdown.* Port Townsend WA: Feral House.

Friis, Harald T. 1946. "A Note on a Simple Transmission Formula." *Proceedings of the I.R.E. and Waves and Electrons.*

Galvan-Tejada, G. M et al. 2022. "Space Weather Observations and HF Transmissions around the 2021 autumn Equinox." *19th International Conference on Electrical Engineering, Computing Science and Automatic Control (CCE)* 1–7.

Giessler, Helmuth. 1961. "The Breakthrough of the 'Scharnhorst' –Some Radio-Technical Details." *IRE Transactions on Military Electronics.*

GPSJAM. https://gpsjam.org/ (geopend 12 28, 2025).

Guard, The National. *107th Lights up the Sky.* Wikimedia Commons.

Haigh, Karen, and Julia Andrusenko. 2021. *Cognitive Electronic Warfare: An Artificial Intelligence Approach.* ARTECH HOUSE.

Hezlet, Arthur. 1975. *The Electron and Sea Power.* Peter Davies Ltd.

Hofstetter, Jeremy, and Adam Wojciechowski. 2020. "Electromagnetic Spectrum Survivability in Large-Scale Combat Operations." *Infantry, 109 (4), 21, 24.*

Imagebank, AEA. *Radiation Check.* Wikimedia Commons.

International Telecommunication Union. sd. https://www.itu.int/en/mediacentre/backgrounders/Pages/itu-r-managing-the-radio-frequency-spectrum-for-the-world.aspx (accessed: January 27, 2026).

Jones, RV. 1978. *Most Secret War: British Scientific Intelligence 1939–1945.* Hamilton.

Kong, Lingmin et al. 2024. "Analysis of Electromagnetic Denial Technology and Characteristics." *Journal of Physics: Conference Series,* 2803(1), 12019.

Mizzima. https://eng.mizzima.com/2025/12/26/29549 (accessed: Januari 7, 2026).

Mulder, R.E. 2024. "How to Think and Act Faster Than the Enemy." *Militaire Spectator.*

NASA. *GPS Satellite (artist's impression).* Wikimedia Commons.

Norin, Lars, Niklas Wellander, and Abhay Devasthale. 2023. "Anomalous Propagation and the Sinking of the Russian Warship Moskva." *Bulletin of the American Meteorological Society,* 104(12), 2286–2304.

Price, Alfred. 1984. *The History of US Electronic Warfare.* Association of Old Crows.

TWZ. https://www.twz.com/news-features/inside-ukraines-fiber-optic-drone-war (accessed: December 28, 2025).

UK Ministry of Defence. *Dragonfire Laser System Test Firing.* Wikimedia Commons.

UK Ministry of Defence. 2015. *Strategic Trends Programme; Future Operating Environment 2035.* London: Ministry of Defence.

Vignoni, David. *Radar Icon Green.* Wikimedia Commons.

Watson-Watt, Robert. 1957. "Battle Scars of Military Electronics–The Scharnhorst Break-Through." *IRE Transactions on Military Electronics.*

Withington, Thomas. 2023. "Jamming JDAM: The Threat to US Munitions from Russian Electronic Warfare." *www.rusi.org.* 6 June. https://www.rusi.org/explore-our-research/publications/commentary/jamming-jdam-threat-us-munitions-russian-electronic-warfare (accessed: November 30, 2025).

CHAPTER 7

Navigation Warfare

Access Denial to Position, Navigation and Timing Information

Barend Lubbers

Abstract

In recent years Global Navigation Satellite Systems (GNSS), such as the American Global Positioning System (GPS), have become essential for military operations. These systems depend heavily on very weak satellite signals, which makes them susceptible to interference. Adversaries can exploit this vulnerability through methods such as GNSS jamming and spoofing. With relatively low power, a jammer can impact large areas, making it an effective technique to disrupt the operations of adversaries. This explains why GNSS jamming is employed in almost all current military conflicts. While the required transmission power is relatively low, the origin of the jamming can still be easily detected and localised, making it possible to neutralise the threat. To counter this, one might turn to spoofing. Although spoofing is often employed to deceive adversaries by transmitting counterfeit GNSS signals, it can also serve as a method of anti-access. Spoofing requires significantly less transmission power than classical jamming, making it much more difficult to detect and localise its source. This could explain the significant rise in spoofing incidents, as can be observed in the media coverage over recent years.

Keywords: Navigation warfare; Global Positioning System; Global Navigation Satellite System; Jamming; Spoofing; Position navigation and timing information

Introduction

Historically, positioning and navigation were specialisations requiring years of intensive training. The introduction of the Global Positioning System (GPS) radically changed this. Nowadays everyone can determine their position with an accuracy an order of magnitude better while much faster than the specialists decades ago. GPS also provides time synchronisation on a par with atomic clocks for a fraction of the price. It is, therefore, no wonder that GPS transformed the use of Position Navigation and Timing (PNT) information in both civil and military applications.

In military operations, GPS proved to be a crucial component supporting functions that extend beyond classical navigation.[1] For example, in targeting and weapon delivery systems, GPS allows for pinpointing coordinates and striking with unprecedented precision. This enhances the effectiveness of weapon systems while minimising collateral damage. Furthermore, GPS allows the coordination of operations across multiple domains, allowing communication and synchronisation among units. Although GPS is the best known and most used satellite navigation system, there are similar systems operational.

Currently there are four satellite navigation systems providing global coverage, commonly known as Global Navigation Satellite Systems (GNSS). The most established and widely used among these is the Global Positioning System (GPS). The other systems are the Russian GLONASS, European Galileo and Chinese BeiDou. All these GNSSs rely on satellites in Medium Earth Orbit (MEO) at an altitude of roughly 21,000 km. This altitude represents a balance between the number of satellites, and the required signal transmission power.[2] Because of the large distance to the satellites, the signals received at the earth surface are extremely weak, typically in the order of 10^{-16} Watt. These weak signals pose a significant vulnerability as the signals are easily overwhelmed by interference, leading to reduced performance of the receiver or even preventing a position fix. GNSS uses spread spectrum techniques to suppress interference, requiring a narrowband interferer to be approximately 40 dB (10,000 times) stronger than the civil GNSS signals, or about 50 dB (100,000 times) stronger than the military signals. Despite the effectiveness of these spread spectrum techniques, interferers still require little power to suppress the desired signals, leaving the receivers vulnerable to interference.

Interference can occur unintentionally, for example due to harmonics from other radio signals or by faulty electronic devices. In some cases, however, interference is intentional with the aim of suppressing GNSS signals. Suppressing GNSS signals is a form of Anti-Access (A2) where the access to the GNSS signals is suppressed, preventing receivers from providing (reliable) PNT information. The suppression of GNSS signals is commonly referred to as GNSS jamming, which is intensively used in conflict areas such as Ukraine, Russia and Israel, amongst others.[3]

Next to jamming there is a more sinister form of interference, called spoofing. With spoofing, counterfeit signals are used to mislead the PNT information provided by GNSS receivers. This allows the attacker to interfere with decision processes. Spoofing, however, can also be used as an A2 technique. Spoofing has a significant benefit over conventional jamming as it requires only a fraction of the power compared to conventional jamming. By transmitting less power, the attacker is more difficult to detect and locate and therefore to eliminate. This may explain the rapidly increasing intensity of GNSS spoofing as observed by monitoring stations[4] and published in several news articles[5] or websites.[6]

A2 of PNT information results in a reduction of operational efficiency of military operations, for example with the use of GPS guided munitions.[7] It may also have a large impact on civil society if employed as part of hybrid warfare. Many critical infrastructures such as (data) communication, energy and financial networks are dependent on the timing provided by GNSS. Also, emergency services, agriculture and transport are avid users of PNT information.[8] A2 of PNT information may, therefore, have a large impact on both military operations and civil society.

In this chapter we explore the use of GNSS jamming and spoofing as a tool for A2 against reliable GNSS based PNT information. First jamming and its effects on a GNSS receiver is explained and quantified. Then GNSS spoofing and its effects are explained. Finally, it is shown by experiments with a GNSS signal simulator and a common GNSS receiver that spoofing requires significantly less power to achieve the same effect as a conventional jammer, explaining the sharp increase in spoofing observed in conflict areas.

Navigation Warfare

The interference with GNSS signals in conflict areas is part of navigation warfare (NavWar). According to STANAG 4621,[9] NavWar is defined as "preventing the hostile use of PNT information while protecting the unimpeded use of the information by NATO forces and preserving peaceful use of this information outside the area of operations". However, NATO's adversaries are unlikely to conform to this definition, particularly the aspect concerning "preserving peaceful use of this information outside the area of operations". For instance, civil aviation frequently experiences GNSS jamming even far from active conflict zones.[10] This restriction on operational use should not be included in a definition; it belongs in doctrine or standard operating procedures instead. Moreover, the NATO definition seems to omit the manipulation of PNT information, which is an important factor that should be included. Therefore, the following revised definition of navigation warfare is proposed and used in this chapter: *"Preventing access to or manipulating/falsifying PNT information of hostile forces while protecting the access and integrity of PNT information of friendly forces."*

PNT information can be subject to access denial and manipulation through various means. For example, an ancient form of NavWar which is still relevant today is the alteration or removal of road signs. For instance, following the invasion of Ukraine, the Ukrainian government removed or modified road signs to confuse the Russian invaders.[11] Although NavWar can target any system providing PNT information, nowadays most NavWar activities focus on GNSS. The reason is two-fold: first GNSS is the most used PNT information provider and secondly GNSS has some inherent vulnerabilities which can easily be exploited.

The first vulnerability is the low power at which the signals are received, as explained in the introduction. The second vulnerability results from the open nature of the civil signals. These civil signals are unencrypted and usually do not have a form of authentication. An exception is the Galileo open service, which features so-called Open Service Navigation Message Authentication (OSNMA) which provides authentication of the navigation message. Authentication of the navigation message, however, is not as secure as encryption of the entire signal,[12] known as Signal Access Control (SAC), which is implemented by the military GPS signals and Galileo Public Regulated Service (PRS). SAC serves a dual purpose: it restricts access to authorised users, and it makes it almost impossible to generate counterfeit signals and thus serving as an Anti-Spoofing (AS) method. The lack of encryption or authentication of the open/civil signals provides adversaries a vulnerability which can be exploited.

NavWar aimed at GNSS includes a wide range of activities, for example electromagnetic warfare (EW) and cyber & electromagnetic activities (CEMA), such as jamming and spoofing. It also includes conventional cyber warfare, for example aimed at the GPS Command and Control (C2) infrastructure. It may even include kinetic attacks aimed at GNSS control and space segment components, such as satellites, control, reference and upload stations. This chapter, however, focuses only on EW and CEMA NavWar A2 methods against reliable PNT information at the user level, in particular jamming and spoofing.

GNSS Jamming

GNSS jamming aims to deny access (A2) to GNSS signals, preventing receivers from providing accurate PNT information. A GNSS jammer emits a Radio Frequency (RF) signal that matches or overlaps with the frequencies of GNSS signals. As a result, this jamming signal interferes with the desired signals, leading to reduced accuracy of the PNT information provided by the receiver, or potentially rendering the receiver completely incapable of providing any PNT information at all.

Jammers can use various signal modulations, including continuous wave (CW), swept continuous wave/chirp, (white) Gaussian noise, phase or frequency modulation, pulsed signals, and matched spectrum or a combination of these.[13] Each of these signal types has its own advantages and drawbacks. For instance, while a CW jammer is relatively simple to implement, it can also be easily countered, for example with a notch filter. As a result, CW jamming is not commonly used. A more frequently used waveform is a chirp signal,[14] where the frequency of a CW is rapidly changing within a certain bandwidth. Although these jammers are more difficult to suppress, advanced GNSS receivers are equipped with adaptive notch filters[15] that estimate the jamming signal and apply a notch filter to mitigate its effects. Another form of jamming uses (white) noise with a certain bandwidth centred on a specific frequency, often referred to as Additive White Gaussian Noise (AWGN) or simply noise jamming. This kind of jamming is more difficult to eliminate and the effect on a generic receiver is usually accurately predictable. Because of this predictability, the effect of noise jamming on a receiver makes a suitable example for analysis.

The effect of a noise jammer can be effectively assessed by its influence on the carrier-to-noise ratio (C/N_0). The C/N_0 is defined as the ratio of the total received power from the desired signal C to the power spectral density of the (thermal) noise N_0. Most GNSS signals have a nominal C/N_0 range between approximately 38 to 50 dB-Hz. The lower number is for satellites positioned at a low elevation, close to the horizon, while the higher figure is for satellites located directly overhead (at zenith). To predict the impact of a noise jammer, the effective C/N_0 or $(C/N_0)_{eff}$ is introduced. This $(C/N_0)_{eff}$ is the carrier-to-noise ratio where the power of the noise is increased by the noise caused by the jammer. This increase in noise power is depending on the received power, the bandwidth and central frequency of the noise jammer. For narrowband jammers, where the bandwidth of the jammer is smaller than that of the desired signal, the increase in noise spectral density can be accurately determined by a model.[16] For broadband noise jamming, where the bandwidth of the jammer is larger than that of the desired signal, the model can be simplified by adding the jammer's noise spectral density at the location of the receiver to the nominal noise spectral density N_0.[17]

A degradation of the $(C/N_0)_{eff}$ can result in three primary effects on the tracking of signals: loss-of-lock, loss-of-acquisition and loss-of-accuracy. When loss-of-lock occurs, the receiver can no longer track the desired signal(s). Typically, the weaker signals, usually the signals from satellites at a low elevation, will be lost first. As the power of the jammer increases, additional signals will also be lost. This reduction of the number of tracked satellites will first cause a decrease in accuracy and finally, once the number of tracked signals drops below 4, the receiver won't be able to provide a 3D position fix at all. The threshold for the $(C/N_0)_{eff}$ where a generic GPS receiver will lose its lock is around 25 dB-Hz. However, this threshold can vary significantly depending on the design of the tracking loops within the receiver. Some receivers may maintain tracking at a $(C/N_0)_{eff}$ of 15 dB-Hz while other receivers might lose lock already at 30 dB-Hz.

The acquisition of GNSS signals typically requires a higher C/N_0[18] compared to tracking the same signals. As a result, receivers in acquisition mode are more vulnerable to jamming than those in tracking mode. The signal acquisition threshold lies approximately between 30 and 35 dB-Hz.[19] Again, these numbers may vary significantly depending on the exact acquisition techniques implemented in the receivers. As acquisition requires a stronger signal, if a receiver loses its lock on a signal due to jamming, the reacquisition of the signal will be impossible while the jammer is present.

Before the receiver loses its lock, a jammer will cause a reduction in the accuracy of the range measurements and thereby the accuracy of the position solution. The additional noise will introduce errors in the range measurements which results in a lower accuracy of the position estimation.[20][21] For example, if the C/N_0 is decreased by 5.5 dB, the standard deviation of the range measurements is increased by a factor of 2. A more significant decrease of 15 dB results in an almost tenfold increase in standard deviation of the range measurements. The actual effect on the position accuracy may even be bigger as, especially for the 15 dB decrease, the signals of some satellite may drop below the tracking threshold, resulting in loss-of-lock and thus a position solution based on fewer measurements with a less favourable geometry.

Considering the three effects described above, loss-of-lock, loss-of-accuracy and loss-of-acquisition, we can identify three types of effective range for a jammer: the no-fix range, loss-of-accuracy range and the no-acquisition range. The no-fix range is the range where the receiver can no longer maintain lock on the minimum number of signals to obtain a position fix. This is the range most referred to; however, it may underestimate the actual effective range of a jammer as it does not account for the fact that users often have specific accuracy requirements on the PNT information. The loss-of-accuracy range takes this accuracy into account and is defined as the range at which the receiver is no longer able to provide the required accuracy. The decrease in accuracy is caused by the two effects described before: the reduction in the number of tracked satellites and the increase of measurement noise. As the reduction of accuracy takes place before loss-of-lock, the loss-of-accuracy range may be significantly larger than the no-fix range.

Finally, the no-acquisition range is the range from the jammer where the receiver is no longer able to acquire signals. This range is of interest in situations where a receiver is frequently turned on and off, for example to save batteries. It is especially relevant for submarines, where the receiver spends most of its time in acquisition because once a position fix is obtained the antenna is retracted below the water to minimise the risk of detection. Typically, a generic receiver requires a minimum C/N_0 of approximately 35 dB-Hz for acquisition,[22] so if the C/N_0 drops below this threshold, successful acquisition is unlikely.

To get a feeling for the numbers, the range and impact of a noise jammer on civil GPS receivers is illustrated in Figure 7.1. This figure shows the various effects described before, and at which distance from the jammer these occur. The figure is based on the equations found in[23] and shows the impact of a noise jammer with a power up to 1 Watt, a bandwidth of 2 MHz, positioned at an altitude of 25 m and using an isotropic antenna. The different coloured areas indicate the effect of the jammer. For example, a jammer of 1 Watt has a no-fix range between 5 and 10 km, a no-acquisition range extending up to 33 km and a loss of accuracy range from 30 to over 60 km for high accuracy requirements. As signals at the GNSS frequencies are restricted to line-of-sight, the range may be limited by the curvature of the earth. This is indicated in the figure by the vertical lines. For example, the jammer has no effect on receivers at an altitude of 10 m beyond approximately 34 km. Also, other obstructions, such as buildings or mountains, may limit the actual range of the jammer. Further, it is important to note that these ranges are indicative and can vary significantly based on several factors, including the characteristics of the receiver.

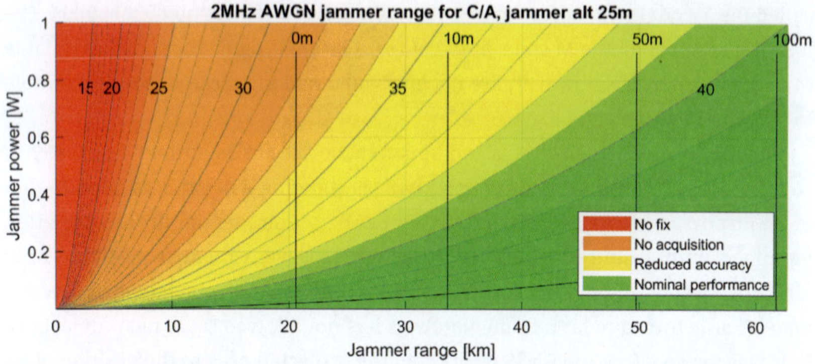

Figure 7.1: Jammer range vs jammer power for a 2 MHz noise jammer with an isotropic antenna and a generic GPS C/A code receiver. The vertical black lines indicate the line-of-sight range for a receiver at heights between 0 and 100 m. The numbers at the range curves indicate the C/N_0 belonging to that curve.

Jamming is clearly observable in the RF spectrum as the signals need to be strong relative to the power of the GNSS signals; typically, about 40 dB (10,000 times) higher for the civil signals. This allows the localisation of the jammer which enables opposing forces to eliminate the jammer. Also, the conspicuous jamming signals allows smart antennas to blank them. Smart antennas such as Controlled Reception Pattern Antennas (CRPA) are composed of multiple antenna elements which allow for detecting and determining the direction of strong signals. As GNSS signals are extremely weak, all strong signals must be interference such as jamming. Once a strong signal is detected and the direction is determined, the signals captured by the individual antenna elements are combined in such a way that the signals from the direction of the jammer are suppressed, which is called null steering.[24]

GNSS Spoofing

In addition to jamming, GNSS receivers can also be attacked by spoofing. Spoofing involves the transmission of counterfeit signals aimed at modifying the PNT information calculated by the targeted GNSS devices. Spoofing attacks can vary in complexity, from basic replay attacks, often referred to as meaconing (masking beacon), to very sophisticated attacks using time-synchronised signals causing a seamless take-over of the target receiver.[25] While the first documented successful development of a spoofer was in 2008,[26] the vulnerability of GNSS signals to spoofing has been recognised since the early development of GPS in the 1970s. To counter this threat, the legacy military P-code signal has been encrypted (after encryption called the Y-code) to prevent adversaries from generating counterfeit signals. The new military signal, the M-code, is, of course, also encrypted. The signals available to the general public, however, are not encrypted making these signals vulnerable to spoofing.

Spoofing can be used to mislead opponents or opposing unmanned systems, such as drones. A common approach involves spoofing a drone's position to that of an airport, as described in a report by C4ADS.[27] Commercial drones are often equipped with a built-in geo-fence designed to prevent them from flying near airports. By spoofing the GNSS receiver of a drone to an airport, the drone won't take off or is forced to land immediately as it is not allowed to fly near an airport. This way, by spoofing the GNSS position in the protected area to the position of an airport, it becomes more difficult to get eyes in the sky using commercial off-the-shelf drones.

Another counter drone method is to spoof the position of a drone to a higher altitude. The drone, attempting to keep its set altitude, will rapidly descend, potentially crashing or flying against obstacles. These attacks can be undirected such that all drones receiving the spoofed signals are affected and their GNSS receiver will provide the same false position. In this case, as the spoofed position is likely off from the actual position of the drone, it will fly off to a certain direction while descending rapidly. These kinds of undirected spoofing attacks can impact large areas as seen on the Black Sea, in or near Russia, and in Syria.[28]

Spoofing attacks, however, can also be aimed at a single specific receiver. For this, a time-synchronised seamless-takeover spoofer is used. This spoofer will exactly replicate the genuine signals at the position of the targeted receiver. For this, the actual position of the target receiver needs to be known. When the counterfeit signals exactly match the genuine signals, the spoofed and true PNT information will align. The spoofer's transmission power is adjusted to slightly above the weak incoming genuine signals. Now, the spoofer can slowly move the spoofed position away from the genuine one, taking control of the GNSS receiver. As the counterfeit signals are specifically generated to target the tracking loops of the target receiver, only this receiver, or receivers very close to the targeted receiver,[29] will be spoofed.

Although spoofing can mislead GNSS users, it can also be used as a form of A2 against PNT information, like conventional jamming. This can be done by obviously spoofing a receiver such that the user knows the position cannot be relied upon. When compared to conventional jamming, spoofing is beneficial considering the required power to achieve its desired effect. As a spoofer mimics the actual signals, it circumvents the anti-interference protection of spread spectrum signals and will, therefore, require significantly less power for an attack. As the spoofer requires less power, it will be more difficult to detect and localise and, therefore, to eliminate. Also, as the interference signal is more difficult to detect, smart antennas, such as CRPAs, might not be able to suppress it.

With a seamless-takeover attack, the counterfeit signals need to be slightly stronger than the genuine signals, as the counterfeit signals are perfectly aligned with the genuine signals. For undirected attacks, the tracking loops will not readily take the counterfeit signals in track. In this situation a knock-off jammer can be used to cause loss-off-lock to the genuine signals. This forces the receiver into re-acquisition where it will find the slightly stronger counterfeit signals and start tracking these. If the spoofed signals are strong enough, though, a receiver may spontaneously start tracking the spoofed signals. This requires more power, but still significantly less power than required for a (knock-off) jammer. It should be noted, however, that for successful access denial, it is not strictly necessary to actually spoof the receiver. A simple loss-of-lock on the genuine signals, or very large errors caused by counterfeit signals is sufficient for anti-access.

The effects of noise jamming can be predicted accurately by models.[30] For access denial spoofing however, no such models are (yet) available. So, to assess the effectiveness of access denial spoofing with respect to noise jamming, some experiments were conducted using a GNSS signal simulator.

Jamming and Spoofing Experiments

Transmitting signals within the GNSS frequency bands is prohibited by law; therefore, over the air experiments are difficult to arrange. GNSS signal simulators allow for jamming and spoofing experiments without transmitting over the air. So, for quick initial assessment on the effects of jamming and spoofing, GNSS simulators are indispensable.

The signal simulator used in the experiments was an Orolia/Safran Broadsim, which is capable of accurately simulating GNSS signals at a user defined time and receiver location. Additionally, the simulator can simulate signals from jammers and spoofers with user-defined locations, power and other signal properties. For the experiments, the simulator was configured with a jammer and spoofer located 1 km south of the receiver position. The jammer transmitted white Gaussian noise with a bandwidth of 2 MHz around the centre frequency of 1575.42 MHz (L1). The spoofer was configured to transmit counterfeit signals to spoof the receiver to the position of the spoofer, so 1000 m from the actual (simulated) receiver position. The antenna gain of the jammer and spoofer in the direction of the receiver was set to 2.5 dBi and the receiver antenna gain to -7 dBi. These numbers are based on a dipole antenna for the jammer and a patch antenna, receiving signals from a low elevation, for the receiver.

The target receiver was a u-blox M8, a receiver commonly found in low-cost commercial off-the-shelf drones, such as the DJI Mavic 3. This is a single frequency multi-constellation GNSS receiver. Although the receiver is multi-constellation capable, the experiments focussed on the civil GPS signals only. The results, however, are representative for all civil GNSS signals.

Each simulation started with 3 minutes without interference to allow the receiver to lock onto the (simulated) real signals and calculate its position. After three minutes the jammer or spoofer was activated while the simulation was continued for 5 minutes. The receiver output was logged using u-blox U-centre software. Al experiments were repeated at least 10 times to assess the repeatability. The described values are the average of the experimental results. It should be noted that only the effects on position information are considered and not the effects on the timing information. The more longer-term effects, those which occur after 5 minutes once the jammer or spoofer was turned on, are also not considered.

First, the observations of the jamming experiment will be discussed. These experiments showed that jamming with a transmission power of 100 mW results in the receiver losing its position fix almost instantly. This matches with the expectations, as can be seen in Figure 7.2. This figure shows the theoretical effect of the simulated jammer as calculated using the model provided by the literature.[31] This figure shows red at a power of 100 mW and a distance to the jammer of 1 km, which indicates that it is unlikely that a receiver can obtain position fixes. A reduction of the jamming power to 40 mW resulted in the receiver struggling to keep position fixing. The resulting position fixes showed very large errors of over 600 m. After approximately a minute the receiver failed to provide a position solution at all. Again, this coincides with expectations based on Figure 7.2, which shows dark orange for this combination of power and distance, which indicates that the receiver will struggle to maintain its fix and it is likely that the fix will be lost sooner or later. Reducing the jammer's transmission power further to 20 mW showed the receiver still struggling but most of the time the receiver was able to provide a position fix. The position, however, showed large errors with an RMS of over 140 m.

With a jammer power of 10 mW, the receiver was able to provide position fixes where the RMS of the error was in the order of 25 m, which is still quite high but for many applications this is sufficient.

In summary, based on the jamming simulations, we can conclude that for this receiver, a noise jammer with a 2 MHz bandwidth requires between 20 to 40 mW of transmission power to reliably deny a u-blox M8 from providing accurate PNT information at a distance of 1 km. The required power of the jammer at the location of the receiver is between 13 thousand up to 26 thousand times higher than the power of the GNSS signals. This results in a jamming to signal ratio (J/S) of between 41 and 44 dB.

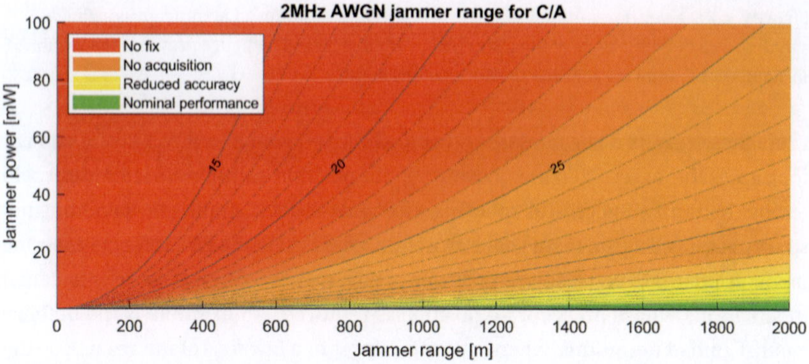

Figure 7.2: Jammer range vs jammer power for a 2 MHz noise jammer with an antenna gain of 2.5 dBi, and a generic GPS C/A code receiver with an antenna gain of -7 dBi. The numbers at the range curves indicate the C/N_0 belonging to that curve.

With the required power for jamming determined we can continue with the discussion of the spoofing experiments. The numbers provided are based on spoofing only, so without the use of a knock-off jammer. The experiments showed that a power of only 0.1 mW was sufficient to consistently spoof the receiver. It took between 2 to 3 minutes for the receiver to lock onto the counterfeit signals. However, as soon as the spoofer was turned on, the position quickly drifted away from the actual position, rendering the position solution unusable. After about 30 seconds up to one minute the receiver lost its position fix, after which it took between 30 seconds up to 2 minutes to get a spoofed position fix. Reducing the power to 0.05 mW proved insufficient to spoof the receiver; however, the position solution still quickly drifted away from the actual position (Figure 7.3). As soon as the spoofer turned on, the position starts to drift away rapidly, reaching an error of 100 m within 45 seconds. After these 45 seconds, the drift rate decreased reaching an error of around 140 m at the end of the simulations at 8 minutes. Although the receiver was occasionally not able to provide a position fix, most of the time a position fix was available.

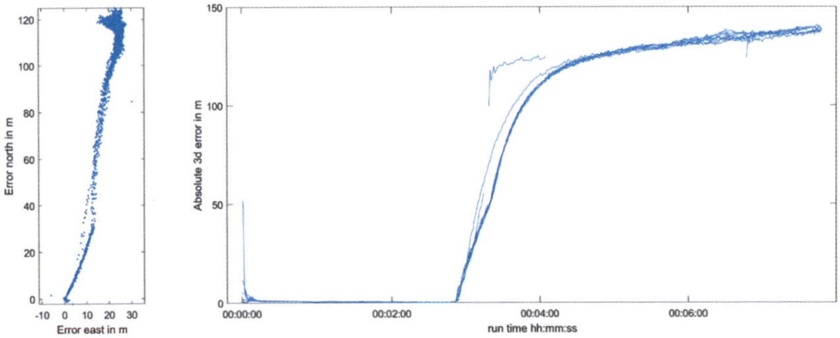

Figure 7.3: Receiver horizontal position error (left) and absolute 3d error (right) during a spoofing attack with 0.05 mW (-13 dBm) from 1 km distance. The figure provides the results of 10 simulations, showing that the results are consistent.

From these results it becomes clear that a spoofing power of only 0.1 mW is enough for successful access denial spoofing, as the position solution became unusable almost instantly. Successful spoofing after loss-of-lock is 'nice to have' but not strictly required as the main goal is to deny access to reliable PNT information. One might argue that even a spoofing power of 0.05 mW is sufficient for access denial in situations where an accurate position is required. An example is the defence against guided munitions, where a position error of a few tens of metres will be the difference between a hit or miss.[32]

From this we can conclude that for this receiver a spoofing attack between 0.05 and 0.1 mW will be successful depending on the goal. The power of the spoofed signals at the location of the receiver is roughly 30 to 65 times stronger than the average power of the genuine signals, which results in a J/S of between 15 to 18 dB, which is between 200 up to 800 times lower than found in the jamming experiment, showing the immense effectiveness of access denial spoofing.

The effectiveness of access denial spoofing can be increased even further. Once the receiver is successfully locked onto the spoofed signals, the power of the spoofer can be reduced. Final experiments showed that once a receiver is locked onto the counterfeit signals, the receiver could maintain a spoofed position fix while reducing the spoofing power to less than 1 μW. The receiver lost its position fix on the spoofed signals once the spoofer power was reduced to between 1 and 0.5 μW. Although the receiver lost lock on the spoofed signals, it could not re-acquire the genuine signals, hence the receiver was still successfully jammed. The power could be lowered even further down to 10 nW while still preventing the receiver from locking onto the genuine signals within 5 minutes. A power of 10 nW results in a received power from the jammer at the location of the receiver of 100 times lower than the average power of the genuine signals.

These last results show that a spoofer, at a distance of 1km, can transmit for a short time with a power of around 0.1 mW to spoof the target receiver, after which the power can be reduced to only 10 nW to prevent the receiver from requiring the genuine signals. This strategy reduces the transmitted power to a minimum, making it almost impossible to detect and localise the spoofer.

In conclusion, the simulator experiments showed that access denial spoofing requires 200 up to 800 times less power than conventional jamming. And, with a smart spoofing strategy, the power can even be reduced to far below the reception power of the genuine signals, making it almost impossible to detect the spoofing attack.

Concluding Remarks and Recommendations

In navigation warfare, jamming is widely used to deny adversaries the access to PNT information. The required power for effective jamming is relatively low, with only 1 Watt, depending on the exact requirements, a range of over 10 km can be achieved (Figure 7.1). Although the required power is low, the power at the receiver is more than 10 to 20 thousand times higher than the GNSS signals themselves. This makes the jammer conspicuous and prone to detection and localisation, and thus a potential target.

As shown by the experiments, spoofing requires significantly less power to achieve similar access denial as conventional jamming. When a receiver is locked onto the counterfeit signals, the jamming power can even be reduced to a power level far below the power level of the genuine signals. As the required power is much lower than for a conventional jammer, the interference signals are less visible in the electromagnetic spectrum, making it more difficult to detect and localise the origin of these signals. As the lower power is more difficult to detect, smart antennas might not be able to detect the interference and therefore supress it, making spoofing a possible way to circumvent smart antennas.

Because of the lower power requirements of GNSS spoofing, it can be expected that spoofing will become more prevalent. This is already noticeable by the sharp increase of reported spoofing around the world and in particular near conflict areas. Spoofing may have become the new jamming!

To effectively counter the effects of GNSS interference on military operations, both doctrine and technology should be adjusted. From a doctrinal perspective, it should be anticipated that in conflict areas GNSS is likely not available due to jamming and/or spoofing by the adversary. Here spoofing may serve two goals: anti-access or manipulation of GNSS-based PNT information. Doctrines should accommodate operations in environments where GNSS is unavailable, even though this may lead to reduced efficiency in achieving operational goals. GNSS should not be deemed a critical requirement for the successful execution of a mission.

During operational planning, and the operational phase itself, it is important to monitor the electromagnetic spectrum for situational awareness on NavWar by detection and localisation of GNSS jammers and spoofers. As outlined in this chapter, (access denial) spoofing incidences are likely to increase due to the advantages of the lower electromagnetic emissions. The development of interference detection and localisation techniques should, therefore, focus on these very low power spoofers.

Next to the detection of jamming and spoofing, we need to ensure maximum availability of PNT information on the battlefield. For this, there are three main technological tracks to follow. The first is to make the GNSS-based system as robust and resilient as possible, for example by using smart antennas and enhanced multi-constellation multi-frequency GNSS receivers. The second is the enhancement and development of GNSS independent PNT sources. And finally, all PNT information should be integrated into one robust PNT system where all individual PNT sources contribute to the best possible PNT solution with enough redundancy such that no individual source, including GNSS, is indispensable.

Within all these developments, it should be anticipated that access denial spoofing will become an increasing threat that should be accounted for. However, instead of seeing access denial spoofing as a threat, we could also see it as an opportunity and exploit it to gain our own advantage on the battlefield of the future.

Notes

1 Defence Science Board, *The Future of the Global Positioning System.*

2 Misra and Enge, *Global Positioning System, Signals, Measurements, and Performance;* Kaplan and Hegarty, *Understanding GPS/GNSS, Principles and Applications.*

3 For current and past GPS jamming, see http://GPSJam.org.

4 GPSPATRON and Gdynia Maritime University, *GNSS Interference Monitoring in the Baltic Sea.*

5 Yerushalmy, "Flying in the New Age of Conflict"; FitzGerald, "GPS Spoofing Surge Adds Headaches"; NOS Nieuws, "Valse gps-signalen misleiden passagiersvliegtuigen boven Midden-Oosten".

6 For current and past GPS jamming and spoofing, see http://gpswise.aero

7 Khushudyan and Horton, "Russian Jamming Leaves Some High-tech U.S. Weapons"; Abdurasulov, "Ukraine's Invisible Battle to Jam Russian Weapons".

8 Capgemini Consulting, *Synthese Rapport, Inventarisatie Kwetsbaarheid Uitval Satellietnavigatie; CGI, S[&]T, NLR. Eindrapport IKUS-II, Inventarisatie Kwetsbaarheden Uitval Satellietnavigatie.*

9 North Atlantic Treaty Organization, *STANAG 4621 C3 (edition 1).*

10 IATA. *Global Navigation Satellite Systems.*

11 Auyezov, "Ukraine Roads Company Removing Road Signs"; Bella, "Ukrainian Agency, Urging Removal of Road Signs".

12 Wang et al., "Practical Spoofing Attacks".

13 Kaplan and Hegarty, *Understanding GPS/GNSS*.

14 Mitch et al., "Signal Characteristics of Civil GPS Jammers".

15 Borio, O'Driscoll and Fortuny, "GNSS Jammers"; Borio, O'Driscoll and Fortuny, "Fast and Flexible Tracking".

16 Lubbers, "GNSS Accuracy Under White Gaussian Noise Jamming"; Kaplan and Hegarty, Understanding GPS/GNSS.

17 Lubbers, Mildner, Oonincx and Scheele, "A Study on the Accuracy of GPS Positioning "

18 Or $(C/N_0)_{eff}$ if a jammer is present. From here on for both $(C/N_0)_{eff}$ and C/N_0 the notation C/N_0 is used as for the receiver there is no difference. So, C/N_0 is either the undisturbed C/N_0 or the $(C/N_0)_{eff}$.

19 Winternitz et al., "Navigator GPS Receiver for Fast Acquisition".

20 Lubbers et al., "A Study on the Accuracy of GPS Positioning".

21 Lubbers, "GNSS Accuracy Under White Gaussian Noise Jamming".

22 Winternitz et al., "Navigator GPS Receiver for Fast Acquisition".

23 Lubbers, "GNSS Accuracy Under White Gaussian Noise Jamming"; Kaplan and Hegarty, *Understanding GPS/GNSS*.

24 Jones, "Anti-jam Technology".

25 Humphreys et al., "Assessing the Spoofing Threat"; Van Der Merwe, "Classification of Spoofing Attack Types".

26 Humphreys et al., "Assessing the Spoofing Threat".

27 C4ADS, *Above Us Only Stars*.

28 Ibid.

29 Depending on the design of the tracking loops, receivers within a radius of several tens of metres are likely to be spoofed.

30 Lubbers, "GNSS Accuracy Under White Gaussian Noise Jamming".

31 Ibid; Kaplan and Hegarty, *Understanding GPS/GNSS*.

32 Khushudyan and Horton, "Russian Jamming Leaves Some High-tech U.S. Weapons".

References

Abdurasulov, Abdujalil. 2023. "Ukraine's Invisible Battle to Jam Russian Weapons," BBC, August 4: https://www.bbc.com/news/world-europe-66279650.

Auyezov, Olzhas. 2022. "Ukraine Roads Company Removing Road Signs to Confuse Russians," Reuters, February 27, https://www.reuters.com/world/europe/ukraine-roads-company-removing-road-signs-confuse-russians-2022-02-27/.

Bella, Timothy. 2022. "Ukrainian Agency, Urging Removal of Road Signs, Posts Fake Photo with a Colorful Message for Russia," The Washington Post, February 26.

Borio, D., C. O'Driscoll, and J. Fortuny. 2012. "GNSS Jammers: Effects and Countermeasures," 6th ESA Workshop on Satellite Navigation Technologies (Navitec 2012) & European Workshop on GNSS Signals and Signal Processing, Noordwijk.

Borio, D., C. O'Driscoll, and J. Fortuny. 2014. "Fast and Flexible Tracking and Mitigating a Jamming Signal with an Adaptive Notch Filter," InsideGNSS, March/April.

C4ADS. 2019. "Above Us Only Stars: Exposing GPS Spoofing in Russia and Syria."

Capgemini Consulting. 2016. Synthese Rapport, Inventarisatie Kwetsbaarheid Uitval Satellietnavigatie. Ministerie van Infrastructuur en Milieu, Den Haag.

CGI, S[&]T, NLR. 2022. Eindrapport IKUS-II, Inventarisatie Kwetsbaarheden Uitval Satellietnavigatie. EGNSS Centre of Excellence.

Defence Science Board. 2025. *The Future of the Global Positioning System.* Washington, D.C.: Office of the Under Secretary of Defense.

FitzGerald, Drew. 2025. "GPS Spoofing Surge Adds Headaches in Middle East for Commercial Airlines," The Wall Street Journal, June 24.

GPSPATRON and Gdynia Maritime University. 2025. GNSS Interference Monitoring in the Baltic Sea: Shipborne Observations near the Kaliningrad Enclave Marine Border. Poland.

Humphreys, T., B. Ledvina, M. Psiaki, B. O'Hanlon, and P. Kintner. 2008. "Assessing the Spoofing Threat: Development of a Portable GPS Civilian Spoofer." Proceedings of the 21st International Technical Meeting of the Satellite Division of The Institute of Navigation (ION GNSS 2008), 2314– 2325. Savannah, GA.

IATA. 2025. *Global Navigation Satellite Systems GNSS Radio Frequency Interference, Safety Risk Assessment.*

Jones, Michael. 2017. "Anti-jam Technology: Demystifying the CRPA", GPS World, April 12: https://www.gpsworld.com/anti-jam-technology-demystifying-the-crpa/

Kaplan, Elliott D. and Christopher J. Hegarty. 2017. *Understanding GPS/GNSS, Principles and Applications* (Third Edition), Boston: Artech House.

Khushudyan, Isabelle and Alex Horton. 2024. "Russian Jamming Leaves Some High-tech U.S. Weapons Ineffective in Ukraine," The Washington Post, May 24: https://www.washingtonpost.com/world/2024/05/24/russia-jamming-us-weapons-ukraine/.

Lubbers, B. 2025. "GNSS Accuracy Under White Gaussian Noise Jamming." *Engineering Proceedings* 88, 26: https://doi.org/10.3390/engproc2025088026.

Lubbers, B., S. Mildner, P. Oonincx, and A. Scheele. 2015. "A Study on the Accuracy of GPS Positioning during Jamming," 2015 International Association of Institutes of Navigation World Congress (IAIN). Prague, Czech Republic.

Merwe, van der J., X. Zubizarreta, I. Lukcin, A. Rügamer and W. Felber. 2018. "Classification of Spoofing Attack Types." 2018 European Navigation Conference (ENC). Gothenburg, Sweden.

Misra, P., and P. Enge. 2012. *Global Positioning System, Signals, Measurements, and Performance,* Revised Second Edition. Lincoln, Massachusetts: Ganga-Jamuna Press.

Mitch, R., R. Dougherty, M. Psiaki, S. Powell, B. O'Hanlon, J. Bhatti, and T. Humphreys. 2011. "Signal Characteristics of Civil GPS Jammers," Proceedings of the 24th International Technical Meeting of the Satellite Division of The Institute of Navigation (ION GNSS), Portland, OR, 1907–1919.

North Atlantic Treaty Organization. 2003. *STANAG 4621 C3 (edition 1) – Navigation Warfare Definition.* Brussels: NATO Standardization Agency (NSA).

NOS Nieuws. 2024. "Valse gps-signalen misleiden passagiersvliegtuigen boven Midden-Oosten," NOS Nieuws, March 16: https://nos.nl/artikel/2512977-valse-gps-signalen-misleiden-passagiersvliegtuigen-boven-midden-oosten.

Wang, H., Y. Zhang, J. Zhu, J. He, S. Zhao, Y. Shen, and X. Jiang. 2025. "Practical Spoofing Attacks on Galileo Open Service Navigation." arXiv:2501.09246, preprint.

Winternitz, L., M. Moreau, G. Boegner Jr., and S. Sirotzky. 2004. "Navigator GPS Receiver for Fast Acquisition and Weak Signal Space Applications," Proceedings of the 17th International Technical Meeting of the Satellite Division of The Institute of Navigation (ION GNSS), 1013 –1026. Long Beach, CA.

Yerushalmy, Jonathan. 2025. "Flying in the New Age of Conflict – the Hotspots Diverting Flights and Leaving Pilots Blind," The Guardian, June 24.

CHAPTER 8

Smart Prioritisation for Detecting, Tracking and Neutralising Targets in Swarming Attacks

Mark Voskuijl, Ronald Nijboer, Martijn Scherpenzeel, Tomas van Summeren, Jory Veldman, and Denise Mollink

Abstract

Attacks with numerous unmanned systems (air, sea and land), cruise missiles and ballistic missiles are common in contemporary warfare. These coordinated swarming attacks are intended to overwhelm the adversary and to saturate defensive systems. With the current developments in artificial intelligence and autonomous systems, it can be expected that these tactics will further evolve and become more effective. To this end, it is essential that effective countermeasures and tactics are in place. Defence systems must be able to detect, track and neutralise as many targets as physically possible in a large scale simultaneous attack. The current research highlights the necessity to prioritise available resources within the different phases (detect, track and neutralise) of the process. Two case studies are presented. The first case study focuses on the search and tracking functions of a phased array multifunction radar. It is demonstrated through the simulation of different combat scenarios that smart prioritisation and effective time allocation is vital to maximise the performance of the radar. In the second case study, the effectiveness of a ground based laser guided rocket system against a swarm of Chinese loitering munitions is analysed. Various swarming scenarios are simulated which illustrate the value of giving priority to the right target(s) when neutralising threats.

Keywords: Swarming; Loitering munitions; Multi-function radar; Prioritisation; Scheduling; Point defence; Threat neutralisation

Introduction

During the war between Ukraine and Russia and during Iran's attack on Israel in 2024, large numbers of drones and/or missiles have been used. Firing these large numbers simultaneously attempts to inflict as much damage as possible.

Russia carried out an attack with many drones and missiles on Ukraine on October 10, 2022.[1] The target of this attack was energy infrastructure. It also hit the capital city of Kyiv. For this attack, Russia used 84 cruise missiles and 24 suicide drones. With multiple waves of attack, Russia inflicted severe damage. There were many casualties due to this attack. In addition, energy distribution was severely disrupted. Over half of the national energy infrastructure was no longer operational according to Ukrainian President Volodymyr Zelenskyy. This left more than 10 million residents without electricity. By December 2024, the numbers of drones and missiles used in the attacks had increased to more than 1,000. By June 10, 2025, a single attack used 315 drones and 7 missiles. So the numbers of drones and missiles used in an attack are increasing.[2]

Ukraine has shown that drone attacks are used not only on land but also on naval vessels. They have carried out multiple attacks with drones and/or missiles on Russian ships. The Admiral Makarov was damaged in an attack on Sevastopol on Oct. 29, 2022. This involved the use of multiple air and sea drones in a coordinated attack. Reports indicate that at least one sea drone successfully hit the ship. In the process, the ship's radar system was disabled. Then on May 24, 2023, the Yury Ivanov-class intelligence ship Ivan Khurs was attacked by three sea drones. At least one of them caused damage to the ship. On August 4, 2023, damage was inflicted on the Olenegorsky Gornyak, a Project 775 Ropucha-class landing ship. This probably happened with a sea drone. Then on Sept. 13, 2023, Ukrainian Su-24 bombers attacked the Kilo-class submarine Rostov-on-Don and the landing ship Minsk with Storm Shadow missiles. The Minsk was visibly destroyed and the Rostov-on-Don was reported to be so badly damaged that it was beyond repair. On Feb. 1, 2024, Ukraine claimed in a video that it sank the Tarantul-class corvette R-334 Ivanovets using MAGURA V5 unmanned surface vehicles (USVs). Soon after, another video on Feb. 14 claimed the sinking of the large landing ship Tsezar Kunikov. This was also attacked with MAGURA V5 sea drones.[3]

Iran's Islamic Revolutionary Guard Corps (IRGC) carried out an attack on Israel on April 13, 2024. The attack targeted multiple military targets. Iran used a combination of drones, cruise missiles and ballistic missiles. In total, more than 300 missiles and drones were fired. According to Iran, it is in retaliation for the airstrike on the Iranian consulate in Damascus. This airstrike appears to have been carried out by Israel and killed a senior Iranian officer. According to the IRGC's Tasnim News Agency, the coordinated attack on Israel began with a wave of hundreds of HESA Shahed 136 loitering munitions with the aim of overwhelming and saturating the defences provided by the Iron Dome and David's Sling systems. This would create a gap in the missile defence systems that a second wave of attacks would then have to take advantage of. The second wave includes cruise and ballistic missiles. In addition, advanced and rapidly flying jet-propelled Shahed 238 loitering munitions were used. This multi-layered approach aimed to maximise the impact on the targeted defences and increase the likelihood of penetrating the protective measures in place.[4]

Swarming, a military tactic involving the simultaneous attack of numerous drones or missiles, aims to overwhelm a superior enemy. This coordinated approach leverages speed, surprise, and overwhelming numbers to create chaos and disrupt the opponent's strategy. Increasingly vital in modern warfare, especially with the proliferation of drones and enhanced coordination technologies, swarming can be executed using a variety of assets, including drones, naval vessels, soldiers, or even cyberattacks.

In an Anti-Access/Area Denial (A2/AD) environment, swarm attacks represent a major threat – one for which current platforms were not originally designed. Maintaining comprehensive and uninterrupted threat visibility is paramount in defending against these attacks, particularly those of significant magnitude. This thorough monitoring is the foundation of a multi-layered defence. After identifying targets, weapons like anti-aircraft missiles are used to eliminate the danger. These weapons depend on radar to guide them precisely. Ideally, using several radar systems to divide the work allows for better threat detection, tracking, and targeting. However, if one Multi-Function Radar (MFR) system handles everything at once, it can easily become 'saturated': it gets overloaded and is unable to keep up with all the threats. In that case it is paramount to prioritise the targets to be tracked and neutralised, while still spending time on searching for new, potentially more dangerous, targets.

In this chapter we investigate how smart prioritisation can increase a unit's detection and neutralisation capability during a swarm attack. This is done by considering two case studies. In the next section we investigate the case of a Multi-Function Radar on an Air Defence and Command Frigate (ADCF) for which we adapt the prioritisation rules in order to be more effective in Searching, Tracking and Targeting a large number of targets. In the subsequent section we investigate the case of the prioritisation of neutralising swarms of targets. Some general conclusions are given in the final section.

Case Study 1: Multi-Function Radar

Introducing Multi-Function Radar Sensor Management

A phased array Multi-Function Radar (MFR) performs many functions, such as surveillance, tracking, and fire control. Phased array radars utilise electronic steering. Instead of physically moving, they manipulate the timing (or 'phase') of signals from multiple antennas. This allows them to instantly redirect their radar beam, making them super responsive. This rapid adaptability makes them perfect for high-speed environments, such as tracking threats in air defence.

The MFR operates with a finite set of resources necessary to carry out its various functions. These resources typically include a time budget, an energy budget, and a processing budget. Among these, the time budget is the most critical constraint and will be considered here. The time budget is a fixed amount of time in which all radar tasks are to be executed. The energy budget corresponds to the amount of power available in a radar beam. This power can, in theory, be divided over multiple targets (i.e. radar beams) at the same time, but this will not be considered here. We assume that at every time interval a single beam is generated with which a single function can be performed at the same time instance. Lastly, it is assumed that the system is not limited by the processing budget.

At its core, a radar system needs to both identify the precise position of objects and continuously track the movement of objects already under surveillance. Faced with time constraints and a multitude of necessary functions, radar systems must strategically determine the most critical tasks to execute first. This system of prioritising different functions in a radar system is known as Radar Resource Management (RRM).

An MFR uses a method called time-sharing to distribute its attention across its antennas. All tasks need to be performed in a fixed time interval. This interval is broken down into a large number of very short time intervals. These short intervals are carefully assigned to different radar tasks, guaranteeing each task gets the required budget to complete its assigned role. However, when a radar system is overloaded, it doesn't have enough time to handle all the demands placed upon it. Consequently, the system must prioritise its tasks, as it can't complete everything. Urgent tasks, like detecting immediate dangers, get done first. Less important tasks might be delayed or skipped altogether, which could mean missing something or reacting slower to smaller threats. Neglecting less important tasks means the radar may not have a comprehensive understanding of the surrounding environment. Therefore, careful time management is paramount for the radar to operate effectively.

A radar system's search and track functions compete for resources, requiring a delicate balance to ensure optimal performance. To successfully track a target, it must first be detected through the search function. Focusing too much on tracking existing targets can seriously hamper a radar's ability to find new threats. If the system gets bogged down tracking every blip it sees, it leaves no time for actively scanning its surroundings. This creates a dangerous blind spot. Smart prioritisation and effective time allocation are vital, ensuring the radar system remains both proactive in its search and reactive to known dangers.

Modelling Multi-Function Radar Scheduling

Effectively allocating the time budget involves prioritising tasks. Levels of priority can be assigned to specific functions by using an Operational Rule Set (ORS). The ORS presents the prioritisation using a slider mechanism, where the width of the slider indicates the range of priorities that can be assigned to each function. The wider the slider, the greater the range of priority levels available for the associated tasks. Usually, these settings remain static and are not adjusted even when the MFR's mission changes.

We investigate the implications of altering the priority ranges of eight radar functions in relation to the MFR's mission. We consider three types of search functions, four types of tracking functions and one weapon guidance function. The Command Aim (CA) and Threat Warnings (TW) provide key information for mission success and danger evaluation. Using these existing feeds, the sensor manager can intelligently re-prioritise radar tasks, making the radar system quicker and more effective.

In naval warfare, battles are primarily fought in three distinct zones: combating underwater vessels (ASW), engaging surface ships (ASuW), and defending against airborne threats (AAW). Each of these warfare areas is specifically designed to neutralise dangers originating from their respective environments: beneath the waves, on the ocean's surface, and within the skies. These three warfare areas determine the Command Aim. The Threat Warnings are determined by the colours red, yellow and white. White means that an attack by a hostile unit is unlikely. Yellow means that an attack from a hostile unit is likely. Red means that an attack from an enemy unit is imminent or is in progress.

The current investigation builds on an existing radar resource management model, extending it with additional radar functions and a refined prioritisation mechanism.[5] The methodology centres around a modular simulation model, divided into five interconnected components that simulate the decision-making process and operation of a multifunction radar system (Figure 8.1). Each module reflects a different stage in the radar's operational cycle, from operator input to task execution. This cycle is run through each time interval, prioritising and distributing tasks over the available time sub-intervals.

The first block initialises the radar system at each time step. Based on intelligence, previously tracked targets, and mission inputs, it sets radar configurations and defines the Command Aim (CA) and Threat Warnings (TWs). It also converts intelligence into virtual targets: search areas which represent expected but undetected threats.

Using the radar settings and both virtual and real targets, the second block generates radar tasks. Virtual targets result in search tasks assigned to functions

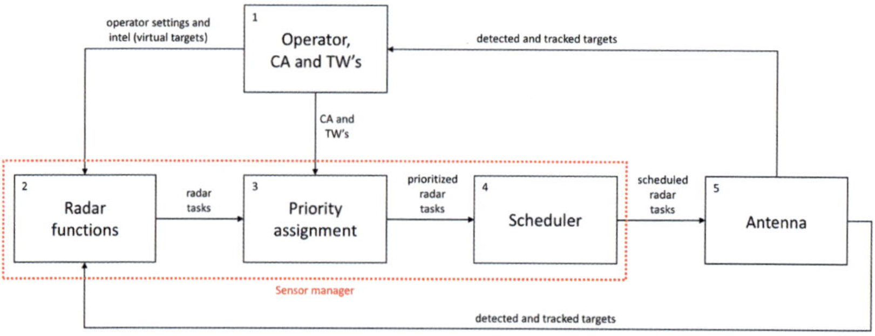

Figure 8.1: Decision-making process and operation of a multifunction radar system

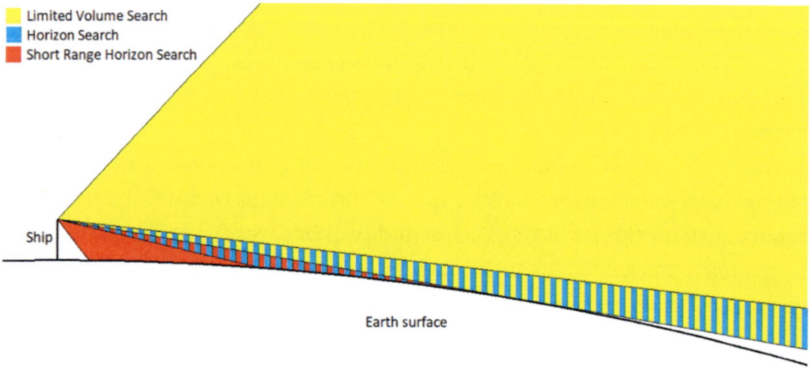

Figure 8.2: Search tasks of multi-function radar

like Horizon Search (HS), Short Range Horizon Search (SRHS), or Limited Volume Search (LVS) (see Figure 8.2).

Real targets are assigned to one of four track functions: Air Surveillance Track (AST), Surface Surveillance Track (SST), Air Weapon Track (AWT) and Surface Weapon Track (SWT), and additional tasks may be created for missile guidance if engagement occurs.

In the third block each radar task is assigned a risk-based priority (value between 0 and 1) reflecting the threat level of its associated target (based on Closest Point of Approach Distance, CPAD). In the original approach by Bolderheij this would be the priority of the task. In the new approach where CA and TW are taken into account this risk-based priority is then mapped onto a priority range specific to the radar function it belongs to, generating the final priority of the task at hand. The radar function priority ranges are derived from the CA and TWs, enabling adaptive, mission-aware scheduling. First, the CA gives priority to one of three warfare areas: ASW, ASuW, or AAW. For each of these areas specific priority ranges are defined for the 8 different functions. Then, depending on the TW, the priority range may be given increased values. If the warfare area ASW is yellow or red, the SRHS function is increased. If the warfare area ASuW is yellow or red, the HS function is increased. If the warfare area AAW is yellow or red, the LVS function is increased. If the TW is yellow, the values are increased by 0.1. If the TW is red, the values are increased by 0.2. Finally, a check is made to ensure that the maximum value of the priority range does not exceed 1. If it does, this value is reset to 1. The higher the risk, the higher the priority of the task becomes. So the objective is to keep the risk to one's ship as low as possible.[6]

Block four checks if the total time required to perform all tasks on each radar face exceeds the available time budget. If it does, a scheduler selects the most important tasks based on their scaled priorities, ensuring optimal system performance. Less critical tasks may be dropped.

All scheduled tasks are executed by the radar antennas in block five. Task results (e.g., target detections, track updates, engagement outcomes) are evaluated and used as input for the next time step. This creates a feedback loop, allowing the radar to continuously adapt to the evolving situation. A target is detected in a certain time step when the target is within Line-of-Sight and within the radar beam and the search task is being executed. Once detected the target goes from status virtual to status real and can be tracked.

Simulations and Results

The simulated scenario is as follows (see also[7]). Tensions are high as Orange Country, a significant regional power, has unilaterally expanded its territorial waters, seizing control of vital offshore oil fields. This move is strongly contested by the global community. To assert international resolve, an Air Defence and Command Frigate (ADCF) is ordered to navigate a pre-defined course just outside Orange Country's original territorial limits, but clearly within the newly claimed zone. The ADCF is equipped with advanced sensors, including a Multi-Function Radar (MFR). Orange Country has responded by deploying two mobile, land-based missile launchers (Site 1 and Site 2) and a small aircraft carrier (FR Revolution), positioned within the contested waters. These assets are armed with subsonic Surface-to-Surface Missiles (SSMs). Launcher 1 and the carrier fire missiles programmed to activate their internal radar systems after a set delay, aiming for a predicted impact point. Launcher 2 utilises SSMs that follow a pre-determined route defined by waypoints before activating their radar. Fortunately, the ADCF has been provided with this crucial intelligence. As the mission begins, the conflict has escalated to full-scale war. As the ADCF enters the range of Launcher 1 (red circle), four Exocet missiles are launched in succession, each 50 seconds apart. Upon entering the range of

Launcher 2, two Harpoon missiles are launched with a much larger interval of 2,300 seconds. Adding to the complexity, three civilian aircraft (2 Airbus A-300s and 1 Boeing 737) are traversing the area along a designated airway (purple line). Furthermore, a fighter aircraft launches from the carrier, completes its mission, and returns.

In addition to this base scenario it is expanded to include a group of UAVs that collectively come flying from the mainland towards the ADCF (scenario 3 and 4), to include a group of USVs that come from all directions towards the ADCF (scenario 5 and 6), and a combination of both (scenario 7 and 8). The scenarios are run twice. The first time no restrictions are placed on the priority range of functions (scenarios 1, 3, 5, 7), while the second time the setting of the priority range is determined using the Command Aim and Threat Warnings (scenarios 2, 4, 6, 8). An overview of the scenarios is given in Table 8.1 and the specific objects that are used are summarised in Table 8.2.

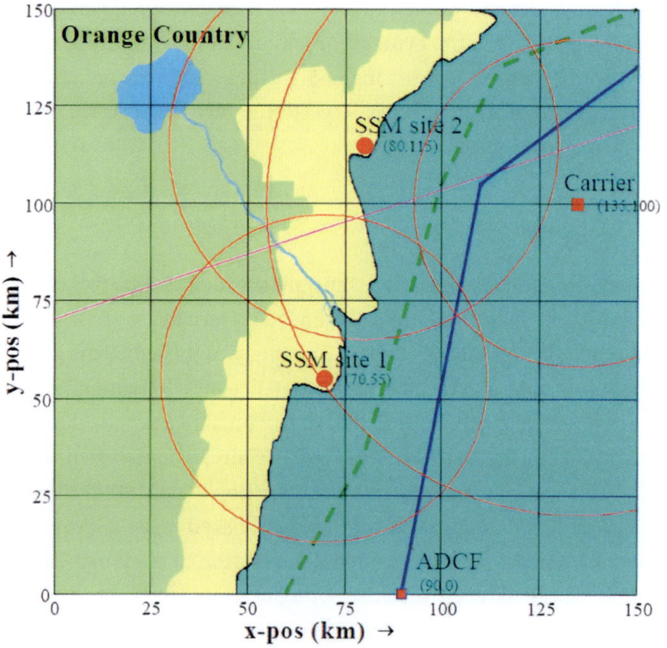

Figure 8.3: Geographic setting of the multi-function radar scenario

Table 8.1: Overview of the scenarios

Scenario	No priority range	With priority range				
			CA	TW: ASW	TW: ASuW	TW: AAW
No drones	1	2	AAW	Yellow	Yellow	Red
Air drones	3	4	AAW	Yellow	Yellow	Red
Sea drones	5	6	ASW	Red	Yellow	Yellow
Air and Sea drones	7	8	AAW	Yellow	Yellow	Red

Table 8.2: Overview of targets and missiles

Object	Type	Domain
SSM Site 1	Missile launcher	Shore
MM38 Exocet	Anti-ship missile	Air
SSM Site 2	Missile launcher	Shore
Harpoon RGM-84A	Anti-ship missile	Air
FR Revolution	Aircraft carrier	Sea
Crusader	Fighter	Air
Airbus A300	Scheduled flight	Air
Boeing 737	Scheduled flight	Air
Shahed 136	UAV	Air
MAGURA V5	USV	Sea

Table 8.3: Overview of detection times and differences

Target	Launch Time	Detection Time											
		1	2	$\Delta 1-2$	3	4	$\Delta 3-4$	5	6	$\Delta 5-6$	7	8	$\Delta 7-8$
FR Revolution	0	14657	10539	4118	14657	10539	4118	14657	10539	4118	14657	10539	4118
Airbus A300	28000	28001	28001	0	28001	28001	0	28001	28001	0	28001	28001	0
Airbus A300	4000	4361	4361	0	4361	4361	0	4361	4361	0	4361	4361	0
Boeing 737	9000	9001	9001	0	9001	9001	0	9001	9001	0	9001	9001	0
Fighter	6000	6319	6002	317	6319	6002	317	6319	6002	317	6319	6002	317
USV 1	4000							6212	4976	1236	6212	4976	1236
USV 2	4100							6298	5063	1235	6298	5063	1235
USV 3	4200							6385	5150	1235	6385	5150	1235
UAV 1	4400				6674	6125	549				6674	6125	549
UAV 2	4450				6712	6162	550				6712	6162	550
UAV 3	4500				6769	6220	549				6769	6220	549
UAV 4	4550				6816	6266	550				6816	6266	550
UAV 5	4600				6874	6325	549				6874	6325	549
UAV 6	4650				6913	6362	551				6913	6362	551
UAV 7	4700				6969	6419	550				6969	6419	550
UAV 8	4750				7016	6466	550				7016	7016	0

Target	Launch Time	Detection Time											
		1	2	Δ1 – 2	3	4	Δ3 – 4	5	6	Δ5 – 6	7	8	Δ7 – 8
Exocet 1	7165	7167	7167	0	7173	7173	0	7167	7167	0	7173	7173	0
Exocet 2	7215	7217	7217	0	7223	7223	0	7217	7217	0	7223	7223	0
Exocet 3	7265	7267	7267	0	7272	7272	0	7267	7267	0	7272	7272	0
Exocet 4	7315	7317	7317	0	7322	7322	0	7317	7317	0	7322	7322	0
Harpoon 1	15327	15414	15414	0	15414	15414	0	15414	15414	0	15414	15414	0
Harpoon 2	17627	17734	17630	104	17734	17630	104	17734	17630	104	17734	17630	104

Launch and detection times of the various targets for the different scenarios are shown in Table 8.3. In all scenarios scheduling using priority ranges improves the detection times of targets. The aircraft carrier is detected 4118 seconds earlier in all scenarios with priority ranges and the Fighter 317 seconds. There is no difference in the detection times of the three airliners in the various scenarios.

Detection of the Exocets competes with the detection of the UAVs. There is no difference in detection times between scenarios with and without priority ranges, but the scenarios including UAVs show later detection of the Exocets. The Harpoons do not suffer from such competition and here the second Harpoon is detected earlier in the priority range scenarios.

In our scenario the detection of USVs does not affect the detection of UAVs and Exocets, except for scenario 8 where the eighth UAV is not detected earlier than in scenario 7, whereas this is the case in scenario 4 versus 3.

More detailed analysis[8] of the simulations reveal that all scenarios show saturation of the MFR. In the scenarios using priority ranges (i.e. 2, 4, 6, 8) more time budget is spent on Search functions and the less requested Search budget is not being executed. This shows that scheduling using priority ranges makes sure that Search functions are given more time budget, while at the same time the simulations showed improved detection times.

Case Study 2: Coordinated Assaults with Loitering Munitions through Swarming Tactics

Introduction

Unmanned aircraft systems (UAS) are deployed in unprecedented numbers in modern warfare.[9] They are used for a variety of roles such as information, surveillance, target acquisition and reconnaissance (ISTAR), electronic warfare, support of artillery and as communication relay. UAS can also be equipped with an internal warhead to operate as a guided munition. This specific category is known under different names such as strike drones, loitering munitions or one way effectors. The Munitions Safety Information Analysis Center uses the following definition for this category.[10]

> *A loitering munition is a remote controlled, lock-on-after-launch operated guided munition, following a non-ballistic and operator influenced trajectory and capable of non/beyond line-of-sight target verification and precision attack, which is destroyed by functioning of its payload.*

Various types of loitering munitions with different configurations and flight performance exist.[11] Mass precision strike effects can be achieved at low cost with UAS if different tasks are assigned to different types of UAS.[12] In essence, UAS with an ISTAR role can be used over the battlefield to find targets and loitering munitions can subsequently be used to strike these targets. Finally, the ISTAR platform is used for battle damage assessment. At present, this hunter-killer approach is being frequently and successfully used in the Russian-Ukraine war.[13] Large scale attacks with numerous unmanned systems are common as well. Russia currently uses Shahed-136/Geran-2 UAS for long range combined attacks on Ukraine on a daily basis.[14] The attacks with hundreds of these UAS include decoys and jet powered unmanned systems but also limited numbers of cruise missiles and rockets. These swarming attacks are intended to overwhelm the adversary and to saturate defensive systems. The phenomenon of swarming is studied in military sciences, engineering and biology. Arquilla and Ronfeldt define swarming from a military perspective as follows.[15]

> *Swarming is seemingly amorphous, but it is a deliberately structured, coordinated, strategic way to strike from all directions, by means of a sustainable pulsing of force and or fire, close-in as well as from stand-off positions.*

With the current developments in artificial intelligence and autonomous systems, it can be expected that these tactics will further evolve and become more effective. A notable development with respect to swarming and autonomy is the deployment of the Russian V2U drone. It is reported that the V2U operates in a group in which the individual UAS actively collaborate and recognise each other by colours painted on the wing. Furthermore, artificial intelligence is used for autonomous target recognition[16]. The V2U has a tandem wing configuration with a wing span of 1.2 m and a warhead of 3.5 kg.[17]

The overall research question addressed with this second case study is the following: “How does smart prioritising increase a unit’s neutralisation capability during a swarm attack?” To answer this question, a single scenario is defined in which a defensive position protected by a Counter UAS system is attacked by an autonomous swarm of loitering munitions. In this scenario, a specific type of counter UAS system is analysed in combination with a specific type of loitering munition and a specific type of swarming algorithm. The variables in the scenario are the prioritisation logic and the geometry of the swarm at the start of the attack (numbers of unmanned systems and starting positions). The types of counter UAS system, loitering munition and swarming algorithm considered are deemed representative for contemporary warfare. Nevertheless, such a highly specific scenario has its limitations in terms of generalisability of the conclusions. To address these limitations, the Counter UAS system is subsequently compared to two other types of counter UAS systems in terms of operational performance and economic costs.

Scenario

In 2020, the People's Liberation Army of China conducted tests with a swarm of 48 loitering munitions launched from a truck.[18] At present it is still unclear what the level of autonomy of that specific loitering munition swarm was. This loitering munition swarm is taken as example use case in this research. The specific type of loitering munition is the CH-901 Feihong developed by the China Aerospace Science and Technology Corporation. Various swarming algorithms can potentially be used.[19] It is assumed that an artificial potential field method is used as a swarming algorithm in a point defence scenario. With the artificial potential field method, artificial forces are modelled which attract UAS to a target and which repel UAS from obstacles and each other. With this relatively simple algorithm, a large swarm can be controlled in an area with obstacles. A detect and avoid functionality is naturally included in this algorithm. The performance of a counter UAS system based on laser guided rockets is considered in this scenario. The overall point defence scenario is schematically displayed in Figure 8.4.[20]

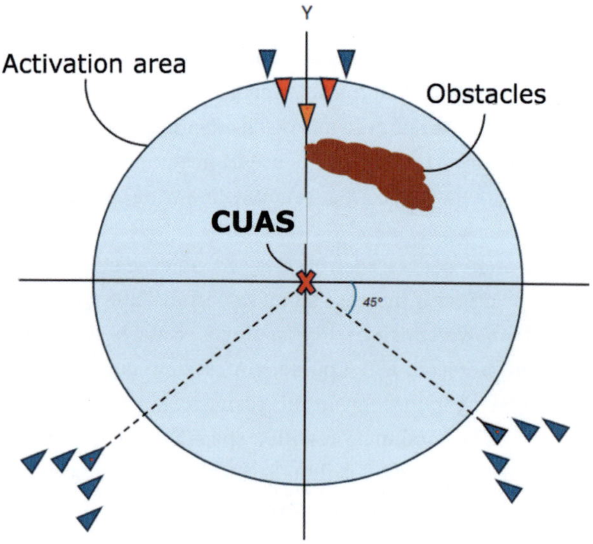

Figure 8.4: Swarming scenario in which groups of loitering munitions approach a defensive position from different directions

Following the point defence scenario, a comparison is made of the baseline counter UAS system with two other types of counter UAS systems in terms of operational capacity and economic costs. The three systems which are compared are:

- Laser guided rocket system
- High speed drone interceptor (race drone)
- Man-Portable Air-Defence System (MANPADS) with infrared guidance

The development of the simulation models of the autonomous loitering munition swarm and the three counter UAS systems is described in the following section.

Simulation Models

Simulation models of the loitering munitions, including the swarming algorithm, and the Counter UAS systems are required for the case study. First, a flight performance simulation model of the CH-901 Feihong is created.[21] The aerodynamic characteristics are analysed using OpenVSP, which is a first order panel method in combination with an empirical parasitic drag estimation method.[22] The geometry of the vehicle required for the aerodynamic analysis is estimated based on pictures in the public domain. The weight of the CH-901 Feihong is known and reported by various sources. The propulsion system characteristics are estimated based on pictures of the propellers and statistical data of batteries and electric motors. Finally, the overall flight performance of the UAS is calculated using a point mass simulation model. Results are summarised in Table 8.4.

Table 8.4: Flight performance analysis of the CH-901 Feihong loitering munition based on open source data[23].

Parameter	Value
Vehicle mass	9 kg
Warhead mass	3.5 kg
Cruise speed	87 km/hr
Endurance	2.3 hours
Maximum climb rate	6.1 m/s
Flight range	174 km
Launch method	Tube (48 in a single vehicle)

Next, simulation models of the three Counter UAS systems are developed. As a representative laser guided rocket system, the Advanced Precision Kill Weapon System is selected. This is a precision guided munition based on the unguided Hydra 70 rocket. Aerodynamic and propulsion system characteristics of the Hydra 70 are available in literature.[24] A flight performance model of the APKWS was developed primarily based on this data.[25] As a guidance method, proportional navigation was used. Before a rocket is launched toward a target, several steps need to be taken. It is assumed that the targets are identified with a radar. Next, a laser designator is pointed at the target. After detection, the launcher for the guided rockets will turn to acquire the correct rotation and elevation angles. The time required for each step is modelled[26].

The SA-7B Grail is selected as a representative MANPADS. Just like the laser guided rocket system and CH-901 Feihong loitering munition, the equations of motion are modelled as a point mass simulation model. Based on data in the public domain, models for the geometry, propulsion system and mass are created. The aerodynamic characteristics are analysed using Missile DATCOM, a semi-empirical method developed by the US Air Force.[27] Proportional navigation is used as a guidance method. All the sub models are combined in a complete flight performance model.[28]

A wide variety of drone interceptors is currently under development worldwide. The RacerX UAS is selected as the interceptor for the current scenario. This a very fast battery-powered racing quadcopter UAS. It is the system of preference due to its speed in combination with the design simplicity and design transparency. The RacerX is minimalistic in design and therefore embodies the essence of a typical quadcopter design; a system for which the performance is completely determined by the four propulsion (motor-propeller) assemblies. A complete flight simulation model is developed for the current research.[29] The core elements of this model are the battery, electronic speed controller, brushless electric motors and propellers. Drone interceptors with a quadcopter configuration can potentially use completely other guidance methods than conventional missiles because there is control over the thrust force of individual propeller motor combinations. In the current research, trajectory optimisation is used to determine the most optimal trajectory towards an objective point in 2D space given certain limitations, such as system dynamics or an arrival pitch angle. The direct collocation method described is used to solve the trajectory optimisation problem.[30]

Simulation of the Scenario

All the simulation models are integrated in an overall simulation of the complete scenario. The point defence scenario is analysed with the laser guided rocket system as the CUAS method. Step by step, the complexity of the coordinated assault is increased. At first, a situation is analysed where the 48 loitering munitions attack from a single direction. Individual UAS approach with time intervals of 10 seconds. In this case there is sufficient time to destroy each loitering munition of the swarm.

If the interval between the approaching loitering munitions is reduced to 5 seconds, the attack becomes much more challenging. In that case, only 15 loitering munitions are destroyed before one reaches the defensive position. The situation worsens when loitering munitions arrive in pairs with 13 loitering munitions destroyed. Next, the complexity is increased by letting the loitering munitions attack from two separate directions with a 60-degree angle. The maximum angular velocity of the launcher is 30 degrees per second. Due to the increased time required to turn the launcher, the effectiveness of the CUAS system is reduced even further. In a final scenario, three separate groups of loitering munitions perform a swarming attack. The realism of this scenario is increased by introducing an obstacle in the North which the swarm will have to avoid. All results are summarised in Table 8.5.

Table 8.5: Swarming simulation results for a point defence scenario with a laser guided rocket CUAS system

Swarming scenario	Number of loitering munitions destroyed
Attack from a single direction, 10 sec interval	48
Attack from a single direction, 5 sec interval	15
Attack from a single direction (pairs flying in parallel)	13
Attack from two directions (60-degree angle)	10
Three groups attacking from different directions	9

The prioritisation logic used in the results presented above works as follows. The loitering munition that will arrive at the target in the least amount of time, which depends on the combination of distance and flight speed, is selected by the counter UAS system. In case the loitering munitions approach from different angles, valuable time is needed to rotate the counter UAS system in the right direction. Hence, if multiple loitering munitions approach from the same direction it could be advantageous to attack multiple loitering munitions in a cluster before rotating and attacking another cluster. To investigate the effect of prioritisation logic, the swarming scenario with three groups attacking from different directions is repeated. Two prioritisation methods are compared. In logic A, the prioritisation is on the loitering munition which will arrive at the target in the least amount of time. Logic B selects the loitering munition for which the smallest rotation is required with the counter UAS system.

Table 8.6: Effect of prioritisation logic

Size of the swarm	prioritisation logic A			prioritisation logic B		
	Number destroyed	Number reached target	Effectiveness	Number destroyed	Number reached target	Effectiveness
15	11	4	73%	9	6	60%
30	12	18	40%	10	20	33%
45	12	33	27%	11	34	24%

It can be observed that there are clear differences in terms of effectiveness. The effectiveness is defined as the number of loitering munitions destroyed as a percentage of the total swarm size. This demonstrates that the prioritisation logic has an effect on the neutralisation capacity. Note that the numbers presented in Table 8.6 depend to a large extent on the maximum rotation velocity of the counter UAS system and the geometry of the attack (location and number of attacking loitering munitions). A more comprehensive investigation and more details can be found in the work by Veldman.[31] Hence, one should not directly conclude that logic A is better than logic B. The results are merely intended to demonstrate that prioritisation logic can have a significant effect on the outcome. It is possible to optimise the prioritisation of targets when both the geometry of the attack and the characteristics of the counter UAS system are known. It is expected that this mathematical optimisation can be performed in real time since the computational effort required for the presented results is limited.

Comparison of Different Counter UAS Systems

The laser guided rocket system is compared to a high speed drone interceptor and a MANPADS with infrared guidance. First the comparison is done in terms of operational capacity and next the comparison is done in terms of economic costs.

Flyout fans are trajectories from a starting location (a ground launch point) at regular angle intervals to form fan-shaped regions, which are used to visualise and analyse weapon engagement zones within those sectors. A separate flyout fan is created for each CUAS method. These flyout fans are presented in Figure 8.5–Figure 8.7.

The flyout fan of the drone interceptor (Figure 8.5) shows that a range of almost 4000 m can be achieved in a flight close to the ground. The range is somewhat less in a vertical trajectory where potential energy is gained. At the range limit, which takes about 55 seconds to reach, the battery only has 5% energy left. This is a maximum speed scenario. In a different strategy, the drone interceptor could also fly

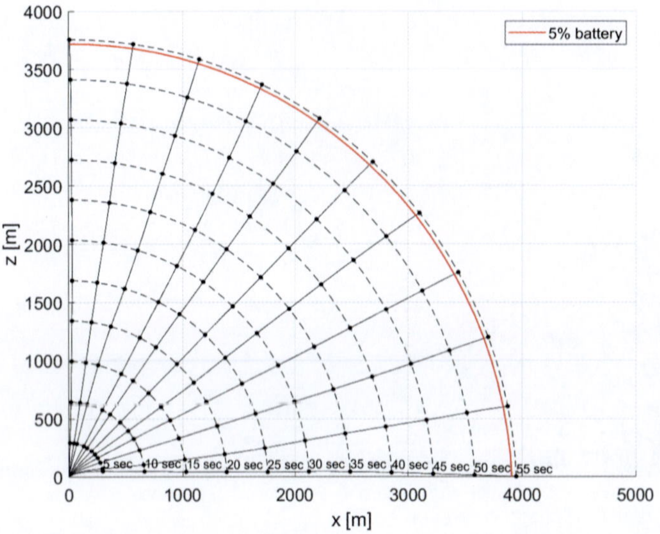

Figure 8.5: Flyout fan of drone interceptor

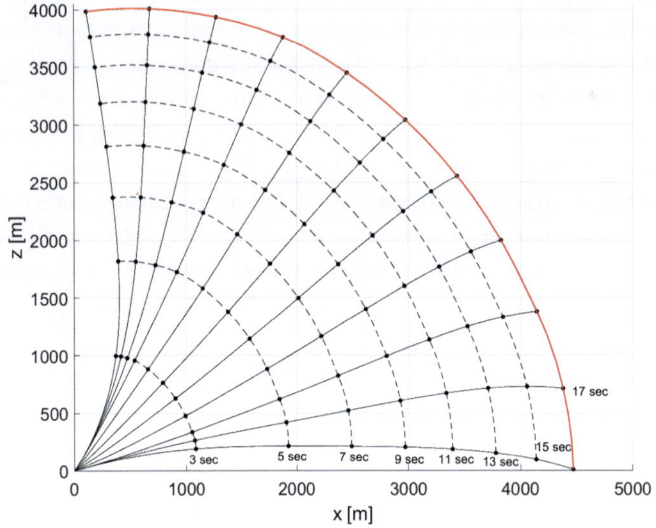

Figure 8.6: Flyout fan chart of MANPAD (infrared guided surface to air missile)

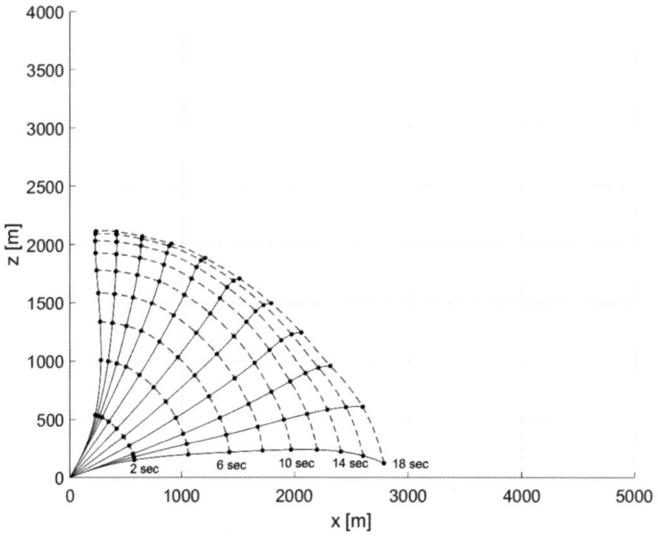

Figure 8.7: Flyout fan of laser guided rocket

much slower towards a location, save battery energy and wait for targets to appear. The MANPADS has a similar range as the drone interceptor (see Figure 8.6). Flight trajectories are however somewhat different due to the nature of the guidance method. The most notable difference of the MANPADS with the drone interceptor is the flight speed. A distance of 4000 m is achieved in approximately 14 seconds and 55 respectively. Hence, the MANPADS is about three times faster. The final flyout fan presented (Figure 8.7) is the laser guided rocket. It has a significantly smaller flight range than the drone interceptor and the MANPADS. The average flight speed is well above that of the drone interceptor but below the speed of the MANPADS. Clearly the three Counter UAS solutions considered have a significantly different operational capacity. This is also reflected in the economic costs of the different solutions.

Determining the exact costs of weapon systems is a complex task in general. Unit costs for weapon systems are typically not publicly available. Furthermore, when a cost assessment is done properly, it should not only include the direct costs but also the costs for training, costs for research and development, personnel costs and maintenance costs. Finally, it should be noted that the development and production of weapon systems can also have positive economic effects for the country in which they are produced. Nevertheless, even a rudimentary comparison of cost estimates of different counter UAS systems can be very insightful since there are very significant differences. The cost of a RacerX is estimated in the price range of \$400 –\$1000. The cost of a typical MANPADS on the other hand ranges in orders of magnitude higher. The SA-18 Grouse, a modern Russian MANPAD which has been used against UAS in Ukraine, costs roughly $70,000. A NATO equivalent MANPAD, the FIM-92 Stringer, costs roughly $119,000 per system.[32] The APKWS can be launched from numerous platforms (air, land and sea). The cost of an APKWS missile is estimated (without launcher) at $27,500.[33]

When both costs and operational capacity are considered, the preferred counter UAS solution against an autonomous loitering munition swarm is a high speed drone interceptor. If economic costs are of less importance and the emphasis is placed purely on operational capacity, then the use of MANPADS is recommended.

Conclusions

In this chapter we investigated how smart prioritisation can increase a unit's detection and neutralisation capability during a swarm attack. This is done by considering two case studies.

The effect of smart prioritisation on detection capability is assessed in the first case study, a Multi-Function Radar on an Air Defence and Command Frigate (ADCF). Several conclusions can be drawn based on the simulation results. Scheduling of an MFR using prioritisation of tasks based on Command Aim (CA) and Threat Warning (TW) leaves more time for Search functions when the system is saturated. At the same time, the scenarios that were considered showed earlier (or equal) detection times for all targets. This way of smart prioritisation depends on the correctness of external intelligence through the setting of the CA and TW. Further research is needed in order to optimise the setting of the priority ranges for scenarios of increasing numbers of drones. Also the optimisation for detection range or arrival time, instead of detection time could be investigated.

The effect of smart prioritisation on a unit's neutralisation capability was investigated in the second case study. Apart from an analysis of operational capacity, also a basic cost effectiveness was included in the research. Coordinated assaults with state of the art loitering munitions which employ swarming tactics were simulated in a point defence scenario. The simulation results clearly demonstrate that coordinated assaults through swarming tactics greatly improve effectiveness compared to uncoordinated assaults. A key factor in relation to the effectiveness is time. The loitering munitions have a flight speed with which they approach a target. The characteristics of the air defence system determine the minimum time required to detect, identify and neutralise a single target. These characteristics encompass the detection range, the fly-out range and flight speed of the kinetic interception measure (drone, missile, rocket, …). In addition, time is required in several steps of the process such as target designation and pointing the Counter UAS system in the correct direction. A coordinated assault with swarming tactics can be designed such that the time required to take out a single target by the Counter UAS system is maximised. By doing so, this reduces the number of loitering munitions required for a successful attack. Analogous to this, Counter UAS systems should use a smart prioritisation logic in the selection of targets to maximise effectiveness. Finally, based on a comparison of different types of Counter UAS systems it is concluded that drone interceptors show great promise. They are relatively slow in terms of flight speed but at the same time they are relatively cheap and thus they can be deployed in large numbers. In addition, the flight range is quite large and drone interceptors can perform completely different manoeuvres than conventional rockets and missiles. A crucial enabler for the drone interceptor is an accurate detection system with sufficient range.

Altogether, the two case studies combined highlight the potential value of smart prioritisation for detection and neutralisation of swarm attacks. The case studies are very specific but the methods used are transferable to other scenarios and weapon systems.

Notes

1 Stepanenko et al., "Russian Offensive Campaign Assessment, October 10".

2 Anokhin, "A Comprehensive Analytical Review".

3 Wertheim, "Ukraine's Magura Naval Drones: Black Sea Equalizers".

4 Ganzeveld et al., "Iran Update, April 13, 2024".

5 The original radar resource management model was developed by: Bolderheij, *Mission-driven Sensor Management.*

6 For a mathematical description of the prioritisation approach, see: Scherpenzeel, *MFR Sensor Management*; Bolderheij, *Mission-driven Sensor Management.*

7 Simulations are based on a scenario that was used in: Bolderheij, *Mission-driven Sensor Management*. The original scenario is expanded with swarms of UAVs, USVs, or both.

8 For a more detailed analysis of the scenarios, see: Scherpenzeel, *MFR Sensor Management*; Bolderheij, *Mission-driven Sensor Management.*

9 Kuchma, "Operational Experience in Ukraine".

10 Breiner and Ferran, "Loitering Munitions S3 Assessment Challenges".

11 A comprehensive overview of the capabilities of loitering munitions is provided in: Voskuijl, "Performance Analysis and Design of Loitering Munitions".

12 Bronk and Watling, "Mass Precision Strike".

13 Sarmat, "Strike Stories from Black Forrest".

14 Slyusar, "Lessons Learned from the Usage of Loitering Munitions".

15 Arquilla and Ronfeldt, "Swarming & the Future of Conflict".

16 Slyusar, "Lessons Learned from the Usage of Loitering Munitions".

17 Anonymous, "Obscure Russian V2U Drone".

18 Trevithick, "China Conducts Test of Massive Suicide Drone".

19 A comprehensive overview of swarming algorithms is presented in: Xu et. al. "Algorithms and Applications of Intelligent Swarm Cooperative Control".
20 Veldman, "De verdediging tegen zwermgedrag bij loitering munition".
21 Mollink, "De prestaties van een FH-901 kamikaze drone".
22 McDonald and Gloudemans, "Open Vehicle Sketch Pad".
23 Mollink, "De prestaties van een FH-901 kamikaze drone".
24 Aerodynamic and propulsion system characteristics were determined based on flight test and wind tunnel experiments. Dahlke and Batiuk, "Hydra 70, MK66 Aerodynamics and Roll Analysis".
25 Vetsuypens, "Defining the Launch Boundary"; Snip, "Performance van APKWS als ground-to-ground missile".
26 Veldman, "De verdediging tegen zwermgedrag bij loitering munition".
27 Rosema, Doyle and Blake, "Missile Data Compendium (DATCOM)".
28 Bras, "The Simulation of the SA-7B Grail"; Wartena, "The Threat of the SA-7B 'Grail'".
29 The complete model development, model analysis and the associated parameters are described in: Summeren, "The Quadcopter Interceptor".
30 Kelly, "An Introduction to Trajectory Optimization".
31 Veldman, "De verdediging tegen zwermgedrag bij loitering munition".
32 Pohasii et al., "Assessing the Economic and Technical Effectiveness".
33 For indications of the costs per missile, see Newdick, "Drone-Killing APKWS Laser-Guided Rockets", and Helfrich, "Laser-Guided Rockets Are Getting New Highly-Versatile Anti-Armor Warhead."

References

Anokhin, I. 2026. *A Comprehensive Analytical Review of Russian Shahed-type UAVs Deployment against Ukraine in 2025*, Institute for Science and International Security.

Anonymous. 2025. Obscure Russian V2U Drone Unraveled By Intelligence: Autonomous Loitering Munition Powered By Nvidia Chip, *Defense Express*, 9 June, 2025.

Arquilla, J., and D. Ronfeldt. 2000. *Swarming & the Future of Conflict*, RAND National Defense Research Institute.

Bolderheij, F. 2007. *Mission-driven Sensor Management, Analysis, Design, Implementation and Simulation.* PhD Thesis, TU Delft.

Bras, P. 2022. *The Simulation of the SA-7B Grail: Research to the Danger of the SA-7B Grail around a Runway for Landing and Departing Aircraft*, Bachelor's Thesis, Netherlands Defence Academy, Den Helder, the Netherlands.

Breiner, J., and M. Ferran. 2024. *Loitering Munitions S3 Assessment Challenges*, Technical report L-297, Munitions Safety Information Analysis Center.

Bronk, J., and J. Watling. 2024. *Mass Precision Strike. Designing UAV Complexes for Land Forces*, Occasional Paper, Royal United Services Institute (RUSI) for Defence and Security Studies.

Dahlke, C.W., and G. Batiuk. 1990. *Hydra 70, MK66 Aerodynamics and Roll Analysis*, Technical Report RD-SS-90-6, US Army Missile Command, Red Stone Arsenal, Alabama.

Ganzeveld, A., A. Braverman, J. Moore, A., Jhaveri, B. Carter, and N. Carl. 2024. *Iran Update, April 13*, Institute for the Study of War.

Helfrich, E. 2025. "Laser-Guided Rockets Are Getting New Highly-Versatile Anti-Armor Warhead", *The War Zone*.

Kelly, M. 2017. "An Introduction to Trajectory Optimization: How to Do Your Own Direct Collocation", *SIAM Review*, 59 (4), 849–904.

Kuchma, Y. 2025. "Operational Experience in Ukraine", *SAE Loitering Munitions Conference*, London, England.

McDonald, R.A., and J.R. Gloudemans. 2022. "Open Vehicle Sketch Pad: An Open Source Parametric Geometry and Analysis Tool for Conceptual Aircraft Design", *AIAA SCITECH 2022 Forum*, San Diego, California & Virtual.

Mollink, D. 2025. *De prestaties van een FH-901 kamikaze drone*, Bachelor's Thesis, Netherlands Defence Academy, Den Helder, The Netherlands.

Newdick, T. 2025. "Drone-Killing APKWS Laser-Guided Rockets Eyed For Eurofighter Typhoon", *The War Zone*.

Pohasii, S., M. Mashchenko, O. Serhiienko, O. Klimenko, T. Andriushchenko, and T. Milevska. 2023. "Assessing the Economic and Technical Effectiveness of Intercepting UAVs using Alternative Methods". *7th International Symposium on Multidisciplinary Studies and Innovative Technologies (ISMSIT)*, Ankara, Turkey.

Rosema, C., J. Doyle and W.B. Blake. 2014. *Missile Data Compendium (DATCOM) User Manual – 2014 Revision*, Technical Report AFRL-RQ-WP-TR-2014-0281, Air Force Research Laboratory, Wright-Patterson Air Force Base, Ohio.

Sarmat. 2025. "Strike Stories from Black Forrest" *SAE Loitering Munitions Conference*, London, England.

Scherpenzeel, M. 2026. *MFR Sensor Management*, Master's Thesis, Netherlands Defence Academy, Den Helder.

Slyusar, V. 2025. "Lessons Learned from the Usage of Loitering Munitions During the War in Ukraine", *SAE Loitering Munitions Conference*, London, England.

Snip, J. 2023. *Performance van APKWS als ground-to-ground missile*, Bachelor's Thesis, Netherlands Defence Academy, Den Helder, The Netherlands.

Stepanenko, K., G. Barros, R. Bailey, A. Howard, and F.W. Kagan. 2022. *Russian Offensive Campaign Assessment, October 10*, Institute for the Study of War.

Summeren, T. 2025. *The Quadcopter Interceptor: The Analysis of a High-Speed Air Defense Quadcopter*, Master's Thesis, Netherlands Defence Academy, Den Helder, The Netherlands.

Trevithick, J. 2020. "China Conducts Test of Massive Suicide Drone Swarm Launched from a Box on a Truck", *The War Zone*, 14 October 2020.

Veldman, J. 2025. *De verdediging tegen zwermgedrag bij loitering munition: een case study met de APKWS en de Feihong 901*, Bachelor's Thesis, Netherlands Defence Academy, Den Helder, The Netherlands.

Vetsuypens, N. 2023. *Defining the launch boundary of the advanced precision kill weapon system (apkws)*, Master's thesis, Netherlands Defence Academy, Den Helder, The Netherlands.

Voskuijl, M. 2022. "Performance Analysis and Design of Loitering Munitions: A Comprehensive Technical Survey of Recent Developments". *Defence Technology*, 18 (3), 325–343.

Wartena, G. 2023. *The threat of the SA-7B "Grail" against commercial aircraft: the development of an infrared detection model for the SA-7B "Grail"*, Bachelor's Thesis, Netherlands Defence Academy, Den Helder, the Netherlands.

Wertheim, E. 2025. "Ukraine's Magura Naval Drones: Black Sea Equalizers", *Proceedings of the U.S. Naval Institute*, 151 (9).

Xu, X., X. Yan, W. Yang, K. An, W. Huang, and Y. Wang. 2022. "Algorithms and Applications of Intelligent Swarm Cooperative Control: A Comprehensive Survey". *Progress in Aerospace Sciences*, 135.

CHAPTER 9

When Models Meet War

OR and AI Robustness for Western Doctrine in A2/AD

Ramon Soesan, Mark Patrick Roeling and Roy Lindelauf

Abstract

Anti-Access and Area Denial (A2/AD) environments demand rapid, resilient, and well-informed decision-making under contested conditions. Both Operations Research (OR) and Artificial Intelligence (AI) methods support decision-making. OR offers analytical frameworks and optimisation techniques, while AI promises data-driven capabilities to process growing sources of information. Recent discourse has emphasised AI's ability to generate and evaluate courses of action, but questions remain about the robustness of these methods in realistic operational contexts. This scoped review systematically evaluates robustness through a three-layer framework, operational, structural, and adversarial, comparing representative AI approaches with selected OR methods used in defensive planning. Our review reveals a gap between academic contributions versus implementation in BMSs. Adversarial robustness of AI methods requires investigation before it can be considered sufficient for A2/AD planning against opponents.

Keywords: Anti-Access/Area denial; Robustness; BMS; Artificial Intelligence; Operations research; Western doctrine

Introduction

In contested environments, implementing Anti-Access and Area Denial (A2/AD) strategies quickly becomes a complex, multi-domain issue.[1] Increasingly interconnected domains of warfare generate large volumes of data and simultaneously create opportunities for coordinating multi-domain operations like A2/AD.[2] However, developing and adapting A2/AD strategies is highly demanding and resource-intensive,[3] fuelling a growing demand for capable Decision Support Systems (DSS) that can operate under uncertainty and time pressure. Traditionally, Operations Research (OR) methods have provided the analytical backbone of DSS in the military decision-making process (MDMP), for example by delivering insights into defensive strategies and breaking down resource allocation problems.[4] Today, advances in Artificial Intelligence (AI) are enabling a new generation of DSS: AI-DSS.

This new generation can integrate more complex data at a vast speed, supporting commanders in developing Courses of Action (COA).[5]

Recent strategic analyses underscore the growing role of AI in military decision-making, particularly in the high-tempo, contested environment characteristics of A2/AD operations. Probasco emphasises the potential for AI-enabled systems to shorten decision cycles, and enhance adaptability in complex operational settings.[6] Pfaff and Hickey highlight AI and machine learning as enablers for accelerating COA development.[7] Several RAND studies display the benefits of using AI, but also stress the importance of human-AI teaming in order to ensure that automation augments rather than replaces human judgement.[8] Complementing these largely conceptual perspectives, Nadibaidze provides an empirically grounded view, examining AI-DSS deployments in Project Maven, the Russia-Ukrainian war, and the Israel-Hamas conflict.[9] They find that such systems can improve speed and situational awareness. Nevertheless, they also raise concerns about over-reliance, ambiguous accountability, and adherence to humanitarian norms. While these reports focus on the 'what' and 'why' of the adoption of AI in DSS, they offer less on the 'how'. In particular, the question remains how state-of-the-art academic AI methods can be used in practice.

This chapter focuses on the readiness of academic AI methods for actual deployment in military DSSs that support A2/AD strategies. Although most DSSs are not specifically designed for A2/AD scenarios, the contested, uncertain, and multi-domain characteristics of the A2/AD environment make structured decision support indispensable. As an interface between algorithms and commanders, Battle Management Systems (BMS) impose concrete constraints that any deployed DSS must satisfy. To make that integration problem concrete, we used WINLAS (an academically documented BMS) as a reference BMS.[10] Focusing on robustness as a central criterion, we developed a three-layer framework (structural, operational, and adversarial) for assessing AI integration readiness. Moreover, the framework is also applied to a selection of OR methods; they serve as an established, well researched benchmark.

Using a scoping literature review, we selected relevant AI methods for assessment. This enabled us to provide an overview of recent academic developments in AI for A2/AD-relevant DSS. To the best of our knowledge, no prior work on robustness (explained in the next section) of AI methods for A2/AD-relevant DSS has been conducted. To address this gap, we focus on the following research questions:

RQ1: What are the current academic developments of AI-DSS for A2/AD scenarios and what are the applications of OR methods?

RQ2: How is robustness handled in AI-DSS methods and how does it compare to OR methods?

To answer these research questions, the remainder of this chapter is constructed as follows. First, the robustness framework and literature search strategies are laid out in the next section. Thereafter, the findings and discussion are presented. To conclude this chapter, practical, policy and doctrinal implications of the findings are discussed in the final section.

Methodology

Robustness Framework

To evaluate the robustness of state-of-the-art academically developed AI and OR methods relevant for A2/AD, we first defined a robustness framework. The term robustness lacks an exact definition,[11] but in general, it reflects on the accuracy of a model under uncertainty in the input data.[12] Since we were interested in the actual integration of AI-DSS into real-world A2/AD operations, we developed a robustness framework that is grounded in both theoretical as well as practical views on robustness.

For the more theoretical part, our main source was Ashmore et al.[13] They argue that performance of any algorithm depends on the completeness of the input domain space. For an input domain space to be complete, three components are specifically put forth as subdomains. The operational subdomain controls the data that the model could expect as input, the failure subdomain entails the set of inputs if failure occurs somewhere in the system, and the adversarial subdomain considers the set of input data if the system is being attacked by an adversary.

To apply those subdomains to defence purposes and more specific to DSS for A2/AD, the terms coined by Ashmore et al. require integration into the military context. So, how could algorithms actually be used in A2/AD scenarios? Implementing any AI-DSS into a military operation needs an interface (frontend) between computational infrastructure (backend) and the military commander. A BMS is a ubiquitous interface used in military operations for this purpose.[14] Trivially, most BMSs are confidential and no details are known. However, some open source BMSs exist. For this chapter, we used WINLAS as an open source BMS reference.[15] With its focus on integrating AI, we were able to extract practical implications for enabling such integration. Despite the relevance of WINLAS, we are aware of the limitations that arise from the choice to only consider one BMS as reference. The risk is that it might depict a too one-sided representation of what a BMS looks like. Nevertheless, we justify this decision to only use WINLAS as a reference system because it is one of the most advanced, transparent BMS and unique in that it focuses on AI integration in particular.[16] WINLAS requires AI modules to operate as adaptive, real-time data processing planners. The AI modules should be able to cope with the ever-changing conditions on the battlefield.

Combining the theoretical and practical aspects of the two above-mentioned sources, we defined the following three-layer robustness framework:

- Structural: design and internal resilience of decision-support system. How the system handles component loss, degradation, redundancy and modular integration.
- Operational: performance under real-world circumstances. Concerns for example partial observability, sensor noise, communication latency or loss, and energy/processing constraints.
- Adversarial: resilience to intelligent, adaptive adversaries and deliberate deception. For example, spoofing, kinetic threats like missiles and mines, jamming, and camouflage.

These layers are interacting with each other in practice. Trivially, a weakness in one layer often produces cascading failures in others. For example, if a method is not robust against adversarial perturbations, it is likely to affect the structure of the method as well: failing nodes could lead to suboptimal structural decisions.[17] Besides the interactive dynamics of the robustness layers, the multi-domain nature of A2/AD environments also underlines the importance of all three layers. In multi-domain environments, a variety of elements is present. In order for this relation between elements to be effective, all should be robust in the broadest sense. Therefore, addressing all three robustness layers is of major importance in order to provide sufficient decision support.

Literature Search Strategy

With the robustness framework in place, we constructed a literature search strategy to retrieve relevant OR and AI methods. This search strategy was not equal for the OR and AI methods. As mentioned before, the OR methods are used as more of an established benchmark, rather than fully evaluated for their novelty. Since the field of OR for military operations predates modern AI research by decades, it has had substantially more time to mature, and many of its developments have already diffused to real-world applications. Therefore, it is interesting to see how robustness is handled in the more well established OR methods compared to the novel AI methods.

In line with this rationale, we searched for a selection of representative papers that described an OR or AI method relevant for an A2/AD scenario. Due to the different purposes of including OR and AI methods, we used two distinct protocols for retrieving relevant literature. To get a selection of the most prominent OR literature, we drew upon an overview paper of Hauskens.[18] He decomposed the field of OR in defence into five different subcategories and offered a selection of the last 50 years of literature in each of them. One pillar that he suggested was committed to protection strategies. With the large overlap between protection and A2/AD, we used the literature selection from the protection subcategory of Hauskens' paper as the core OR selection. This selection consisted of five papers (see Appendix).

For selecting the AI literature, we followed the PRISMA-ScR protocol.[19] The search scope focuses on AI-DSS for A2/AD scenarios. We made use of the PCC (population, concepts, and context) search term selection strategy. Regarding the population search terms, the following terms were utilised: decision support, battle management, course of action, COA, mission planning, and resource allocation. For the concepts, we used: artificial intelligence, reinforcement learning, and large-language models. Lastly, the context search terms were: defence, military, A2/AD, anti-access, area denial, command-and-control, and C2. The final search string was constructed such that it always contained one term out of each of the categories.

To identify potentially relevant papers, we conducted an electronic search (in September 2025) in the Netherlands Defence Academy (NLDA) and Naval Postgraduate School (NPS) databases, covering the period 2020 –2025. Both have a vast collection of databases, amongst others, IEEE, ProQuest, Springer. Only English-language, peer-reviewed articles were included in the search. However, we noticed that due to the novelty of LLM focused research relevant in the military context, not many peer-reviewed papers exist. Therefore, for the LLM methods we did not require them to be published in a peer-reviewed article.

The initial search identified 1218 papers. First, we removed 65 duplicates and 5 retracted papers. Next, we screened the remaining papers for relevance based on their titles and abstracts; 919 papers were excluded in this step, mainly because they lacked military relevance or did not propose any new method: 240 papers were sought for retrieval, 238 were actually retrieved. For those papers, the eligibility criteria displayed in Table 9.1 were used. Since the main focus of this book, NLARMS 2026, is on Dutch defence research, we decided to only include Dutch and NATO member state related papers. On top of that, the other two inclusion criteria enabled us to apply the robustness framework and to only focus on A2/AD scenarios. This resulted in the exclusion of 218 papers and a final selection of 20 papers. Table 9.1 presents a visual overview of the selection procedure.

Table 9.1: Inclusion and exclusion criteria for paper selection

Inclusion criteria	Exclusion criteria
First author affiliation in NATO member state	Not originating from NATO member state
Technical details of method described	No details
A2/AD relevant	Not relevant for A2/AD

Robustness Scoring

For the final selection of 25 papers (20 AI related, 5 OR related), we extracted robustness aspects along the line of the framework that we proposed earlier in this section. To answer RQ1 and RQ2, we coded each included paper for method class, domain/COA role, structural/operational/adversarial robustness. For the robustness evaluation, each layer could score four different values: strong, medium, weak, NA. For every layer, we flagged whether a layer was mentioned, modelled, simulated, and if real-world data/systems were used for evaluation. The value 'strong' was noted when a robustness layer scored positively on all flags, 'medium' when only the mentioned and modelled or simulated flag was positive, and weak if only robustness considerations were mentioned. A score of NA was given when that specific robustness layer was irrelevant for the paper's goal. To be able to compare robustness scores, the scores were converted to ordinal data: from strong/medium/weak/NA to 3/2/1/0. An important remark for this scoring is that the scores are not linearly related with each other. In other words, the difference between scores of 1 and 2 is not the same as the difference between 2 and 3: actual testing on real data or hardware is a significant bigger step than going from mentioning to modelling or simulating.

Since the search strategy focused on a military context, it could be expected that the majority of papers mentioned adversarial elements and somehow integrated adversarial agents/actions in their methods. Therefore, we added an extra robustness flag for the adversarial layer to provide more depth to the layer. Next to the aforementioned scoring system, we evaluated whether the adversarial agents/actions were modelled or simulated in an intelligent or adaptive way.

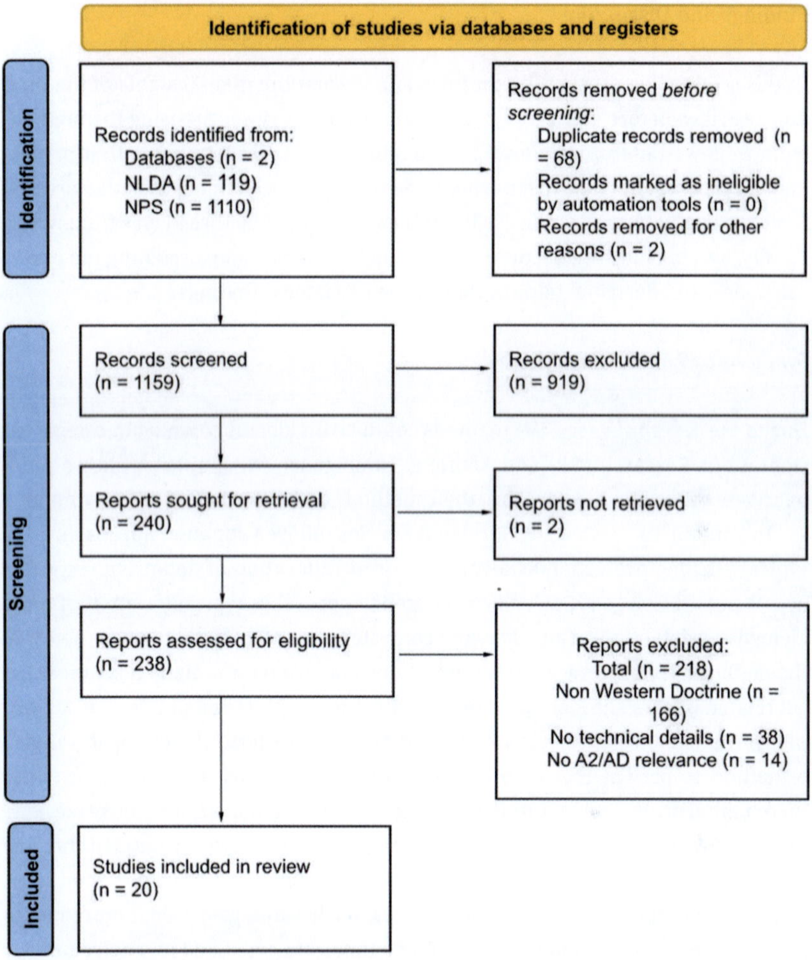

Figure 9.1 Flow chart of AI papers selection procedure

Findings and Discussion

In this section, we present the findings and discuss the main insights of the literature review. Before going into detail about robustness, synthesising the methods from reviewed literature allowed us to address RQ1. Therefore, we first present an overview about what specific methods were proposed in the included articles. Next, we evaluate and synthesise the robustness along the three-layer framework. Finally, we conclude this section with a dedicated subsection discussing the practical, policy, and doctrinal implications of the robustness findings.

Synthesis by Method Class (RQ1)

Due to the search strategy, the methods could be divided into two main categories: Operations Research (OR) and Artificial Intelligence (AI). Table 9.2 shows a full overview of the included articles, their methods and a short application summary.

In general, the reviewed OR related articles follow a common approach. They typically frame protection as a game-theoretic allocation of defensive resources across possible targets. Such game-theoretic approaches typically consist of three elements: a defender, an attacker, and contested assets. The main nuances between the methods are the underlying assumptions made about the elements. Across the OR related papers, the key variations are illustrative: Zhuang and Bier developed a model that balances strategic attacks and natural disasters.[20] Bier et al. developed a method to predict the marginal reduction in expected damage achieved by incremental protection investment.[21] Dong et al. extended this line by incorporating uncertainty in the attacker behaviour, and Chen et al. further extended this line by modelling dynamic target valuations.[22] Guikema et al. critiqued the usual static attacker assumptions of game-theoretic approaches and proposed a more intelligent attacker.[23] All the above mentioned methods are evaluated in a deterministic setting, in which the outcomes are analytically solved. Translated to A2/AD, these models are useful for considering where to invest defences in the light of a variety of threats.

In the AI related literature, the variety of use cases stood out. In the AI method class, a subdivision could be made in three subcategories: Large Language Models (LLM), Reinforcement Learning (RL), and a residual set of approaches (like Genetic Algorithms, Bayesian Inference and Multi Criteria Decision Analysis). The three articles in the LLM subcategory all focused on tools to aid specifically in the strategic end of the military decision-making process. Goeks and Waytowich developed an LLM-based COA generation framework evaluated in the real-time strategy (RTS) game environment called StarCraft II.[24] Lamparth et al. compared LLM and expert behaviour in wargame simulations and identified escalation tendencies and sensitivity differences.[25] Rivera et al. investigated LLM sensitivity to contextual and prompt changes in strategic decision tasks, and found that LLMs tend towards escalation.[26] Overall, these studies highlight LLMs' possible value for rapid and flexible COA drafting, which is an important part of constructing A2/AD strategies.

In contrast to the LLM methods, which tend to focus on the strategic level of the decision-making process, the RL related articles typically have a narrower scope and application-driven approach. Out of the six selected RL papers, three proposed a UAV related method; Wan et al. incorporated a surveillance mission method with honey drones as a decoy.[27] Allen et al. proposed an RL based pursuit game with uncertainty about the target information.[28] Gadiraju et al. developed a coordination algorithm for UAVs in dynamic military environments.[29] The remaining RL related works targeted broader COA or mission-planning functionality, but were still only evaluated for very specific situations.[30] In relation to A2/AD scenarios, the RL methods are useful for a variety of steps in the decision-making process. For example, surveillance by drones can be efficiently planned and executed by specific RL methods, or tactical COA planning can be optimised by means of RL.

In the residual set, various methods are discussed. However, a categorisation by use case could be made. Six out of eleven articles concerned applications for U(A) Vs. The issue to optimally coordinate a group of U(A)Vs under different constraints (e.g. vehicle availability, adversarial threat, or battery capacity) has received ample attention. Ulas et al. and Silva et al. both proposed a mission planning application for UAVs that aimed for strong network stability, respectively by a Genetic Algorithm and a hybrid method.[31] Babel proposed a similar application, but with missiles as UAV variant and RRT* (Rapidly Exploring Random Tree star algorithm) as methodological foundation.[32] Li and Epireanu and Chirala et al. developed UV planning methods (ABM and stochastic programming) with vehicle availability constraints, focusing respectively on vehicle status and fuel capacity.[33] Lee et al. leveraged the use of a Digital Twin to perform efficient route planning for UAVs.[34] The use cases discussed in those papers align very well with A2/AD: the U(A)Vs are mostly used for surveillance or protecting a specific area from penetration.

The four remaining articles proposed a variety of methods that can be used in A2/AD strategies. Wielgosz and Malysko looked specifically at a vessel detection MCDA method to protect maritime infrastructure.[35] Zigulic et al. came up with an image detection method to enhance military decision-making.[36] In contrast to the technical applications of the aforementioned methods, Kohn et al. and Munir et al. considered implementation related issues.[37] Kohn et al. focused on decision making for the use of lethal autonomous weapons, Munir et al. focused on the level of trust that commanders can give to situational awareness tools.

Taken together, the reviewed OR and AI studies provide a diverse set of methodological building blocks for A2/AD decision support. The OR contributes rigorous, strategic level reasoning on where to allocate protection, while AI literature supplies flexible, simulation-driven building blocks for COA drafting.

Table 9.2: Overview of the included papers in the review. The Method column displays the main methodology used, and the Application column shows a brief summary of the application.

Author/Date	Method	Application
Ulaş, Sezgin, and Boyaci (2025)	Genetic Algorithms	Dynamic drone networks for real-time usage.
Wielgosz and Malyszko (2025)	Multi Criteria Decision Analysis	Detection of vessels potentially threatening critical maritime infrastructure.
Silva, Reis, and Sargento (2023)	Domain Specific Language Modelling	Management and control of fleets of UAVs.
Dimitriu, Michaletzky, Remeli, et al. (2024)	Reinforcement Learning	Military simulation in strategy game.
Li and Epureanu (2020)	Agent Based Model	Competition of autonomous vehicle fleets with application to fleet modularity.
Goecks and Waytowich (2024)	Large Language Model	COA generating in strategy game.
Babel (2023)	RRT*	Flight path planning for a fleet of missiles.
Lee, Van Bossuyt, and Bickford (2021)	Digital Twin	Route planning for UAVs under contested environments.
Rivera, Mukobi, Reuel, et al. (2024)	Large Language Model	Military decision-making simulation.
Gadiraju, Karmakar, Shah, et al. (2024)	Reinforcement Learning	Coordinated UAV control in dynamic military environments.
Lampart, Corso, Ganz, et al. (2024)	Large Language Model	Behaviour in wargame simulation: humans vs. LLM.
Moreira, Velez-Bedoya, and Arango-López (2025)	Deep Neural Network	Recommendation systems for strategic decisions in strategy game.
Žigulić, Glučina, Lorencin, et al. (2024)	YOLOv5	Image detection for enhancing MDMP.
Wan, Cho, Zhu, et al. (2024)	Reinforcement Learning	Optimising drone surveillance missions via honey drones.
Allen, Schmidt, and Gabriel (2022)	Reinforcement Learning	Speed-overmatched pursuit games with uncertain target information.

Author/Date	Method	Application
Liles, Robbins, Lunday (2023)	Reinforcement Learning	Air battle management through contextual decomposition.
Altun, Furkan Ceran, Kutay Metin, et al. (2025)	Reinforcement Learning	Tackling complex tactical simulations with multi-agent training.
Kohn, Cohen, Johnson, et al. (2024)	Bayesian Decision Model	Supporting ethical decision-making for lethal autonomous weapons.
Munir, Aved, Pham, et al. (2024)	Link-weighted Trust Scoring Model	Quantifying trustworthiness of situational awareness systems.
Chirala, Venkatachalama, and Smereka (2023)	Two-Stage Stochastic Recourse	UV mission planning under uncertainty in vehicle availability.
Guikema and Aven (2009)	Game Theoretic Model	Perspective on approaches for assessing risk from intelligent attacks.
Zhuang and Bier (2007)	Game Theoretic Model	Defensive strategies with endogenous attacker effort.
Chen, Dong, Hunt, et al. (2023)	Game Theoretic Model	Defensive resource allocation with dynamic target valuations.
Dong, Chen, Hunt, et al. (2021)	Game Theoretic Model	The roles of forecast information and risk control on resource allocation.
Bier and Gutfraind (2019)	Game Theoretic Model	Defensibility of critical systems.

Abbreviations: YOLOv5; You Only Look Once version 5, which is a well-known open source computer vision model for object detection and classification. RRT*; Rapidly Exploring Random Tree star algorithm for trajectory generation.

Synthesis by Robustness (RQ2)

This section discusses robustness and we remind the reader that robustness scores of strong/medium/weak/NA were converted to ordinal data 3/2/1/0. Note that NA scores were excluded for mean computation. In this section, we synthesise the scores at high-over level, highlighting patterns and explain how method-level assumptions drive the observed strengths and gaps. A full per-paper robustness scoring is available in the Appendix.

At the highest-over level, we compared robustness between AI and OR methods. Table 9.3 shows the robustness scores for both categories. The OR methods scored higher across all robustness layers. Regarding the structural layer, the average values of 2 (standard deviation of 0.47) and 2.75 (0.38) for AI and OR respectively indicate that OR methods showed stronger structural robustness, whereas AI methods tended toward medium-level evidence. For the operational layer, the difference is smaller: an overall score of 1.7 (0.56) for the AI methods versus 1.75 (0.75) for OR. Again, for the adversarial layer, OR methods had a higher mean compared to the AI methods; 2.25 (0.38) versus 1.88 (0.52). To discover the mechanisms behind these differences, we next examined the categories in closer detail.

In the AI category, clear differences appear on the structural axis. LLM based methods are inherently black box, making structural implications for the outputted decision-support difficult to audit. For example, Lamparth et al. and Rivera et al. evaluated specific war-like scenarios with LLMs via prompts. The COAs given by the LLMs were purely textual and relations between structural elements remained unclear. Consequently, LLM studies scored low on structural robustness with a mean of 1.33 (0.44).[38] By contrast, RL studies often made the structure of COA elements an explicit modelling subject. This focus is reflected in the RL structural mean of 2.5 (0.5), which lies between medium and strong under our rubric. In the residual category, the structural robustness is moderate, considering the mean score of 1.86 (0.49). However, due to the diverse decision support roles, it could be useful to look at the different applications. The U(A)Vs related papers had a large focus on structural aspects, which is reflected in a structural robustness score of 2.2 (0.32). For the remaining methods, the structural score was 1.75 (0.75), remarkably lower.

Along the operational axis, the scores of the AI subcategories were situated closer to each other. Again, LLM based methods scored lowest, with a mean of 1.33 (0.44). The prompt evaluation setting and black box origin provided little visibility into operational implications of the suggested decision support. For the RL studies, a mean score of 2 (0) shows homogeneous medium operational robustness. Operational constraints, like energy consumption and battery capacity, were often explicitly mentioned and simulated in the RL methods but not evaluated using real-world data or hardware. For the residual group, the split between U(A)Vs and other applications does not seem to influence the scores: 1.67 (0.67) and 1.8 (0.69).

Regarding the adversarial axis, LLM (2 (0)), RL (1.83 (0.28)) and residual group (2 (0.5)) cluster near the medium robustness band. This was expected: since our search criteria targeted papers with a military or contested-context focus, most studies at least mentioned and simulated adversarial conditions. In this light, a score of around 2 shows that real-world adversarial tests remain rare. Interestingly, only one paper tested its method on real-world data: Wielgosz and Malyszko leveraged real maritime data to evaluate their vessel identification method.[39] Due to the sensitive nature of real-world adversarial data, and the subsequent scarcity of it in academic literature, the adaptive/intelligent flag could be insightful. Only Li and Epireanu satisfied this rubric by integrating and documenting an intelligent adversary[40]. Several other papers also included adversarial elements that may be adaptive in practice, but their black-box simulation reporting prevented us from confidently classifying them as intelligent. For example, multiple papers used real-time strategy (RTS) games for evaluating their proposed method.[41] Although those RTS games incorporate adversarial elements and actions, the adversarial behaviour was not transparent and it remained unclear how they influenced the proposed methods exactly. Therefore, adversarial robustness was scored as medium (2) for those methods.

For the OR methods, structural robustness scores relatively high with a mean of 2.75 (0.38). Chen et al., Bier et al. and Dong et al. all scored strong on structural robustness.[42] This reflects that these OR papers both construct explicit analytical/structural models (topologies, cascade or dependency formulations) and validate them by using real-world data. On the operational axis, OR methods score lower, 1.75 (0.56). In the OR methods, operational focus is typically not further reaching than assuming common information across all agents. The adversarial score of 2.25 (0.38) was the result of a similar pattern as the structural robustness score; adversarial assumptions were modelled and tested with real world data.

Overall, OR and AI methods display complementary but distinct robustness profiles. OR methods score higher on structural robustness: they typically start from explicit analytical models and frequently validate those models against empirical or case data, which produces clearer system-level provenance. By contrast, AI work is more simulation- and component-driven: LLM studies are black-box and weak on structural provenance, while RL methods often make COA structure explicit and therefore score higher on the structural axis. On the operational axis, the gap between OR and AI narrows: methods in both categories rely on assumptions and simulations rather than actual real-world representations. For the adversarial axis, many AI studies simulate adversarial behaviour, but limited auditability due to black-box and unclear evaluation settings lead to medium scores. OR approaches often model adversaries analytically and typically validate the model by evaluating a real-world case study. Taken together, the reviewed methods offer complementary strengths: formal structural modelling from OR and flexible, scenario-level testing from AI. However, few articles combine structural, operational and adversarial robustness aspects.

Table 9.3: This table shows the averages and standard deviations for AI methods versus OR method across structural, operational and adversarial robustness categories. The values indicate that across all axes the OR methods score stronger on robustness compared to the AI methods.

		Structural	*Operational*	*Adversarial*
AI methods	*LLM, n=3*	1.33 *(0.44)*	1.33 *(0.44)*	2 *(0)*
	RL, n=6	2.5 *(0.5)*	2 *(0)*	1.83 *(0.28)*
	Residual, n=11	1.86 *(0.49)*	1.64 *(0.69)*	1.88 *(0.59)*
	Overall, n=20	2 *(0.47)* *n_zero=3*	1.7 *(0.56)* *n_zero=0*	1.88 *(0.52)* *n_zero=2*
OR methods	*n=5*	2.75 *(0.38),* *n_zero=1*	1.75 *(0.75),* *n_zero=1*	2.25 *(0.38),* *n_zero=1*

Conclusion

In this study, we explored the current opportunities for AI decision support in A2/AD execution. Through a semi-systematic literature search, we identified 20 papers that proposed relevant AI methods. Moreover, we selected 5 OR A2/AD related methods as a comparative corpus. Opportunities were assessed twofold. We reviewed the latest developments of AI methods and applications in academics that could aid in A2/AD strategies. We applied a three-layer – operational, structural, and adversarial – robustness framework on those methods to evaluate the readiness for actual integration into BMSs. Overall, our synthesis showed that OR methods mainly focus on resource allocation problems through game-theoretic modelling, while the reviewed AI methods proposed a wider variety of applications and methods. Those can be used as building blocks for constructing A2/AD strategies.

Regarding the robustness of those methods, eminent gaps emerged across all layers. AI methods often focused on one of the layers, rarely on all three. For example, articles proposing a U(A)V related use case often make structural robustness a key element. On top of that, scoring black box methods on robustness proved challenging: exact functioning of such algorithms remained unclear. The reviewed OR methods scored stronger regarding robustness. Overall, these methods described the assumptions in their models clearer and validated on real-world data.

Doctrinal, Policy, and Practical Implications

What are the implications of our findings for decision-makers, policymakers, and military planners?

On the doctrinal level, our findings show that AI methods could be of use in a variety of applications. Where OR methods could be used for a broader perspective on A2/AD strategy development, the reviewed AI methods could serve as integrated building blocks for setting up and managing A2/AD. However, the integration of AI-DSS in future doctrines is only possible when a clear definition of, and requirements for, A2/AD are established. The reviewed methods were not specifically developed for A2/AD, and this leads to a possible gap between academic methods and military reality. To close this gap, developing clear, structured A2/AD concepts in the doctrines is of major importance. Subsequently, because a clear A2/AD definition is not in place, policymakers cannot decide on which methods to prioritise. Based on the reviewed literature and its connection to A2/AD, the following definition is proposed to guide DSS prioritisation: "A2/AD denotes the coordinated employment of multi-domain capabilities intended to restrict an adversary's ability to deploy, manoeuvre, and sustain forces within a theatre of operations."

We showed that the gap between medium and strong robustness scores is present across all layers. Given the nonlinear trend in the scoring – upgrading from medium to strong requires considerably more effort compared to upgrading from weak to medium – it would be sensible to prioritise the most promising methods and applications first. Focusing on increasing the robustness of those, and especially against adversarial threats, will strengthen the use of AI-DSS in A2/AD scenarios.

The robustness gaps show that military planners need to be very careful when using AI tools in planning and executing A2/AD strategies. Although the state-of-the-art academic AI methods propose relevant use cases, the current lack of strong robustness limits the possibilities for instant integration. Moreover, the black box nature of AI methods like LLMs and RL challenges the successful cooperation between humans and machines. Non transparent and therefore hard to interpret recommendations might not be sufficient for complex, contested environments like A2/AD.

When connecting the proposed A2/AD definition to the robustness assessment, the implications of the identified robustness gaps become clearer. The coordinated employment of multi-domain capabilities relates directly to structural robustness across interconnected systems. Restricting an adversary's ability to deploy and manoeuvre can only be effective when methods are resilient against adversarial adaptation. Moreover, sustaining denial effects under contested conditions requires operational robustness. Therefore, the identified robustness gaps indicate not only technical limitations of the reviewed methods, but also potential constraints for the integration and use of AI-DSS within A2/AD strategies.

Limitations and Future Research

Since we conducted a multidisciplinary literature search, selecting all relevant literature was seemingly impossible. Although we constructed our search terms with as much accuracy as possible, we might have missed possible relevant papers. By focusing on Western doctrine, we narrowed the selection significantly: the majority of relevant papers did not originate from Western doctrines (see Figure 9.1). However, since the scope of this chapter is mainly focused on the robustness of DSS tools for A2/AD rather than a fully exhaustive overview of all possible methods, we are convinced that the results are representative.

Future research should focus on a quantitative approach on scoring robustness for AI-DSS. A benchmark for scoring across the three-layer robustness framework is interesting and could be very effective for assessing AI-DSS tools.

Appendix

Table 9.4: Overview of included papers, the main method class (AI or OR) of those papers, and for the AI articles a method subcategory layer. Robustness scores per paper. Table is ranked on the sum score across the robustness layers. Str., Op., and Adv. stand for structural, operational and adversarial, respectively.

Author/date	Method Cat.	Method Subcat.	Str.	Opl.	Adv.
Munir, Aved, Pham, et al. (2024)	AI	Other	3	3	2
Chen, Dong, Hunt, et al. (2023)	OR	OR	3	3	2
Dong, Chen, Hunt, et al. (2021)	OR	OR	3	2	3
Silva, Reis, and Sargento (2023)	AI	Other	3	3	0
Wan, Cho, Zhu, et al. (2024)	AI	RL	3	2	2
Gadiraju, Karmakar, Shah, et al. (2024)	AI	RL	3	2	1
Bier and Gutfraind (2019)	OR	OR	3	1	2
Li and Epureanu (2020)	AI	Other	2	1	3
Wielgosz and Malyszko (2025)	AI	Other	0	2	3
Dimitriu, Michaletzky, Remeli, et al. (2024)	AI	RL	2	2	2
Liles, Robbins, and Lunday (2023)	AI	RL	2	2	2
Altun, Furkan Ceran, Kutay Metin, et al. (2025)	AI	RL	2	2	2
Chirala, Venkatachalama, and Smereka (2023)	AI	Other	2	2	1
Rivera, Mukobi, Reuel, et al. (2024)	AI	LLM	1	2	2
Goecks and Waytowich (2024)	AI	LLM	2	1	2
Babel (2023)	AI	Other	2	1	2
Moreira, Velez-Bedoya, and Arango-López (2025)	AI	Other	2	1	2
Zhuang and Bier (2007)	OR	OR	2	1	2
Žigulić, Glučina, Lorencin, et al. (2024)	AI	Other	1	2	1
Allen, Schmidt, and Gabriel (2022)	AI	RL	0	2	2
Ulaş, Sezgin, and Boyaci (2025)	AI	Other	2	1	1
Lampart, Corso, Ganz, et al. (2024)	AI	LLM	1	1	2
Kohn, Cohen, Johnson, et al. (2024)	AI	Other	1	1	1
Lee, Van Bossuyt, and Bickford (2021)	AI	Other	0	1	2
Guikema and Aven (2009)	OR	OR	0	0	0

Notes

1 Iv and Matsumura, *The Army's Role in Overcoming Anti-Access.*
2 Burmaoglu et al., "Defense 4.0: Internet of Things in Military".
3 Krepinevich et al., *Meeting the Anti-Access and Area-Denial Challenge.*
4 Yıldırım, "Military Operations Research".
5 Schubert, *Artificial Intelligence for Decision Support.*
6 Probasco, *AI for Military Decision-Making.*
7 Pfaff and Hickey, "Integrating Artificial Intelligence".
8 Scholl, "Understanding the Limits of Artificial Intelligence"; Burdette et al., *An AI Revolution in Military Affairs?.*
9 Nadibaidze, "AI in Military Decision Support Systems".
10 Testón et al., "The Role of Automated Planning".
11 Bradley, "Robustness?".
12 Ben-Tal et al., *Robust Optimization.*
13 Ashmore et al., "Assuring the Machine Learning Lifecycle".
14 Seizovic et al., "Emergent Behaviour in the Battle Management System".
15 Testón et al., "The Role of Automated Planning".
16 Ibid.
17 Wan et al., "Optimizing Effectiveness and Defense".
18 Hausken, "Fifty Years of Operations Research in Defense".
19 Tricco et al., "PRISMA Extension for Scoping Reviews (PRISMA-ScR)".
20 Zhuang and Bier, "Balancing Terrorism and Natural Disasters".
21 Bier and Gutfraind, "Risk Analysis beyond Vulnerability and Resilience".
22 Dong et al., "Defensive Resource Allocation"; Chen et al., "Counterterrorism Resource Allocation".
23 Guikema and Aven, "Assessing Risk from Intelligent Attacks".
24 Goecks and Waytowich, "COA-GPT: Generative Pre-Trained Transformers".
25 Lamparth et al., "Human vs. Machine".
26 Rivera et al., "Escalation Risks from Language Models".
27 Wan et al., "Optimizing Effectiveness and Defense".
28 Allen et al., "Reinforcement Learning Approach".
29 Gadiraju et al., «GLIDE: Multi-Agent Deep Reinforcement».
30 Altun et al., "Strategic Implementation of Super-Agents"; Liles et al., "Solving Nonstationary Markov Decision"; Dimitriu et al., "A Reinforcement Learning Approach".
31 Ulaş et al., "A Hybrid Optimization Framework"; Silva et al., "A Mission Planning Framework".
32 Babel, "Coordinated Flight Path Planning".
33 Li and Epureanu, "AI-Based Competition of Autonomous Vehicle Fleets".; Chirala et al., "UV Mission Planning Under Uncertainty in Vehicles' Availability".
34 Lee et al., "Digital Twin-Enabled Decision Support".
35 Wielgosz and Malyszko, "A Method for Early Identification of Vessels".
36 Žigulić et al., "Military Decision-Making Process".
37 Kohn et al., "Supporting Ethical Decision-Making"; Munir et al., "Trustworthiness of Situational Awareness".
38 Lamparth et al., "Human vs. Machine"; Rivera et al., "Escalation Risks from Language Models".
39 Wielgosz and Malyszko, "A Method for Early Identification".
40 Li and Epureanu, "AI-Based Competition of Autonomous Vehicle Fleets".

[41] Dimitriu et al., "A Reinforcement Learning Approach"; Goecks and Waytowich, "COA-GPT: Generative Pre-Trained Transformers"; Moreira et al., "Integration of Causal Models";

[42] Bier and Gutfraind, "Risk Analysis beyond Vulnerability and Resilience"; Dong et al., "Defensive Resource Allocation"; Chen et al., "Counterterrorism Resource Allocation".

References

Allen, Julia Filiberti, Steve Schmidt, and Steven A. Gabriel. 2022. "Reinforcement Learning Approach to Speed-Overmatched Pursuit Games with Uncertain Target Information". *Military Operations Research (Alexandria, Va.)* 27 (4), 37–50. https://doi.org/10.5711/1082598327437.

Altun, Huseyin Oktay, Huseyin Furkan Ceran, Korkut Kutay Metin, Tolga Erol, and Emre Fisne. 2025. "Strategic Implementation of Super-Agents in Heterogeneous Multi-Agent Training for Advanced Military Simulation Adaptability". *IEEE Access* (Piscataway) 13, 96544–63. https://doi.org/10.1109/ACCESS.2025.3573419.

Ashmore, Rob, Radu Calinescu, and Colin Paterson. 2021. "Assuring the Machine Learning Lifecycle: Desiderata, Methods, and Challenges". *ACM Comput. Surv.* 54 (5), 111:1–111:39. https://doi.org/10.1145/3453444.

Babel, Luitpold. 2023. "Coordinated Flight Path Planning for a Fleet of Missiles in High-Risk Areas". *Robotica* (Cambridge, UK) 41 (5), 1568–89. https://doi.org/10.1017/S0263574722001886.

Ben-Tal, Aharon, Arkadi Nemirovski, and Laurent El Ghaoui. 2009. *Robust Optimization*, 1–576. https://www.torrossa.com/en/resources/an/5576051.

Bier, Vicki, and Alexander Gutfraind. 2019. "Risk Analysis beyond Vulnerability and Resilience – Characterizing the Defensibility of Critical Systems". *European Journal of Operational Research* 276 (2), 626–36. https://doi.org/10.1016/j.ejor.2019.01.011.

Bradley, James V. "Robustness?" 1978. *British Journal of Mathematical and Statistical Psychology* 31 (2), 144–52. https://doi.org/10.1111/j.2044-8317.1978.tb00581.x.

Burdette, Zachary, Dwight Phillips, Jacob L Heim, et al. n.d. *An AI Revolution in Military Affairs? How Artificial Intelligence Could Reshape Future Warfare*. RAND.org.

Burmaoglu, Serhat, Ozcan Saritas, and Haydar Yalcin. 2019. "Defense 4.0: Internet of Things in Military". In *Emerging Technologies for Economic Development*, edited by Dirk Meissner, Leonid Gokhberg, and Ozcan Saritas. Springer International Publishing. https://doi.org/10.1007/978-3-030-04370-4_14.

Caballero Testón, J., Olaya Pérez-Mon, and María D. R-Moreno. 2025. "The Role of Automated Planning in Battle Management Systems for Military Tactics". *Expert Systems with Applications* 297 (February): 129259. https://doi.org/10.1016/j.eswa.2025.129259.

Chen, Xia, Yucheng Dong, Kyle Hunt, and Jun Zhuang. 2023. "Counterterrorism Resource Allocation during a Pandemic: The Effects of Dynamic Target Valuations When Facing a Strategic Terrorist". *Risk Analysis* 43 (6), 1235–53. https://doi.org/10.1111/risa.13992.

Chirala, Venkata Sirimuvva, Saravanan Venkatachalam, and Jonathon M. Smereka. 2023. "UV Mission Planning Under Uncertainty in Vehicles' Availability". *Journal of Intelligent & Robotic Systems* (Dordrecht) 108 (2), 26. https://doi.org/10.1007/s10846-023-01860-z.

Dimitriu, Adonisz, Tamás V. Michaletzky, Viktor Remeli, and Viktor R. Tihanyi. 2024. "A Reinforcement Learning Approach to Military Simulations in Command: Modern Operations". *IEEE Access* 12, 77501–13. https://doi.org/10.1109/ACCESS.2024.3406148.

Dong, Yucheng, Xia Chen, Kyle Hunt, and Jun Zhuang. 2021. "Defensive Resource Allocation: The Roles of Forecast Information and Risk Control". *Risk Analysis* 41 (8), 1304–22. https://doi.org/10.1111/risa.13626.

Gadiraju, Divija Swetha, Prasenjit Karmakar, Vijay K. Shah, and Vaneet Aggarwal. 2024. "GLIDE: Multi-Agent Deep Reinforcement Learning for Coordinated UAV Control in Dynamic Military Environments". *Information (Basel)* (Basel) 15 (8), 477. https://doi.org/10.3390/info15080477.

Goecks, Vinicius G., and Nicholas Waytowich. 2024. "COA-GPT: Generative Pre-Trained Transformers for Accelerated Course of Action Development in Military Operations". *2024 International Conference on Military Communication and Information Systems (ICMCIS)*, April, 01–10. https://doi.org/10.1109/ICMCIS61231.2024.10540749.

Guikema, Seth D., and Terje Aven. 2010. "Assessing Risk from Intelligent Attacks: A Perspective on Approaches". *Reliability Engineering & System Safety* 95 (5), 478–83. https://doi.org/10.1016/j.ress.2009.12.001.

Hausken, Kjell. 2024. "Fifty Years of Operations Research in Defense". *European Journal of Operational Research* 318 (2), 355–68. https://doi.org/10.1016/j.ejor.2023.12.023.

Iv, John Gordon, and John Matsumura. n.d. *The Army's Role in Overcoming Anti-Access and Area Denial Challenges*. RAND.org

Kohn, Spencer, Marvin Cohen, Athena Johnson, Mikhail Terman, Gershon Weltman, and Joseph Lyons. 2024. "Supporting Ethical Decision-Making for Lethal Autonomous Weapons". *Journal of Military Ethics* 23 (1), 12–31. https://doi.org/10.1080/15027570.2024.2366094.

Krepinevich, Andrew, Barry Watts, and Robert Work. n.d. *Meeting the Anti-Access and Area-Denial Challenge*. CSBA.

Lamparth, Max, Anthony Corso, Jacob Ganz, Oriana Skylar Mastro, Jacquelyn Schneider, and Harold Trinkunas. 2024. "Human vs. Machine: Behavioral Differences between Expert Humans and Language Models in Wargame Simulations". *Proceedings of the AAAI/ACM Conference on AI, Ethics, and Society* 7 (1), 1. https://doi.org/10.1609/aies.v7i1.31681.

Lee, Eugene Boon Kien, Douglas L. Van Bossuyt, and Jason F. Bickford. 2021. "Digital Twin-Enabled Decision Support in Mission Engineering and Route Planning". *Systems (Basel)* (Basel) 9 (4), 82. https://doi.org/10.3390/systems9040082.

Li, Xingyu, and Bogdan I. Epureanu. 2020. "AI-Based Competition of Autonomous Vehicle Fleets with Application to Fleet Modularity". *European Journal of Operational Research* 287 (3), 856–74. https://doi.org/10.1016/j.ejor.2020.05.020.

Liles, Joseph M., Matthew J. Robbins, and Brian J. Lunday. 2023. "Solving Nonstationary Markov Decision Processes via Contextual Decomposition: A Military Air Battle Management Application". *Expert Systems With Applications* 233. Elsevier. https://doi.org/10.1016/j.eswa.2023.120949.

Moreira, Fernando, Jairo Ivan Velez-Bedoya, and Jeferson Arango-López. 2025. "Integration of Causal Models and Deep Neural Networks for Recommendation Systems in Dynamic Environments: A Case Study in StarCraft II". *Applied Sciences* (Basel) 15 (8), 4263. https://doi.org/10.3390/app15084263.

Munir, Arslan, Alexander Aved, Khanh Pham, and Joonho Kong. 2024. "Trustworthiness of Situational Awareness: Significance and Quantification". *Journal of Cybersecurity and Privacy* (Washington) 4 (2), 223–40. https://doi.org/10.3390/jcp4020011.

Nadibaidze, Anna. 2024. "AI in Military Decision Support Systems: A Review of Developments and Debates". ResearchGate, 10 January. https://www.researchgate.net/publication/385551984_AI_in_Military_Decision_Support_Systems_A_Review_of_Developments_and_Debates.

Pfaff, C., and Christopher Hickey. 2025. "Integrating Artificial Intelligence and Machine Learning Technologies into Common Operating Picture and Course of Action Development". *Books, Monographs & Collaborative Studies*, 15 July. https://press.armywarcollege.edu/monographs/980.

Probasco, Emelia. 2025. *AI for Military Decision-Making*. Center for Security and Emerging Technology, Georgetown University.

Rivera, Juan-Pablo, Gabriel Mukobi, Anka Reuel, Max Lamparth, Chandler Smith, and Jacquelyn Schneider. 2024. “Escalation Risks from Language Models in Military and Diplomatic Decision-Making”. *Proceedings of the 2024 ACM Conference on Fairness, Accountability, and Transparency* (New York, NY, USA), FAccT ’24, 5 June, 836–98. https://doi.org/10.1145/3630106.3658942.

Scholl, Keller. 2025. “Understanding the Limits of Artificial Intelligence for Warfighters”. Volume 5, Mission Planning | RAND. Accessed 28 July. https://www.rand.org/pubs/research_reports/RRA1722-5.html.

Schubert, Johan. 2018. *Artificial Intelligence for Decision Support in Command and Control Systems*. 1 January. Conference: 23rd International Command and Control Research & Technology Symposium

Seizovic, Aleksandar, David Thorpe, and Steven Goh. 2022. “Emergent Behavior in the Battle Management System”. *Applied Artificial Intelligence* 36 (1), 2151183. https://doi.org/10.1080/08839514.2022.2151183.

Silva, Margarida, André Reis, and Susana Sargento. 2023. “A Mission Planning Framework for Fleets of Connected UAVs”. *Journal of Intelligent & Robotic Systems* (Dordrecht) 108, (1) 2. https://doi.org/10.1007/s10846-023-01820-7.

Tricco, Andrea C., Erin Lillie, Wasifa Zarin, et al. 2018. “PRISMA Extension for Scoping Reviews (PRISMA-ScR): Checklist and Explanation”. *Annals of Internal Medicine* 169 (7), 467–73. https://doi.org/10.7326/M18-0850.

Wan, Zelin, Jin-Hee Cho, Mu Zhu, Ahmed Anwar, Charles Kamhoua, and Munindar Singh. 2024. “Optimizing Effectiveness and Defense of Drone Surveillance Missions via Honey Drones”. *ACM Transactions on Internet Technology* (New York, NY) 24 (4), 1–26. https://doi.org/10.1145/3701233.

Wielgosz, Miroslaw, and Marzena Malyszko. 2025. “A Method for Early Identification of Vessels Potentially Threatening Critical Maritime Infrastructure”. *Applied Sciences* (Basel) 15 (15), 8716. https://doi.org/10.3390/app15158716.

Yıldırım, Ugur. 2021. “Military Operations Research: Origins, Recent Past, and Present”. *Türk Savaş Çalışmaları Dergisi* 2 (1), 1–16. https://doi.org/10.52792/tws.916617.

Zhuang, Jun, and Vicki M. Bier. 2007. “Balancing Terrorism and Natural Disasters – Defensive Strategy with Endogenous Attacker Effort”. *Operations Research* 55 (5), 976–91. https://doi.org/10.1287/opre.1070.0434.

Žigulić, Nikola, Matko Glučina, Ivan Lorencin, and Dario Matika. 2024. «Military Decision-Making Process Enhanced by Image Detection». *Information (Basel)* (Basel) 15 (1), 11. https://doi.org/10.3390/info15010011.

CHAPTER 10

Navigating Contested Airspace

Multi-Objective Path Optimisation for UAVs in A2/AD Environments – Decision-support through Pareto-based Trade-off Analysis

Alexander van Oers, Jochem van Dam, and Relinde Jurrius

Abstract

The ongoing war in Ukraine, often called the first drone war, illustrates how inexpensive drones can shift the battlefield. Unmanned Aerial Vehicles (UAVs) have become indispensable assets for reconnaissance and strike missions, operating in environments where Anti-Access/Area Denial (A2/AD) strategies rely heavily on radar networks to restrict freedom of action. The operational challenge is that UAVs must simultaneously minimise flight distance and probability of detection. These objectives are inherently conflicting: detours can reduce exposure but increase flight time. Conventional shortest-path algorithms do not adequately capture these trade-offs, limiting their utility for mission planning in denied airspace. This chapter presents a method for multi-objective path optimisation that integrates probabilistic radar detection modelling with a metaheuristic optimisation procedure. Our method combines Particle Swarm Optimisation (PSO) with Random Geometric Graphs (RGGs) to generate feasible paths. Rather than a single route, the method produces a Pareto front of candidate paths, offering decision-makers and autonomous systems a spectrum of trade-offs between efficiency and survivability. From an A2/AD perspective, this method enables evaluation of radar coverage effectiveness and provides a tool for UAV route selection under contested conditions.

Keywords: UAV route planning; Pareto optimisation; Radar detection modelling; Particle swarm optimisation; Decision support

Introduction

The ongoing conflict in Ukraine has highlighted the transformative impact of Unmanned Aerial Vehicles (UAVs) on modern warfare. Often described as the first 'drone war,' this conflict illustrates how relatively inexpensive UAVs can provide reconnaissance, surveillance, and strike capabilities. UAVs are particularly valuable in environments where Anti-Access/Area Denial (A2/AD) strategies are employed:

their smaller size, lower cost, and flexibility allow them to operate in contested airspace that would be too risky or costly for crewed aircraft. Nevertheless, adversaries increasingly rely on integrated sensor networks, particularly radar systems, to restrict freedom of manoeuvre and deny access to critical areas. Even for UAVs, operating under such conditions involves a trade-off: minimising flight distance to complete missions efficiently while simultaneously minimising the risk of detection by hostile sensors.

Traditional path planning approaches, such as shortest-path algorithms or deterministic graph searches, are ill-suited to this problem. They fail to capture the probabilistic nature of detection and do not readily account for operational trade-offs between survivability and mission efficiency. Many existing UAV path-planning methods focus on obstacle avoidance or coverage, without explicitly modelling threat exposure or detection risk as a primary optimisation criterion.[1] For example, in Huang and Fei (2018),[2] UAVs were not forbidden to enter a radar detection zone, but the penalty associated with doing so was constant. In Zhang et al. (2022)[3] and Zhang et al. (2023)[4] path planning for stealth-UAVs was done by enforcing safety relative to a fixed detection threshold.

Our approach adopts an a posteriori multi-objective optimisation framework, in which no single 'best' trajectory is prescribed. Instead, we generate a Pareto front of candidate UAV paths, each representing a different balance between two competing mission criteria: flight distance and cumulative detection risk. This Pareto set reflects the reality that operational preferences depend on context: whether path length (travel time) or detection risk (acceptable exposure) is the dominant constraint. By presenting this spectrum of non-dominated options, our method allows mission planners or autonomous systems to evaluate how much additional risk is incurred by shortening a route, or how much extra distance is required to remain below a particular risk threshold.

From a doctrinal and operational perspective, our method acts as a decision-support tool. It supports contingency planning and facilitates discussion of acceptable risk, enabling more resilient UAV employment in contested airspace. In practice, commanders can evaluate route options under varying adversary sensor postures, anticipate the consequences of radar activation or degradation, and select paths that balance survivability, timeliness, and resource constraints.

The remainder of this chapter is organised as follows. The second section introduces the probabilistic radar-detection model and discusses its assumptions. The third section formulates the multi-objective path-planning problem and details the hybrid PSO–RGG optimisation method. The fourth section presents simulation results, including Pareto fronts and comparative simulations. The fifth section discusses the doctrinal and operational implications for employing UAVs in A2/AD environments.

Radar Detection Model

In this section, we describe the radar detection model used to evaluate UAV exposure in an A2/AD environment. We first introduce the formulas that relate radar system parameters to detection performance and combine them into a range–detection probability curve. This relationship is illustrated using representative parameters for a counter-UAS radar. Subsequently, we define how detection probability is used at the mission-planning level to evaluate cumulative detection risk along UAV trajectories.

The resulting range-detection probability model forms the basis for the cumulative detection risk metric used in the optimisation framework described in the third section.

Albersheim's Equation

A widely used approximation[5] in radar detection theory relates the required signal-to-noise ratio (SNR) to a specified probability of detection P_d and false-alarm probability P_{fa}. The SNR quantifies the strength of the received radar signal relative to noise. In general, effective detection requires sufficiently high SNR to achieve a high P_d, while maintaining a low P_{fa}.

Albersheim's equation provides a compact relation between these quantities and can be used to compute the required SNR for given values of P_d and P_{fa},

$$SNR = A + 0.12\,A\,B + 1.7\,B, \tag{1}$$

where

$$A = \log\left(\frac{0.62}{P_{fa}}\right), \text{ and } B = \log\left(\frac{P_d}{1 - P_d}\right).$$

Proposed by Albersheim (1981)[6] and subsequently examined in detail by Tufts and Cann (1983),[7] this approximation has become a standard tool for system-level radar performance modelling and mission-level analysis. It captures radar detection performance with low computational cost and small error over the operationally relevant ranges of P_d and P_{fa}.

In this work, P_d is interpreted as the probability of detection during a single radar sweep (or dwell) at a given range. The model represents an instantaneous detection assessment at a fixed relative position between UAV and radar. Detailed temporal integration across multiple sweeps is not modelled explicitly, as the focus here is on comparative route planning.

Radar Equation

To relate the signal-to-noise ratio SNR to the distance between the radar and the UAV (range R) we use the radar equation,[8]

$$SNR = \frac{P_t G^2 \lambda^2 \sigma}{(4\pi)^3 R^4 k T_s B_n L_s}, \tag{2}$$

where P_t is the transmitted power, G the antenna gain, λ the wavelength, σ the radar cross section of the UAV, k the Boltzmann constant, T_s the system noise temperature, B_n the receiver noise bandwidth and L_s the system losses. The radar cross section is treated as a fixed parameter representative of a small UAV; distinctions between UAV classes are outside the scope of this mission-level study. All radar and target dependent parameters are lumped into a single effective constant *K*, yielding a simplified relation SNR = K/R^4. Different values of *K* are used to represent short-, medium-, and long-range radar systems. The model is therefore parametric and scenario-level, rather than tied to specific UAV classes or platform characteristics.

Equations (1) and (2) together produce a *range–detection probability curve*: for a specified false-alarm probability P_{fa}, Albersheim's equation determines the required SNR for a desired detection probability, while the radar equation maps this SNR to a corresponding detection range. This relationship is shown in Figure 10.1, which illustrates the detection probability P_d as a function of range *R* (and SNR) for false-alarm probabilities P_{fa}.

Using representative parameters for a short-range X-band counter-UAS radar (e.g. operating at 10 GHz with a receiver bandwidth of 200 MHz), the resulting detection curve predicts that a typical small UAV becomes detectable out to approximately 2.2 km when the false-alarm probability is set to 10^{-5}. Higher false-alarm tolerances increase the effective detection range.

A maximum detection range R_{max} can be identified, beyond which the probability of detection becomes negligibly small without generating too many false alarms. Targets beyond this range are considered undetectable by the radar. The maximum range corresponds to the minimum SNR required for detection.

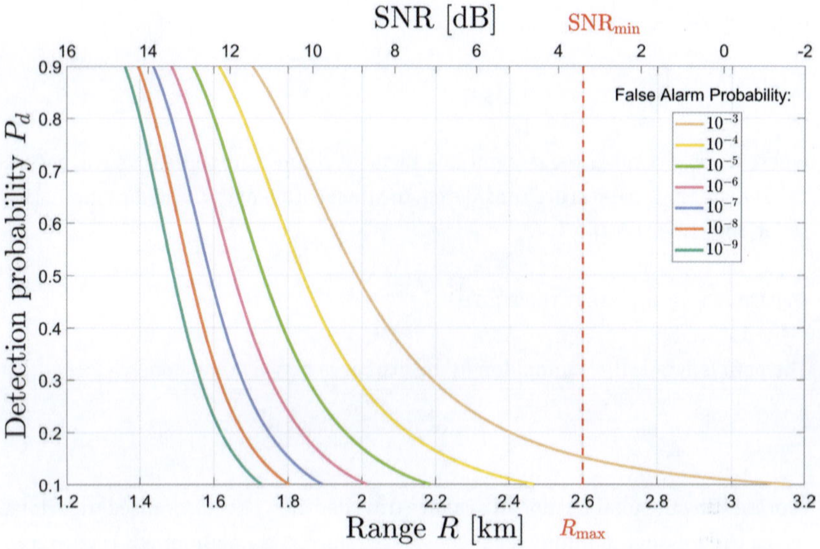

Figure 10.1: Relation between signal-to-noise ratio (SNR), range R, false-alarm probability P_{fa}, and single-sweep detection probability P_d for a representative short-range counter-UAS radar. The curves illustrate how detection probability decreases with range for fixed false-alarm settings, defining an effective detection range R_{max}.

Optimisation Method and Multi-Objective Philosophy

UAV trajectory planning in A2/AD environments is multi-objective: minimal path length and minimal radar detection risk cannot be achieved simultaneously. Any attempt to collapse these conflicting objectives into a single scalar metric risks obscuring the operational trade-offs that military planners must understand. For this reason, we adopt an *a posteriori multi-objective optimisation philosophy*, where the aim is not to produce a single 'best' trajectory but instead to generate a *set of Pareto-optimal paths*. This Pareto set is then presented to planners and commanders as a *decision-support tool* and a *platform for discussion.*

Multi-Objective Optimisation Formulation

In our formulation, a UAV trajectory is discretised into a sequence of 2D nodes (or waypoints)

$$\mathbf{x}_i = (x_i, y_i), \qquad i = 0, \ldots, N,$$

with path segments (edges) as

$$\boldsymbol{e}_i = [\boldsymbol{x}_i, \boldsymbol{x}_{i+1}], \qquad i = 0, \ldots, N-1,$$

where $[\boldsymbol{x}_i, \boldsymbol{x}_{i+1}]$ denotes the straight-line segment connecting successive waypoints.

Two mission criteria are considered: the total path length $\boldsymbol{L}$ and the cumulative radar detection risk $\boldsymbol{D}$.

Objective 1 – Minimise path length

The path length is the summation of the distances between the nodes,

$$L = \sum_{i=0}^{N-1} \|\boldsymbol{x}_{i+1} - \boldsymbol{x}_i\|, \tag{3}$$

representing the total Euclidean distance travelled along the discretised trajectory.

At the mission-planning level considered here, total path length is used as a proxy for mission efficiency, encompassing mission duration and resource expenditure. For UAVs operating at constant cruise speeds, path length relates directly to mission duration and to energy (or fuel) expenditure under nominal conditions. Using distance as an objective keeps the formulation platform-agnostic and avoids assumptions about specific vehicle dynamics.

Objective 2 – Minimise cumulative detection risk

For each path segment (each edge), the radar detection model described in the second section yields a probability of detection $P_d(\boldsymbol{e}_i)$, interpreted as the probability of detection during a single radar sweep at a representative range for that segment. Detection events on different segments are assumed statistically independent.

The probability of remaining undetected over the entire path is given by the product of the non-detection probabilities over all path segments,

$$\prod_{i=0}^{N-1} (1 - \boldsymbol{P}_d(\boldsymbol{e}_i)). \tag{4}$$

In practice, $\boldsymbol{P}_d(\boldsymbol{e}_i)$ is evaluated by sampling the detection model at a small number of representative points along the segment $\boldsymbol{e}_i$ and combining these samples into a single segment-level detection probability.

The cumulative detection risk D, defined as the probability that the UAV is detected at least once along the path, is

$$D = 1 - \prod_{i=0}^{N-1} (1 - \boldsymbol{P}_d(\boldsymbol{e}_i)). \tag{5}$$

These two objectives, shorter distance and lower cumulative detection risk, typically pull the solution in opposite directions. Shorter paths tend to cross high radar exposure regions, while safer paths often require lengthy detours. The central question is therefore not "What is the optimal path?", but rather: "Which trade-off between distance and detection best fits the mission scenario?"

Pareto Front

In multi-objective optimisation, no single solution is universally optimal. Instead, a spectrum of Pareto-optimal solutions exists, each representing a different balance between the competing objectives. One path may be very short but moderately risky; another may be extremely safe but operationally slow; a third may be a robust compromise that remains feasible under fuel or endurance constraints.

Following the established classification in the multi-objective optimisation,[9] our approach belongs to the a posteriori class. We do not ask the decision-maker to specify preference weights in advance, nor do we reduce the multi-objective problem to a single weighted sum. Instead, we first compute a representative set of feasible solutions, from which the Pareto front can be identified (see Figure 10.2). Planners or autonomous mission systems can then compare and select among the alternatives based on the operational context.

This a posteriori approach aligns closely with doctrinal realities: in contested airspace, planners rarely know beforehand what level of radar exposure is tolerable, or how much additional flight time is acceptable. The Pareto front visualises the operational trade-off: for example, how much additional detection risk is incurred to reduce flight time, or how much extra flight distance is required to remain below a given risk threshold. In this work, the Pareto front is not an end result in itself, but a decision-support tool that exposes feasible operational trade-offs under A2/AD conditions.

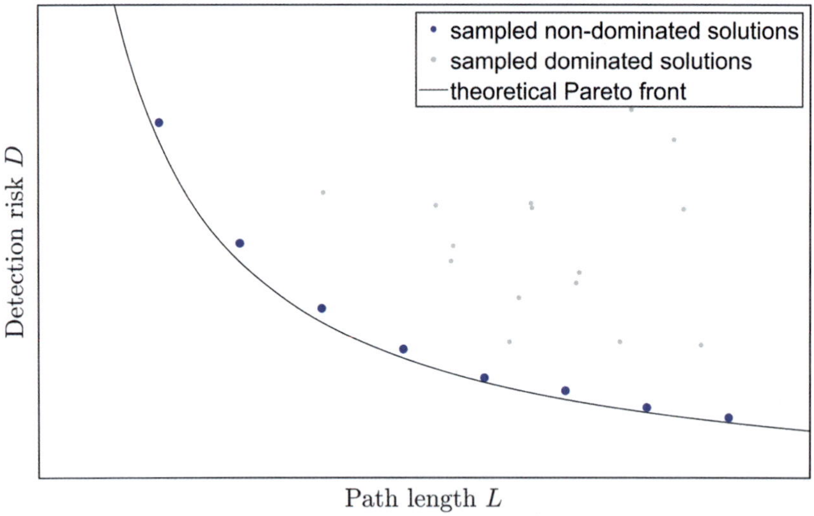

Figure 10.2: Conceptual illustration of a Pareto front for UAV route planning in contested airspace. Each point represents a feasible UAV trajectory characterised by its total path length L and cumulative detection risk D. Blue points indicate sampled non-dominated solutions, grey points indicate sampled dominated solutions, and the black curve represents the theoretical (ideal) Pareto front that bounds the set of achievable trade-offs between the two objectives.

Particle Swarm Optimisation (PSO)

To explore the space of feasible UAV trajectories, we employ Particle Swarm Optimisation (PSO), an evolutionary optimisation method well suited to problems defined over high-dimensional search spaces. PSO maintains a population ('swarm') of candidate trajectories, each iteratively adjusting its position based on its own performance and that of its neighbours. This population-based nature makes PSO well suited for multi-objective problems: a single run of PSO produces a set of solutions that can be used to approximate the Pareto front.

In the present formulation, each particle represents a discrete waypoint-based trajectory. PSO refines candidate solutions by adjusting the positions of the waypoints while preserving their number and connectivity. This allows smooth geometric refinement of paths, such as shifting a segment to reduce radar exposure, without altering the underlying path topology inherited from the initial RGG-based construction.

By enabling gradual, localised adjustments to trajectory geometry, PSO effectively explores trade-offs between path length and cumulative detection risk. The resulting collection of refined trajectories forms the basis for the Pareto analysis and serves as the decision-support output for mission planners operating in contested A2/AD environments.

Random Geometric Graph (RGG) Initialisation

While PSO is effective at refining candidate solutions, evolutionary algorithms often struggle with initial feasibility: many randomly generated trajectories are either invalid, disconnected, or trapped in geographic dead ends (no feasible path due to hard obstacles). To mitigate this, we employ a Random Geometric Graph (RGG) as a structured initialisation mechanism. Nodes are sampled in the operational region and connected based on local geometric proximity, producing a graph that encodes feasible movement through the environment.

The RGG serves several purposes. First, it ensures that all initial trajectories between start and target locations are feasible, avoiding wasted computational effort on swarm particles that do not represent valid paths. Second, it reduces search dimensionality by providing a coarse structure that PSO subsequently refines, preventing the algorithm from wandering arbitrarily through the environment.

In practice, a separate small RGG is generated for each particle in the swarm, and the shortest path on each graph is computed using Dijkstra's algorithm, thereby ensuring diversity in the initial set of feasible trajectories. These graph-based paths are then used to initialise the PSO swarm. PSO subsequently refines the geometry of these trajectories by adjusting waypoint positions while preserving their connectivity, allowing continuous improvement of path length and detection risk without altering the underlying route topology.

This combination, an RGG providing structural diversity and PSO refining continuous path geometry, produces a more robust and comprehensive approximation of the Pareto front than either approach could achieve alone.

Novelty of the Hybrid PSO–RGG Approach

Hybrid methods combining Particle Swarm Optimisation (PSO) with Random Geometric Graphs (RGGs) have appeared in earlier UAV path planning research,[10] but these applications typically involve 'hard' obstacle environments, terrain, buildings, or no-fly zones, that must be strictly avoided. In such settings, the RGG provides feasible corridor-like structures, and PSO refines trajectories within those hard constraints.

In contrast, the A2/AD context addressed in this chapter involves 'soft' obstacles: radar detection fields where higher exposure is undesirable but not forbidden. This changes the nature of the optimisation problem. The objective is not simply to avoid an infeasible region but to reason about graded detection risk, and trade-offs between survivability and efficiency. As a result, the PSO–RGG hybrid acquires a new role: the RGG still provides structurally diverse initial trajectories, but PSO no longer refines paths toward a binary feasible/infeasible boundary. Instead,

PSO continuously adapts trajectories to exploit variations in the radar detection function, skirting threat zones, minimising cumulative detection probability, or trading a short exposure for substantial length reduction.

To our knowledge, this is the first application of a PSO–RGG approach to UAV routing in probabilistic radar environments typical of A2/AD operations, where the optimisation landscape is continuous, soft-constrained, and multi-objective. The hybrid approach therefore goes beyond simple avoidance path generation by revealing a range of operationally meaningful trade-offs, supporting informed decision-making in contested airspace.

Limitations and Scope

The proposed method is intended as a mission-level decision-support tool for UAV route planning in A2/AD environments, rather than as a high-fidelity radar or flight dynamics model. Radar detection is modelled at the level of single sweep detection probability, and cumulative detection risk is accumulated discretely along path segments. Temporal radar effects, detailed sensor behaviour, and track formation are therefore outside the scope of this work.

UAV trajectories are represented as discrete waypoint sequences, with PSO refining waypoint positions while preserving their connectivity. Vehicle dynamics, manoeuvre constraints, and flight-control effects are not considered. The simulations further assume static radar configurations and nominal operating conditions, without modelling adversary adaptation or real-time replanning.

These limitations are intentional: the goal is not to predict exact detection events, but to support transparent, option-based planning by exposing trade-offs between mission efficiency and survivability under A2/AD conditions. Extensions toward higher-fidelity sensing or dynamic environments are left for future work.

Simulations

Figure 10.3 illustrates the UAV path-planning process for a scenario in which the vehicle travels from a starting position near the bottom-left corner of the domain to a target position near the top-right corner. The radar layout is chosen as a representative A2/AD configuration to illustrate trade-offs between efficiency and detection risk. Aggregated radar detection intensity is indicated by the background colours, ranging from no coverage (black) to high coverage (yellow). The optimisation seeks to minimise two competing objectives: the total path length (Equation (3)) and the cumulative detection risk (Equation (5)).

In all simulations, PSO was run with a fixed swarm size of 30 particles and a maximum of 50 iterations, which was sufficient to achieve convergence for all tested risk strategies. The inertia and acceleration coefficients were selected following standard PSO practice and kept constant across experiments to ensure comparability between runs. While the specific numerical values of these parameters influence convergence speed, they were found to have negligible impact on the qualitative structure of the resulting Pareto fronts. The presented results are therefore representative of the underlying trade-off behaviour rather than fine-tuned algorithmic settings.

Two representative optimisation runs are shown. The left column corresponds to a balanced trade-off strategy, in which path length and detection risk are weighted comparably. The right column corresponds to a low risk strategy, in which minimising detection risk is prioritised over path length.

Before initiating the Particle Swarm Optimisation (PSO) procedure, the swarm was initialised using shortest paths computed on multiple small Random Geometric Graphs (RGGs), as described earlier. In Figure 10.3, the first row (iteration 0) shows these initial trajectories as blue lines, while the red line indicates the best-so-far trajectory for the corresponding operational risk strategy. Here, 'best-so-far' denotes the trajectory selected according to the applied risk emphasis.

Starting from this set of feasible initial solutions, PSO iteratively refines the particle trajectories by adjusting the positions of the waypoints, while preserving the waypoint count inherited from the RGG-based initialisation. As the iteration count increases, the spread of candidate paths decreases, indicating convergence of the swarm. By iteration 50, the particles have converged toward a stable solution, with distinct converged routes emerging for the two operational risk strategies.

The resulting set of trajectories obtained across all PSO runs is summarised in the Pareto plot shown in Figure 10.4, which aggregates the final solutions and visualises the trade-off between path length and cumulative detection risk.

Solutions associated with the balanced trade-off strategy in Figure 10.3 cluster around $L \approx 11.75$ km and $D \approx 0.2$, representing moderate increases in path length in exchange for reduced detection risk. Low-risk solutions are located near $L \approx 14.75$ km and $D \approx 0$, corresponding to routes that strongly prioritise survivability. In contrast, distance-minimising (high-risk) solutions appear near $L \approx 10.25$ km and $D \approx 0.45$, reflecting short paths with substantial exposure to radar detection. Together, these clusters illustrate the range of feasible trade-offs between mission efficiency and survivability under identical A2/AD conditions.

Notably, the Pareto front shown in Figure 10.4 was generated without eliciting explicit preference weights from military planners or commanders. Instead of assuming a priori how detection risk should be traded against path length, the simulations explored a range of relative weightings between the two objectives, spanning low to high emphasis on detection risk. These weightings were used solely as a sampling mechanism to populate the solution space; they do not

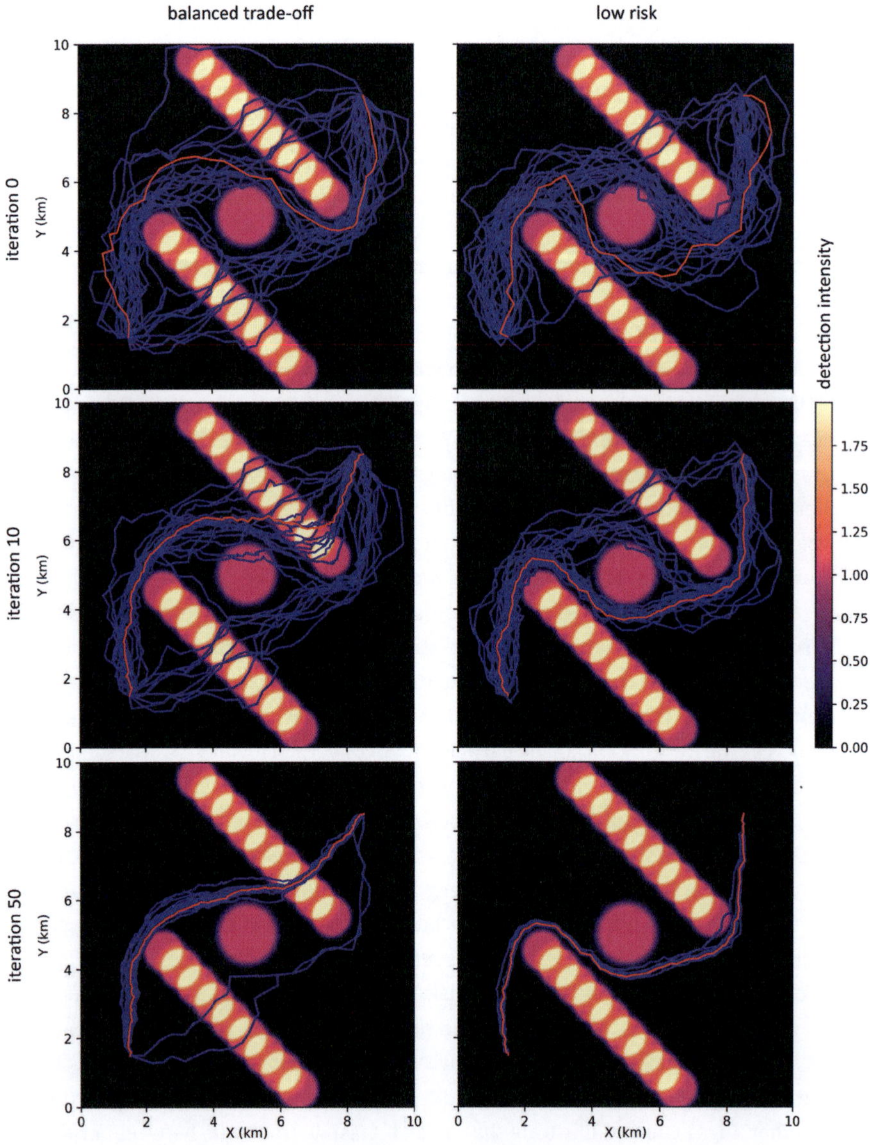

Figure 10.3: PSO convergence for two operational risk strategies. Each panel shows 30 candidate UAV trajectories (blue) and the best-so-far trajectory (red) at the indicated iteration. Left column: balanced trade-off strategy; Right column: low risk strategy. Rows show iterations 0, 10, and 50. Radars are arranged in two linear arrays of short-range radars, with a single medium-range radar at the centre of the domain. The background indicates aggregated radar detection intensity for visualisation. The figure illustrates both within-run convergence (shrinking spread of blue trajectories) and risk strategy-dependent route selection (different converged paths between columns).

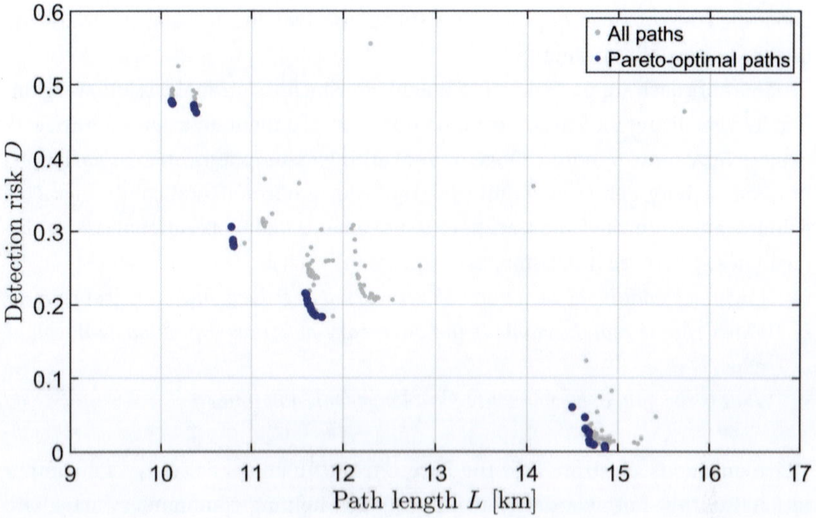

Figure 10.4: Pareto plot of UAV trajectories obtained after 50 PSO iterations (corresponding to the final row in Figure 10.3). Each point represents a complete UAV trajectory characterised by its total path length L and cumulative detection risk D. Grey points indicate dominated solutions, while blue points denote Pareto-optimal (non-dominated) trajectories. The Pareto-optimal set delineates the trade-off boundary between mission efficiency (path length) and survivability (detection risk).

represent prescribed operational preferences. The resulting Pareto front is therefore intended as a tool for discussion, presenting planners with a set of feasible options from which mission-appropriate choices can be made.

Doctrinal Implications for A2/AD Environments

Operating UAVs in A2/AD environments requires commanders to balance survivability, mission timeliness, and resource constraints under uncertainty. A doctrinal implication of the proposed method is the shift from deterministic route planning to option-based planning, enabled by the generation of a Pareto front of feasible UAV paths. Instead of presenting planners with a single 'best' route, the method provides a spectrum of non-dominated alternatives, each representing a different trade-off between detection risk and operational efficiency.

The Pareto front as observed earlier, organises candidate routes into qualitatively distinct regimes, ranging from short, high-risk trajectories to longer, near-zero-risk alternatives, with intermediate compromise solutions in between. These regimes provide a basis for discussion, allowing planners to reason explicitly about the consequences of prioritising speed, survivability, or balanced trade-offs under identical threat conditions.

This approach supports decision-making during mission analysis and wargaming. By visualising the Pareto front, planners and commanders can interactively explore how route options change under shifting assumptions, such as variations in radar activity, communication constraints, or environmental conditions. This enables more informed and transparent judgements about operational risk. It also facilitates structured discussion, using questions such as:

- *How much additional detection risk is incurred to reduce time-on-target by 15%?*
- *Which routes remain viable if the adversary activates a previously dormant radar?*
- *What is the gain in survivability for a longer mission time?*

These questions illustrate how the Pareto front functions not only as a computational output but as a decision-support tool, helping commanders articulate acceptable risk and compare courses of action. The availability of multiple pre-vetted paths also enhances robustness: if the adversary changes their sensor posture, the system can switch to an alternative route without violating constraints on fuel, endurance, or mission timing.

From a doctrinal perspective, the findings support a gradual shift toward multi-option UAV planning, placing greater emphasis on quantifying risk in contested airspace, and tighter integration of decision-support tools into command-and-control processes. The method therefore contributes not only to improved UAV survivability, but also to more adaptive and resilient operational planning in A2/AD environments.

Notes

1 Liu et al., "RGG-PSO+"; Nayeem et al., "UAV Path Planning with an Adaptive Hybrid PSO"; Wu et al., "A Novel Hybrid Enhanced Particle"; Yu et al., "A Novel Hybrid Particle Swarm".
2 Huang and Fei, "UAV Path Planning Based on Particle Swarm Optimization".
3 Zhang et al., "Optimal Path Planning with Modified A-Star Algorithm".
4 Zhang et al., "Efficient and Optimal Penetration Path Planning".
5 Skolnik, *Introduction to Radar Systems*.
6 Albersheim, "A Closed-form Approximation to Robertson's Detection Characteristics".
7 Tufts and Cann, "On Albersheim's Detection Equation".
8 Skolnik, *Introduction to Radar Systems*.
9 Miettinen, *Nonlinear Multiobjective Optimization*.
10 Liu et al., "RGG-PSO+".

References

Albersheim, W. J. 1981. "A Closed-Form Approximation to Robertson's Detection Characteristics". *Proceedings of the IEEE*, 69 (7), 839–839.

Huang, C., and J., Fei. 2018. "UAV Path Planning Based on Particle Swarm Optimization with Global Best Path Competition". *International Journal of Pattern Recognition and Artificial Intelligence*, 32 (6), 1859008. https://doi.org/10.1142/S0218001418590085

Liu, Y., X. Zhu, X.Y. Zhang et al. 2024. "RGG-PSO+: Random Geometric Graphs Based Particle Swarm Optimization Method for UAV Path Planning". *International Journal of Computational Intelligence Systems*, 17, 127. https://doi.org/10.1007/s44196-024-00511-x

Miettinen, K. 1999. *Nonlinear Multiobjective Optimization*. Springer.

Nayeem, G.M., M. Fan, G.M. Daiyan and K.S. Fahad. 2023. "UAV Path Planning with an Adaptive Hybrid PSO". *Proceedings of the International Conference on Information and Communication Technology for Sustainable Development (ICICT4SD)* (139–143). IEEE. https://doi.org/10.1109/ICICT4SD59951.2023.10303551

Skolnik, M. I. 2001. *Introduction to Radar Systems* (3rd ed.). McGraw-Hill.

Tufts, D.W. and A.J. Cann. 1983. "On Albersheim's Detection Equation". *IEEE Transactions on Aerospace and Electronic Systems*, 19 (4), 643–646.

Wu, J., Y. Sun, J. Bi and C. Chen. 2025. "A Novel Hybrid Enhanced Particle Swarm Optimization for UAV Path Planning". *IEEE Transactions on Vehicular Technology*, 74 (8) 11806–11819, https://doi.org/10.1109/TVT.2025.3555285.

Yu, Z., Z. Si, X. Li, D. Wang and H. Song. 2022. "A Novel Hybrid Particle Swarm Optimization Algorithm for UAV Path Planning of UAVs". *IEEE Internet of Things Journal*, 9 (22), 22547–22558, https://doi.org/10.1109/JIOT.2022.3182798.

Zhang, Z., J. Wu, J. Dai and C. He. 2022. "Optimal Path Planning with Modified A-Star Algorithm for Stealth Unmanned Aerial Vehicles in 3D Network Radar Environment". *Proceedings of the Institution of Mechanical Engineers, Part G: Journal of Aerospace Engineering*, 236 (1), 72–81, https://doi.org/10.1177/09544100211007381.

Zhang, Z., J. Wu, J. Dai and C. He. 2023. "Efficient and Optimal Penetration Path Planning for Stealth Unmanned Aerial Vehicles Using Minimal Radar Cross-Section Tactics and Modified A-Star Algorithm". *ISA Transactions*, 134, 42–57, https://doi.org/10.1016/j.isatra.2022.07.032.

CHAPTER 11

Hydraulic Warfare

Low-Tech Area Denial in High-Tech Conflicts

Sebastiaan Leertouwer, Alexander Schmets, Danny Janssen, Dennis Krabbenborg and Edwin Dado

Abstract

Hydraulic warfare, particularly deliberate inundation, is a low-tech way to shape land operations by manipulating water, wet terrain, and water infrastructure to control movement and sustainment. The viability of inundation as an anti-access/area-denial (A2/AD) instrument in contemporary and future high-tech conflicts is assessed using a land-domain obstacle framework. Prepared inundations, improvised inundations, and long-term hydrological modification are compared in terms of effectiveness, resources, time dynamics, and constraints. Hydraulic A2/AD is most credible where water availability and retention enable tactically meaningful flooding and where terrain funnels movement into predictable crossings. Key limitations include environmental dependence, vulnerability of control infrastructure, uncertainty in improvised measures, and risks of self-denial.

Keywords: Anti-Access/Area Denial (A2/AD); Hydraulic warfare; Inundation; Counter-mobility; Sustainment interdiction; Prepared and improvised inundation

Introduction

Throughout history, controlling an adversary's freedom of movement and constraining access to essential resources has been a cornerstone of military strategy. An early example is the emergence of fortified cities in the Middle Ages, where walls and defensive works regulated access and constrained hostile movement. In the 20th century, trench warfare, especially during World War I, constituted a form of terrain manipulation aimed at controlling manoeuvre. Trenches, earthworks, and barbed wire formed static but highly effective obstacles that constrained an adversary's freedom of action, chiefly by restricting freedom of movement. In contemporary doctrine,[1] this approach is termed Anti-Access/Area Denial (A2/AD), which refers to strategies designed to delay or prevent successful offensives within contested areas by imposing prohibitive costs on such efforts.

Another World War I example of deliberate terrain shaping was the 1914 IJzer inundation in Belgium.[2] By opening sluices, Belgian forces deliberately flooded low-lying areas to halt German advance and retain the last unoccupied Belgian territory. Employing water to shape the battlespace is not unique to World War I. The intentional use of hydraulic means, for example by exploiting existing surface-water bodies as a physical barrier has a long history. In the 17th century, Dutch engineers constructed the Water Line to deny access through controlled inundation.[3] More recently, Ukrainian forces employed water-control measures during the defence of Kyiv, including the 26 February 2022 breaching of the Kozarovychi (Irpin) dam.[4] The resulting inundation created a significant area-denial effect by constraining Russian manoeuvre and contributed materially to the overall success of the defence of Kyiv.

This chapter examines whether hydraulic warfare, focused on inundation, remains a viable land-domain A2/AD instrument. The next section defines A2/AD as an operational effect. The following section develops the land-domain obstacle logic and comparison parameters. These are then applied to hydraulic means by defining hydraulic warfare, outlining obstacle types and enabling conditions, and making the time dimension and planning metric explicit. Next, the main hydraulic forms are evaluated against a common set of criteria and the conditions synthesised under which hydraulic A2/AD is credible. The final section concludes with doctrinal, practical, and policy recommendations.

Defining Anti-Access and Area-Denial (A2/AD)

Building on the preceding historical examples, a definition of A2/AD is now provided to elucidate its core mechanisms and temporal dynamics. Anti-Access and Area-Denial (A2/AD) is best understood as a set of effects rather than a specific class of technologies or weapons. In its most general form, A2/AD refers to actions and capabilities intended to prevent an adversary from entering an operational area (anti-access) and to limit its freedom of action once inside (area denial). This is typically achieved by increasing the attacker's risk, time requirements, and sustainment burden.[5] This definition highlights that A2/AD seeks to shape the operational environment in ways that disadvantage an opponent before and during the conflict.[6]

Anti-Access effects constrain or delay the entry of military forces into the operational area. Such effects may be generated by long-range fires, precision artillery, missile systems, electromagnetic warfare, or ISR-enabled strike capabilities[7] that make approach routes prohibitively costly or risky. Area-Denial effects arise once the adversary has entered the operational area, including reductions in mobility, canalisation into predictable corridors, and attrition through delay. These outcomes are commonly produced by obstacles, minefields, engineered terrain features, or concentrations of short-range fires that limit tactical freedom of action. Across both dimensions, A2/AD mechanisms impose escalating time and sustainment burdens, forcing the attacker to allocate additional engineering units, breaching systems, logistics, and capacity to maintain operational momentum.

As noted above, A2/AD systems are intended to prevent an adversary from entering the operational theatre or tactical area. Gaining access therefore requires either degrading the system by exhausting its resources or conducting a breaching operation. Breach operations are typically synchronised, combined-arms actions directed by a manoeuvre commander and are among the most demanding combat tasks a force can face. They follow a systematic five-phase procedure: Suppress, Obscure, Secure, Reduce and Assault (SOSRA). This procedure is designed to overcome obstacles as safely and efficiently as possible. These fundamentals always apply, although their execution will vary according to specific battlespace conditions.[8]

Time is central to the manifestation of A2/AD. A key moment occurs when a peaceful situation, in which disputes are resolved by diplomacy, shifts to one in which the use of force becomes possible. At that point armed forces begin to mobilise, and staff start planning for a range of scenarios related to the emerging crisis. This phase, which may develop into armed conflict but does not necessarily do so, is known as the grey zone.[9] From this moment onward initial activities such as the emplacement of obstacles begin, and the anticipated presence or visibility of denial measures may already contribute to deterrence. The start of the conflict phase is denoted here as t_0; the grey zone therefore occupies an interval before that point ($t < t_0$), often without a sharply defined boundary. During conflict ($t > t_0$) attackers experience obstacles as constraints on movement that demand substantial time and resources to overcome. Importantly, A2/AD effects evolve over operational and seasonal timescales, since terrain, weather and environmental conditions can enhance or reduce the effectiveness of denial measures.

A2/AD in the Land Domain

Modern A2/AD environments combine mutually reinforcing capabilities with classic static obstacles. These capabilities include long-range precision fires, layered air-defence systems, drone-enabled ISR,[10] electronic warfare, cyber operations and deception measures.[11] Each element contributes in different ways to limiting an adversary's access, constraining manoeuvre, or degrading command and control.

Although multidomain A2/AD measures can impose significant operational constraints, they are vulnerable to countermeasures and to the attrition of key enablers.[12] Many rely on detectable, high-value assets that are increasingly exposed to pervasive ISR and precision attack, including drones, and their effectiveness depends on supply-chain robustness, electronic resilience and rapid adaptation cycles on both sides. Given this broad and technologically dynamic landscape, the scope here is limited to static, terrain-based A2/AD measures in the land domain. These physical obstacles derive their effect from shaping the surface battlespace rather than from complex and potentially fragile technological architectures and therefore provide a stable starting point for further analysis. This focus also enables a systematic assessment of hydraulic means, which operate primarily as A2/AD instruments within the surface battlespace.[13]

Obstacles in the Land Domain: Types, Characteristics and Effect

Land-domain obstacles include minefields, trenches, ditches, berms, anti-tank barriers, fortifications, as well as hydraulic obstacles. Their primary effect is to influence and control the adversary's freedom of movement. They shape approach routes, determine engagement geometry, and canalise movement into predictable corridors. Their value lies not only in restricting movement but also in forcing adversary units into disadvantageous engagement areas. In his analysis of the 1973 Yom Kippur War, van Creveld suggested that in future wars minefields, anti-vehicle ditches, and other artificial obstacles would be used "not so much to stop movement as to channel it into preselected 'killing grounds.'"[14] Such observations underscore that land-domain obstacles are employed not primarily to stop movement outright, but to shape it in ways that serve specific operational aims. Within this land-domain focus, the first step is to clarify the objectives that obstacles are designed to achieve.

Objectives of Land-domain Obstacles

Land-domain obstacles are employed to shape an adversary's freedom of movement in support of friendly manoeuvre, fires, and force protection. In offensive operations, obstacle employment aims to preserve friendly freedom of action by improving security, facilitating economy of force, and limiting the adversary's ability to reinforce, counterattack, or reposition on time. In defensive operations, obstacles aim to degrade adversary manoeuvre by controlling approach routes and engagement geometry, increasing delay and exposure to friendly fires, supporting economy of force, and helping to retain key terrain of tactical, operational, or strategic value. In both cases, obstacles are most effective when integrated with observation and fires, because their primary contribution is to shape where and when the adversary can move and under what risk and tempo constraints.[15] These objectives are pursued through different obstacle types and mobility constraints, as classified in Figure 11.1.

Types of Land-domain Obstacles

A classification of land-domain obstacle types and mobility constraints is shown in Figure 11.1.[16] It distinguishes movement constraints that arise from pre-existing terrain and water-related conditions, such as rivers and marshes, from engineered obstacles intended to control manoeuvre. The mobility constraints can be strongly seasonal. For example, the spring *rasputitsa* mud season[17] in Ukraine can sharply reduce off-road trafficability and canalise mechanised movement onto paved routes.[18] The engineered obstacles category includes earthworks and excavations, anti-vehicle barriers and fortifications, minefields and other explosive obstacles, and demolitions used for route denial. Hydrographic features and hydraulic infrastructure are also part of the land-domain obstacle landscape. Deliberate hydraulic manipulation is treated separately in the following section, but hydraulic obstacles[19] are included in Table 11.1 to enable consistent comparison across the various obstacle types.

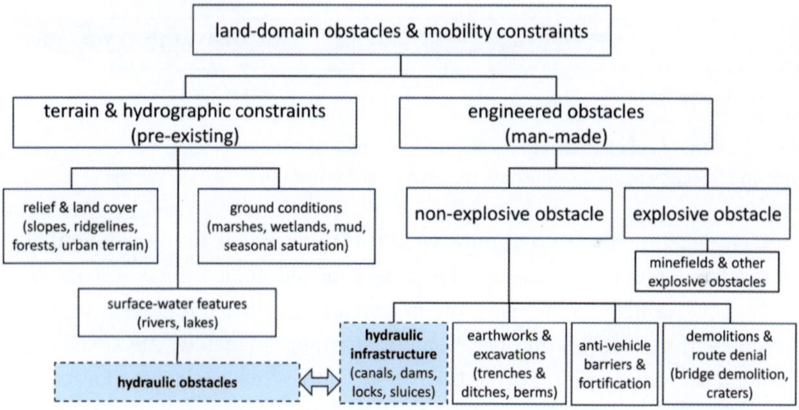

Figure 11.1: Types of static land-domain obstacles and mobility constraints. Deliberate hydraulic manipulation is treated separately later in this chapter.

This classification provides a basis for linking obstacle types to intended movement-control effects. It allows comparison of their characteristic features and of the time and resources needed for emplacement by friendly forces or for breach operations by adversary forces.

Intended Effects on Movement and Engagement

Once obstacle aims are defined and obstacle types identified, the next step is to specify the intended effects on adversary movement and the resulting engagement geometry. Doctrinally, these intended effects provide a shared vocabulary for synchronising obstacle employment with observation and fires, and for linking movement control to the broader scheme of manoeuvre.[20] Figure 11.2 summarises these movement-control effects. It maps each intended movement-control effect to an indicative resource-planning factor, denoting the relative effort and time normally required to realise that effect.[21]

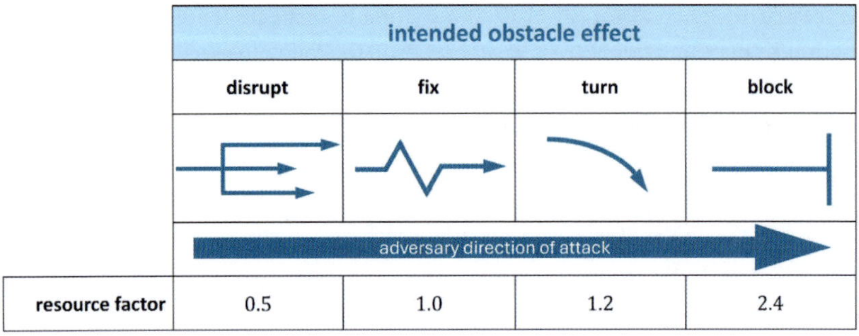

Figure 11.2: Intended obstacle effects on adversary movement along an avenue of approach, with an indicative resource-planning shown for each effect.

Obstacles exert a spectrum of movement-control effects that are commonly categorised as disrupt, fix, turn, and block, as depicted in Figure 11.2. A *block effect* renders an avenue of approach (AA), or mobility corridor (MC), effectively inaccessible within the relevant operational timeframe, compelling the adversary to abandon that axis. A *fix effect* holds adversary forces in a designated location for a specified duration, generating delay and increased exposure without implying permanent denial. A *disrupt effect* interferes with coherent movement by inducing disorder and friction, typically degrading operational tempo and coordination over a shorter period. Finally, a *turn effect* denies access to selected routes at decision points and coerces movement onto a more favourable axis, commonly canalising the adversary toward preselected engagement areas. Across this obstacle effect taxonomy, block corresponds to denial, turn to canalisation, fix to delay or fixation, and disrupt to tempo reduction or disorder.[22] For planning and resourcing, these effects can be translated into an *initial obstacle requirement* using a doctrinal resource factor (RF), which links the selected effect to the required linear obstacle effort (LOE) across an enemy avenue of approach or mobility corridor,[23] typically expressed as LOE = RF × corridor width.[24]

These movement effects are best understood as intents whose operational significance depends on combined-arms coverage. The same physical obstacle can produce markedly different outcomes depending on fire support, observation, and the time the adversary is willing or able to invest in route assessment, bypass, and breaching. RF-based LOE estimates therefore support early sizing of obstacle belts and groups, but they do not by themselves guarantee the intended effect, since realised delay, canalisation, and denial outcomes depend on siting, combined-arms coverage, and the adversary's breaching capacity, and must be refined through reconnaissance and wargaming. This linkage is central for land-domain A2/AD, where denial outcomes are often achieved through constrained mobility, canalisation, and time loss, rather than through absolute physical blockage alone.

Together, these effects and their translation into initial LOE requirements via RF underpin the comparison framework summarised in Table 11.1.

Comparison Parameters for Land-domain Obstacles

Having defined obstacle effects on movement and engagement, the next step is to specify comparison parameters applicable across land-domain obstacles and, later, hydraulic means. Obstacle characteristics, defender and attacker resource requirements, and a provisional metric for obstacle effect are defined, and Table 11.1 synthesises these elements for later comparison.[25]

Obstacle characteristics refer to the physical and operationally relevant properties that govern how an obstacle constrains movement and shapes engagement. These include geometry and depth, concealability and detectability, persistence under weathering and combat degradation, repairability and maintainability, and on combined-arms coverage for achieving the intended outcome. Such characteristics help explain why obstacles with broadly similar intent can generate different operational consequences in different terrain, seasons, and threat environments.[26]

All obstacle types require resources to be constructed (defender perspective) or overcome (attacker perspective). Resources are treated as a combination of manpower, assets and materiel (construction materials, water, earth-moving equipment, explosives, breaching and bridging assets), and time. Because resource demand is strongly shaped by the intended effect, the comparative framework treats effort and time as effect-dependent, consistent with the effect-linked resource factor values shown in Figure 11.2. Defender resource requirements include the effort to emplace an obstacle and, where relevant, to sustain or repair it. Obstacles can also constrain friendly manoeuvre and may require protection and maintenance to preserve their intended effect. Time to completion is decisive because obstacles offer limited effect before emplacement is complete and may be vulnerable to disruption during construction.

Attacker resource requirements include the personnel and assets needed to conduct obstacle reconnaissance and to reduce, bypass, or pass an obstacle while sustaining tempo under threat. These demands can increase sharply with obstacle density, depth, and integration. From the attacker perspective, breaching and passage are typically organised through structured doctrinal frameworks, such as SOSRA.[27]

To support structured comparison, this chapter introduces a provisional metric, Obstacle Effect (OE), defined as the intended movement-control outcome imposed on the adversary. For the purposes of Table 11.1, OE is treated as an ordinal classification aligned with the intended effects taxonomy: block corresponds to denial; turn to canalisation; fix to delay or fixation; and disrupt to tempo reduction or disorder. OE will be refined later through a small number of observable parameters, especially for hydraulic obstacles, to support comparison across obstacle families.[28]

Obstacle performance is inherently time dependent. Prior to the conflict phase ($t < t_0$), obstacles may impose planning burdens and contribute to deterrence by shaping expected avenues of approach. After the conflict begins ($t > t_0$), their effects materialise through realised delay, canalisation, and exposure to fires. In contemporary precision- and ISR-rich environments, the sustainment dimension becomes prominent, because the ability to keep a breach or crossing, and the vulnerability of supporting equipment and logistics, can determine whether an obstacle remains degraded or rapidly regains effectiveness. For example, breach lanes can be re-seeded with scatterable mines, and bridging or breaching equipment can be targeted by precision fires, reducing throughput across the obstacle and allowing it to regain effectiveness even after an initial passage. This temporal and sustainment framing is captured qualitatively in Table 11.1.

Table 11.1: Land-domain obstacle families with indicative characteristics, defender and attacker resource and time drivers, and persistence and sustainment. Typical intended effects are listed in ascending order of effect intensity (D = disrupt; F = fix; T = turn; B = block). Effects are indicative and context-dependent; realised effects depend on siting, observation, fires, and breaching conditions.

Obstacle type/ typical effects	Indicative characteristics	Resources and time drivers		Persistence and sustainment
		Defender	attacker	
Minefields D / F / T / B	Hidden or observable; point or area; scalable density and depth; can be integrated with sensors and fires.	Mines; laying systems; planning time; marking and control; reseeding or relaying.	Detection; proofing or clearance systems; engineer assets; explosive breaching charges; covering fires, time under threat.	Breach lanes can be reseeded; ISR and precision fires can reduce throughput; effects can be restored after passage.
Trenches & ditches F / T	Linear, terrain-bound; constrain vehicle and personnel; creates predictable crossing sites.	Excavators and earth-moving equipment; manpower; time to excavate and shape; spoil management.	Bridging or filling; assault engineer teams; mobility assets; smoke; suppressive fires; assault crossing under fire.	Persistent once emplaced; crossing sites become predictable and targetable.
Anti-vehicle barriers & fortifications F / T / B	Hard obstacles; channel movement; strongpoints; often combined with fires.	Materials (concrete, steel, prefabricated barriers); engineer-ing equipment; time for construction and siting.	Demolition; heavy breaching; bypass engineering; covering fires; time under observation.	Most effective when integrated with observation and fires; otherwise by-passable at cost.
Earthworks & berms D / T	Terrain shaping and line-of-sight modification; can conceal or expose; can impede movement.	Construction equipment; soil; time; camouflage; siting to exploit micro-terrain.	Bypass or regrading equipment; demolition if hardened; time to reroute.	Often used to shape engagement geometry and survivability; effect depends on siting and maintenance.

Obstacle type/ typical effects	Indicative characteristics	Resources and time drivers		Persistence and sustainment
		Defender	attacker	
Demolitions for route denial T / B	Targeted destruction of bridges, culverts, road cuts, chokepoints.	Explosives; initiation and firing systems; access to structure; planning and timing.	Gap crossing equip-ment; bridging; engineer repair; route clearance; time and logistics.	Sensitive to precision fires against bridging and repair assets.
Hydraulic obstacles D / F / T / B	Inundated terrain; saturated soils & reduced trafficability; rivers, canals, wetlands; controlled water levels via hydraulic infrastructure.	Access to water source; control infrastructure; earthworks and preparation; time to achieve target water levels; monitoring and maintenance.	Route reconnaissance and rerouting; bridg-ing and crossing support; dewatering or ground improve-ment; engineer assets; fire support; operations under threat.	Persistent and scalable with sustained water control; sensitive to drainage, drought, and freezing; crossings become predictable and targetable; reflooding can restore effects.

Hydraulic Warfare for A2/AD in the Land Domain

Hydraulic warfare is understood here as the deliberate use or manipulation of water, wet terrain, and water infrastructure to achieve military advantage, especially by controlling movement, shaping engagement geometry, or degrading sustainment. More broadly, water can shape armed conflict either as a contested resource or through deliberate manipulation as a weapon in an otherwise non-hydrological conflict.[29] Hydraulic warfare sits within water weaponisation, the deliberate manipulation of water to produce military, the broader category of political, or coercive effects. Recent conceptual work frames this as using water as 'physical arms' to harm, coerce, or gain leverage over an adversary.[30]

Within this framing, two principal forms of hydraulic warfare are particularly relevant for land operations: deprivation and inundation. Deprivation targets water availability or quality. It can obstruct supply, disable pumps, or damage distribution and treatment infrastructure. Deprivation is an important form of water weaponisation but lies outside the scope of this analysis. The focus here is on inundation, defined here as the release or redirection of a mass of water. This is typically achieved by operating flood control works or breaching storage and control structures. Inundation can disrupt movement, interdict routes, or impose broader coercion. Inundation can be employed defensively or offensively. Defensive applications aim to deny access, slow an advance, and protect key terrain by leveraging local hydrology and infrastructure. Offensive applications aim to collapse mobility and disrupt sustainment. They can also force operational diversion by manipulating water systems in time and space to the attacker's advantage.

This dual-use logic also supports the chapter's title: hydraulic measures often rely on comparatively low-tech actions, such as opening gates, breaching embankments, or exploiting wet-ground dynamics, yet they can generate wide-area effects that interact directly with modern targeting and strike systems by canalising movement, concentrating crossings, and shaping exposure during breach and passage.[31] At the same time, historical practice and recent scholarship suggest normative and political constraints on intentional large-scale flooding or attacks on water control systems, sometimes described as a 'water taboo', which can affect feasibility, escalation dynamics, and choice of method.[32]

The analysis below applies the land-domain obstacle logic introduced earlier to hydraulic means, using the intended effects and comparison parameters summarised in Table 11.1. Because hydraulic effects vary strongly in setup time, scalability, and persistence, Figure 11.3 and Table 11.2 later in this section make these temporal dependencies explicit for planning and comparison.

Types and Characteristics of Hydraulic Obstacles

Hydraulic obstacles can be grouped into four common types. First, *prepared inundations*[33] are planned water-level manipulations that use controlled infrastructure, such as dams, sluices, weirs, canals, levees, and pumps, to achieve a desired depth and extent at a chosen time.[34] Second, *improvised inundations*[35] are ad hoc floods created through breaches, demolitions, diversions, or deliberate damage to control works, typically with less control over water depth, footprint, and duration.[36] Third, *long-term hydrological modification* aims to alter terrain wetness and trafficability over longer horizons, for example by changing drainage regimes, raising groundwater levels, or allowing wetland conditions to expand (peat formation or peatification).[37] Fourth, rivers and canals can function as *hydrographic barriers* whose obstacle effect can be increased by controlling crossings, shaping approaches, and integrating banks and adjacent wet terrain into a broader defensive scheme.

Within the obstacle taxonomy introduced earlier, these four forms constitute a water-based subset of land-domain obstacles. Compared with other obstacle types in Table 11.1, their performance depends on water availability, environmental losses (for example drainage and evaporation), and the extent to which water control can be prepared and sustained.

Across these forms, the primary function is movement control. Hydraulic obstacles can deny mobility, canalise movement, or impose delays that increase exposure to observation and fires. They can therefore be used to block, turn, fix, or disrupt, often across broader frontages when water effects can be scaled in area. Hydraulic obstacles are most often associated with area denial because they increase time, risk, and resource burdens within a defended region. In some settings they can also contribute to anti-access by closing key approaches or crossings before entry, or by making entry routes operationally unattractive through predictable exposure.

Their interaction with modern targeting is particularly important. When practical crossing sites are limited and predictable, enabling assets such as bridging, pumping, earth-moving, and route maintenance become targetable. A defender can then sustain or reconstitute obstacle effects by degrading passage capacity and throughput, strengthening the link between hydraulic obstacles and sustainment interdiction beyond the obstacle itself.

Resources and Preconditions for Hydraulic Obstacles

Hydraulic obstacles impose resource demands and enabling conditions that differ from most dry obstacles. They can deny access by preventing or delaying breach and passage, thereby forcing bypasses and route changes. They can also degrade the attacker's ability to sustain operations beyond the obstacle by reducing throughput and disrupting logistics across constrained crossing points. In contemporary environments, long-range fires and improved targeting make the enabling means of passage, such as bridging, pumping, earth-moving, and route maintenance, increasingly targetable. By attacking these enablers, a defender can reduce passage capacity and thereby reconstitute the obstacle's functional effect over time, even after a first crossing attempt.

Water availability is a necessary precondition for inundation-based obstacles. It may derive from reservoirs and regulated rivers, proximity to lakes and floodplains, sustained rainfall or snowmelt, or high groundwater levels. Terrain morphology, soil permeability, and drainage conditions determine whether water can be retained at tactically meaningful scales and at achievable water depths, and for how long, under expected seasonal variability. When these conditions align with the intended obstacle effect, hydraulic means can burden the attacker in ways that differ qualitatively from many dry obstacles. The passage problem becomes a combined challenge of mobility, timing, and exposure at predictable locations. These preconditions should be identified early in planning and treated as part of intelligence preparation of the operational environment, so that water sources, controllable infrastructure, soil and drainage conditions, and seasonal variability are assessed alongside classic mobility corridors and engagement areas.[38] Climate dependencies can both enable and negate effects. Drought can reduce achievable water depth and duration. Excess rainfall can shift the problem from controlled inundation to uncontrolled flooding. Freezing conditions can convert a water obstacle into a temporary mobility corridor or bridgeable surface, depending on ice thickness and load-bearing capacity.

Infrastructure and time requirements for the emplacement of hydraulic obstacles follow from the intended effect and the degree of control required. Hydraulic obstacles typically rely on either existing infrastructure to control water or deliberate changes to terrain and control works to redirect, retain, or release it. Inundation depends on controllable inlets, sluices, dikes, canals, levees, pumps, and drainage systems, or on the deliberate demolition of such structures to produce the desired flow and water depth. A central planning distinction relative to many conventional obstacles is that hydraulic measures often have substantial setup and arming times, yet can generate large-area effects that are difficult to reduce quickly once established.

Tactical inundation can be achieved rapidly by opening floodgates or breaching control works. Longer-horizon approaches shape terrain wetness and persistence through drainage control or by allowing wetland conditions to expand. In some settings, this can include peat formation (peatification), a long-timescale process in persistently waterlogged terrain that requires planning well in advance.[39] Historic prepared inundation systems illustrate that reliable effects typically require long-term investment in controlled works and supporting terrain preparation, for example in the New Dutch Waterline and the IJssel Line.[40] Improvised inundations can be created more rapidly through breaches, demolitions, or diversions, but they usually offer less control over depth, footprint, and duration. Long-term hydrological modification is therefore most relevant when there is time to shape terrain before imminent conflict, and contemporary programmes have described wetland development as part of broader counter-mobility schemes.[41]

Across these approaches, preparation and timing are often decisive. Hydraulic measures are most reliable when enabling works are in place before hostilities become imminent, or when a force controls the relevant infrastructure and can coordinate release with observation and fires.

Timing, Persistence, and Recovery of Hydraulic Obstacle Effects

The impact of obstacles becomes noticeable and gradually diminishes with time. Even when the intended movement-control outcome is the same, obstacles differ in how quickly they generate a tactically relevant effect, how long that effect persists, and how easily it can be sustained or regained under attack. For hydraulic obstacles, this time dependence is intrinsic. Water must accumulate to sufficient depth and extent to constrain mobility, and it is continuously lost or transformed through drainage, infiltration, evaporation, precipitation, and temperature. As a result, hydraulic obstacle effects can range from near-immediate but uncertain disruption to slow-forming but persistent area denial, depending on method and context.

Table 11.2: Temporal profiles of obstacle effects by obstacle type, highlighting setup time, persistence, and key environmental dependencies

Type of obstacle	Effort to create effect	Effects in time
Peatification	Developing peat-forming marsh and wet woodland conditions that reduce trafficability and hinder military penetration. This typically takes years, though existing marshes can be reinforced over months.	Peat formation is slow and primarily constrained by sustained water availability. Once established, peat-forming wetlands can persist throughout a conflict.
Prepared inundation	Prepared obstacle using inlets, sluices, and dikes, for example the IJssel Line, to fill inundation basins within a defined period to a target depth (30–60 cm) sufficient for a blocking effect.	Arming the obstacle typically takes several days to weeks. Over time, water in the inundated area may decline through soil infiltration.
Improvised inundation	Release of a body of water into a selected area to obstruct adversary movement at a specific location.	The effect can be near-immediate, depending on the water source, but its footprint, depth, and duration remain more uncertain because of the obstacle's improvised nature. Persistence depends on drainage, infiltration, and continued water availability.
Mines	Quick to deploy and capable of targeted effects, though usually over a limited area. No water source is required.	Depending on mine type, the obstacle may require maintenance every 30– 40 days.[42] Rainfall, animal activity, and combat disturbance can degrade effectiveness over time.
Trenches / AT ditches	Relatively time-intensive to construct, but buildable with civilian engineering assets to create a physical obstacle. Their effect can be strengthened by water.	Trenches and anti-tank ditches erode through rainfall, wind, and use. Deterioration is strongly weather-dependent and may unfold over weeks to months.

Table 11.2 provides a qualitative comparison of hydraulic and non-hydraulic obstacles, focusing on planning-relevant drivers rather than precise performance. The entries are indicative and context-dependent, since obstacle effects vary with terrain, season, and adversary capabilities. The key comparison is the temporal profile. Prepared inundations typically require lead time to arm the obstacle through controlled water-level management but can provide broad and persistent effects once established if control and water availability can be sustained. Improvised inundations may generate rapid effects, but with greater uncertainty in footprint, depth, and duration. Long-horizon wetness creation, including peatification, is slow by design and becomes relevant mainly when shaping the environment well in advance is feasible. For context, the table also includes conventional obstacles such as mines and trenches, which can generate useful effects quickly but may degrade, be reduced, or become bypassable unless maintained and integrated with observation and fires.

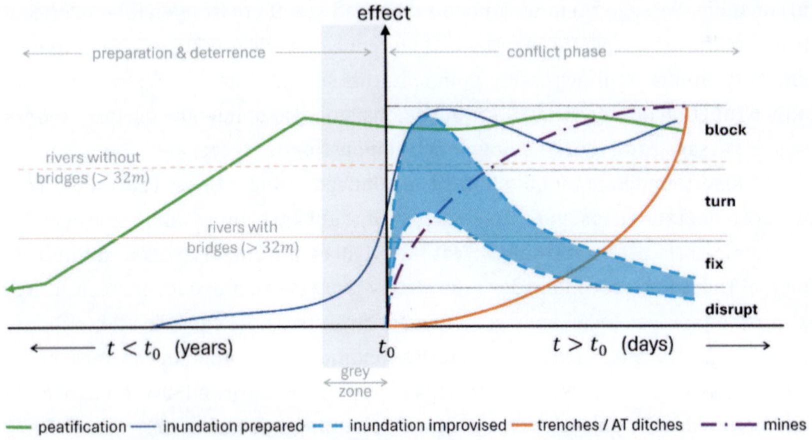

Figure 11.3: Conceptual trendlines of expected obstacle effects over time for selected land-domain obstacle types (schematic). Note that time axis is asymmetric, i.e., years for $t < 0$ and days for $t > 0$.

Figure 11.3 complements Table 11.2 by visualising expected obstacle effect as a function of time compared to the onset of active operations (t_0), including a pre-conflict phase in which preparation, deterrence, and 'grey zone' shaping may occur. The curves are deliberately schematic and not calibrated functions, and neither axis is intended for direct measurement. Their purpose is to support comparison of relative timing, scalability, and persistence across obstacle families. The figure highlights that some obstacles generate useful effects quickly but decay unless sustained, while others require preparation but provide prolonged effects once established. This framing also motivates operationalising temporal obstacle profiles for planning, including time-to-effect, decay under environmental and adversary action, and maintenance or renewal requirements needed to sustain the intended movement-control effect.

For comparison and planning, the effect of a hydraulic obstacle can be expressed as an *obstacle effect value,*[43] defined as the reduction it imposes on the attacker's mobility and throughput over a specified frontage and time window. A practical first operationalisation uses a small set of observable parameters: the depth, width, and duration of wetting or inundation, and the resulting trafficability and throughput available to the attacker. Even shallow water can be decisive when it saturates soils and reduces bearing capacity. Moreover, it obscures visible surface features. Greater depths shift the passage problem toward bridging or ferries, which are slower to establish and typically constrain throughput.

These parameters also capture how hydraulic obstacles concentrate effort. As width and duration increase, viable crossing sites often become fewer and more predictable. Passage then depends on a limited set of crossing enablers concentrated at those sites, so throughput can be reduced by degrading passage capacity and sustainment at the crossing points. In this sense, obstacle effect value is not only a function of physical water dimensions, but also of how the obstacle shapes where passage can occur and how much can pass under threat.

Beyond immediate mobility effects, prolonged inundation and sustained wetness can degrade roads and infrastructure, disrupt agriculture, and impose wider economic costs.[44] They can also create downstream humanitarian and environmental risks. Such second-order effects may increase coercive leverage, but they also raise political, legal, and escalation constraints. These considerations should therefore be treated as feasibility and risk modifiers when applying obstacle effect value in planning and assessment. This supports comparison between hydraulic and non-hydraulic obstacles on a common, planning-relevant scale.

Contribution of Hydraulic Obstacles to Land-domain A2/AD

Hydraulic obstacles contribute to land-domain anti-access and area denial by restricting entry routes, constraining manoeuvre inside a defended region, and increasing the time, risk, and resources required for breach, passage, and sustainment. In a single domain, the distinction between anti-access and area denial is often one of scale and continuity. Long, continuous barrier effects across key approaches can shift a primarily area-denial posture toward a land-domain anti-access effect. In a hydraulic context, the IJssel Line was designed to support such a theatre-level barrier function by enabling controlled inundation to delay or prevent a major ground advance by Warsaw Pact Forces.[45]

Hydraulic obstacles generate A2/AD effects by converting movement into a predictable passage problem. By expanding wet or inundated areas, they can reduce route options and concentrate movement onto limited crossings and causeways. This increases delay and canalisation, and it raises the attacker's resource burden by forcing reliance on specialised crossing enablers and sustained throughput across a small number of sites.

Persistent aerial observation and shortened targeting cycles further amplify this effect. Unmanned aerial systems, including FPV drones, can detect breaching and crossing activity and cue fires against critical engineering assets, contributing to stalled breach attempts.[46] Recent observations from Ukraine illustrate how breach and clearance operations can be tracked, forward engineering assets engaged, and follow-on elements fixed in exposed, constrained terrain with little room to manoeuvre.[47] Where hydraulic obstacles force crossings into predictable locations, the attacker's ability to sustain passage and logistics becomes a central vulnerability. Even after an initial passage, degrading bridging, pumping, earth-moving, and route maintenance can reduce throughput and allow the obstacle's functional effect to recover over time.

Hydraulic obstacles also impose distinct limitations and risks. Their effectiveness depends on control of enabling works and on sufficient time to prepare and to synchronise release with the wider scheme of manoeuvre. They can also constrain friendly mobility and sustainment, particularly if the inundated area expands beyond the intended footprint or if manoeuvre corridors become too few to support friendly repositioning.

Large-scale inundation can generate significant second-order effects and escalation risks. Damage to infrastructure and agriculture, displacement of civilians, and environmental harm may increase coercive leverage, but they also raise political, legal, and legitimacy constraints and can impose long-term costs on the defender and local population. These factors should be considered when judging whether hydraulic obstacles are practical to use, and what risks they bring, as part of an A2/AD approach.

Framework and Evaluation of Hydraulic A2/AD

Hydraulic warfare, particularly inundation, is here as a land-domain A2/AD method in current and future conflict with the aim of identifying the essential preconditions for success. Using the obstacle logic developed earlier, a small set of planning criteria is applied to judge when hydraulic A2/AD is credible and when it is unlikely to succeed.

Evaluation Criteria for Hydraulic A2/AD

Hydraulic obstacles can be assessed as A2/AD instruments by translating the earlier movement-effect logic into criteria that can be applied consistently across hydraulic forms.

First, *effectiveness* concerns whether the measure delivers the intended movement-control outcome (block, turn, fix, disrupt) and produces meaningful delay, canalisation, and throughput reduction. For hydraulic obstacles, this also includes sustainment pressure when passage and logistics are forced through a limited set of predictable crossing sites.

Second, *resource requirements* capture defender effort to prepare, activate, and sustain the effect, and attacker effort to reduce, bypass, or pass while preserving tempo. For hydraulic means, these demands hinge on water availability, control of enabling works, and the capacity to manage flows and losses over time.

Third, *time dynamics* address onset and persistence. The key distinction is between preparation and activation before hostilities ($t < t_0$) and realised effects during conflict ($t > t_0$), including arming time, decay under environmental losses, and the maintenance or renewal effort required to sustain the intended effect.

Fourth, *reversibility and sustainment* focus on how readily mobility and throughput can be restored after an initial passage, and whether the defender can allow the obstacle effect to recover by degrading passage capacity and sustainment at constrained crossing sites.

Fifth, *constraints and risks* define what is feasible and acceptable in practice, including own-force mobility constraints and humanitarian, environmental, legal, and political consequences that shape feasibility and escalation risk.

These criteria are used next to examine the main hydraulic forms.

Applying the Evaluation Criteria to Hydraulic A2/AD

Applying these criteria across the main hydraulic forms shows where each can credibly generate A2/AD effects and where practical constraints are likely to dominate.

Prepared inundations offer the most reliable combination of scale and persistence once activated, provided water control can be maintained. Their A2/AD value is greatest when flooding leaves the attacker only a few usable crossings and causeways, forcing movement and resupply through narrow choke points that can be kept under pressure over time. The trade-off is preparation time: prepared systems require controlled works and supporting terrain preparation, and their performance remains sensitive to environmental losses and seasonal variability.

Where these conditions hold, they can deliver wide-area movement control that many dry obstacles cannot easily match.

Improvised inundations can deliver fast disruption, but with greater uncertainty in footprint, depth, and duration. They are most useful for short-notice delay, local denial, or operational diversion when time is tight, and bypass options are limited. Their main limitations are reduced controllability and a higher risk of unintended consequences. In practice, their value depends on rapid exploitation, and on whether the defender can prolong the denial effect by suppressing passage capacity and logistics at the constrained crossing sites.

Long-term hydrological modification builds slowly, so it matters most when terrain can be shaped well before a crisis. Its strength is durability: it can leave ground persistently soft and hard to cross, even without visible standing water, for example where peat forms in saturated areas. For that reason, it is best treated as a strategic preparation measure that supports deterrence and defence readiness rather than a responsive wartime obstacle option.

Conditions for Viable Hydraulic A2/AD

In practice, hydraulic A2/AD is most credible when several conditions are met. First, water availability must be sufficient to create tactically meaningful depth and extent. Environmental losses must also be manageable under expected drainage, infiltration, evaporation, precipitation, and temperature. Second, the defender must have access to, and control of, enabling works where sustained effects are needed, or the ability to generate effects through deliberate breach or diversion. Third, there must be sufficient time for preparation and synchronisation, either through pre-conflict investment (prepared inundations and long-term hydrological modification) or through an ability to rapidly generate and exploit disruption (improvised inundations). Fourth, terrain must translate hydraulic effects into predictable movement constraints, typically by forcing passage into a limited number of crossing sites where throughput can be controlled over time.

Hydraulic A2/AD is least practical when these conditions are absent. This includes environments with insufficient water availability, rapidly draining ground, seasonal variability that negates persistence, or freezing conditions that convert inundated terrain into temporary mobility corridors or bridgeable surfaces. It is also less practical when enabling works cannot be secured or sustained, when friendly mobility is constrained more than the adversary's, or when humanitarian and escalation costs outweigh operational gains. Under favourable conditions, hydraulic measures credibly complement conventional obstacles by producing wide-area movement control through comparatively simple interventions.

Conclusions and Recommendations

Hydraulic warfare, particularly inundation, is a practical land-domain A2/AD method, most credible in territorial defence and in shaping manoeuvre in water-rich environments where preparation time exists. Its distinctive advantage is wide-area movement control and sustainment pressure: by forcing movement and resupply through predictable corridors and crossing sites, it turns manoeuvre into a throughput problem that can be constrained over time. Its limits are equally decisive: strong dependence on environmental conditions, reduced controllability for improvised measures, vulnerability of control infrastructure, constraints on friendly mobility, and significant collateral and escalation risks.

Successful application depends on a small set of preconditions: favourable hydrographic and hydrological conditions; secure, defendable control infrastructure where sustained effects are needed; sufficient preparation time for engineered options; and integration with observation and fires to convert constrained mobility into operational advantage. Where these conditions hold, hydraulic measures can credibly complement conventional obstacle systems within modern land-domain A2/AD.

Doctrinal, Practical, and Policy Implications

A practical implication of this study is that hydraulic obstacles should be treated as a standard part of land-domain obstacle doctrine, planned with the same movement-effect vocabulary (block, turn, fix, disrupt) and assessed using the same criteria (effect, resources, time, sustainment, constraints). Doctrine should also make the time dimension explicit by separating preparation and activation before hostilities from sustainment and recovery during conflict. Finally, it should place throughput at the centre of obstacle planning because hydraulic denial often works by squeezing passage capacity at predictable crossing sites rather than by creating permanent physical blockage.

Beyond doctrine, hydrology should become a routine input to planning and reconnaissance, alongside terrain and enemy capabilities. This includes early identification of water availability, controllable works, soil permeability and drainage, seasonal mobility baselines, and likely crossing sites. Key control works, and the assets that enable passage (bridging, pumping, earth-moving, route maintenance), should be treated as operationally decisive, with protection, redundancy, and repair planned. Training and exercises should rehearse activation, sustainment, and recovery of hydraulic effects, including attacker attempts to restore passage and defender actions to reduce throughput after a first crossing. Finally, plans should prevent self-denial by defining friendly manoeuvre corridors and contingencies and by managing the risk of uncontrolled spread.

Finally, hydraulic measures should be treated as high-impact decisions, not just engineering options. Because they can trigger major downstream effects, their use should be guided by clear criteria on feasibility, proportionality, escalation risk, and recovery burden. Roles and civil-military coordination for manipulating water infrastructure should be agreed in advance. Dual-use investments (flood control, water storage, drainage, wetland restoration) may offer defence benefits, but only where humanitarian and environmental risks are acceptable and escalation implications are understood.

Acknowledgement

This work forms part of the Netherlands Defence Academy's joint research effort on Anti-Access and Area Denial, as well as the Interreg BONSAI project, in which the Dutch Ministry of Defence is a partner.

Notes

1 NATO Standardization Office, *AJP-3*.

2 Vandenbohede, "The Hydrogeology of the Military Inundation".

3 Van Wieringen and De Zee, "Inundatiewerken"; De Kraker, "Flooding in River Mouths".

4 Moriarty and Thebault, "Satellite Images Show Flooding North of Kyiv".

5 Bonds et al., "What Role Can Land-Based"; Biddle, *Military Power*; Krepinevich et al., "Meeting the Anti-Access and Area-Denial Challenge"; NATO Standardization Office, *AJP-3*.

6 Here we define conflict as beginning at the moment competitive rivalry ends, which occurs when one belligerent moves military assets onto the territory of its opponent.

7 ISR denotes Intelligence, Surveillance, and Reconnaissance. ISR-enabled strike capabilities refer to the ability to detect, identify, track and target forces in near-real time, and to deliver fires rapidly on the basis of that information, linking sensors and decision-making to precision weapons in order to strike effectively at range.

8 Department of the Army, *FM 3-34.2*.

9 The grey zone describes the phase in which diplomacy erodes and sub-threshold coercion becomes routine; it begins when disputes cease to be resolved solely by political means and culminates at the instant kinetic force is used ($t = 0$). During this period, readiness levels are raised and measures are undertaken to shape the battlespace without triggering full war.

10 See (7). In this context, drone-enabled ISR refers to the use of unmanned aerial systems to detect, identify, track, and target adversary forces.

11 Zabrodskyi et al., *Preliminary Lessons in Conventional Warfighting.*

12 Kofman, *Assessing Russian Military Adaptation.*

13 Ibid.

[14] van Creveld, *Military Lessons of the Yom Kippur War.*

[15] Joint Chiefs of Staff, *Joint Publication 3-15.*

[16] This typology is inspired by *Joint Publication 3-15.*

[17] Dunlop, "Dictionary Lemma".

[18] Hambling, "Mud Season in Ukraine".

[19] In this chapter, hydrographic refers to the configuration and location of surface-water features, such as rivers, lakes, and canals. Hydrological refers to water regimes and processes, including soil moisture, groundwater levels, drainage and seasonal soil saturation that affect trafficability. Hydraulic infrastructure refers to the engineered control of water levels and flows, and the forces associated with them, using structures and operational measures such as dams, sluices, locks, canals, pumps, and levees, including deliberate interventions that manipulate water for operational effect.

[20] Department of the Army, *Combined Arms.*

[21] *Joint Publication 3-15.*

[22] Schadd and de Reus, "Intelligent Operational Decision Support".

[23] U.S. Department of the Army, Field Manual 90-7; U.S. Department of the Army. FM 20-32.

[24] Worked example of the use of resource factor (RF) and linear obstacle effort (LOE). Suppose an enemy with AA width = 2 . 0 km. Assume the desired effect is block. Use RF = 2.4, as provided in Figure 11.2. Then LOE = 2.4 × 2.0 km = 4.8 km. Using a common planning assumption that a standard *block obstacle* segment spans about 500 m of frontage, meaning it extends roughly 500 m from left to right across the avenue of approach (perpendicular to the direction of advance), the required 4.8 km LOE corresponds to 4.8 ÷ 0.5 = 9.6, rounded up to 10 segments. So, one can plan 10 standard obstacle segments to achieve the desired block effect. This does not mean ten literal minefields: the same requirement can be met with a mix of obstacles, for example ant-vehicle barriers, ditches or barbed wire entanglements, provided their combined effort delivers the required LOE.

[25] Here, and in the following paragraphs, reference is made to Joint Publication *3-15.*

[26] Department of the Army, *Planning and Design of Roads.*

[27] Department of the Army Headquarters, *Engineer Operations.*

[28] Department of the Army Headquarters, *Combined Arms Obstacle Integration.*

[29] King, "The Weaponization of Water in Syria and Iraq".

[30] Grech-Madin, "Murky Waters of War".

[31] Illustrative examples are: defensive inundation by breaching dikes during the Siege of Alkmaar (1573) and the relief of the Siege of Leiden (1574). Deliberate operational inundation through water-control works at Nieuwpoort during the Battle of the Yser (1914). Large-scale coercive flooding by the intentional destruction of levees on the Yellow River in Huayuankou, China (1938). Offensive inundation by the British air force through breaching sea dikes on Walcheren (1944). A contemporary case is the 6 June 2023 breach of the Kakhovka Dam, which caused extensive flooding along the lower Dnipro River and illustrates how hydraulic effects can constrain routes, and stress sustainment in ISR- and precision-rich environments.

[32] Grech-Madin, "Water and Warfare".

[33] Bruijstens, *Waterlinie, stroming tegen doorstroming*; Schieving, *Water als hindernis*; Smeets, *Het nut van militaire inundaties.*

[34] An example of this is the IJssel Line, constructed in the 1950s in The Netherlands.

[35] Schmets et al., "Novel Emergency Response Intervention for Flood Resilience"; Zoomers, *Onderzoek naar het explosief doorsteken van dijken*; Mekencamp, *Gecontroleerde bresinitiatie met explosieven als noodmaatregel bij hoogwatercrises*; Steneker, *Bresinitiatie als noodmaatregel*; Mossinkoff, *Controlled Explosive Breach Initiation of Levees.*

[36] For example the Kakhovka Dam breaching (2023).

[37] Middel, "Meer moeras is goed voor verdediging Europa én het klimaat"; Schauenberg, *How Rewetting Peatlands Could Help.*

[38] Joint Chiefs of Staff, *Joint Intelligence Preparation.*

[39] Kentucky Geological Survey, *Peat and Peatification.*

[40] Schilt and Beekmans, *Drijvende stuwen voor de landsverdediging.*

[41] Gaszewski, *Poland's East Shield.*

[42] This applies to the ATMF-62 mines currently in service with the Dutch Army. See also: Fortuin, *Gebiedsontzegging in het landdomein.*

[43] In Dutch: hinderniswaarde.

[44] Smeets, *Het nut van militaire inundaties.*

[45] Gaszewski, Poland's East Shield.

[46] Van Dalen, "River Crossing Operation"; Newton, *How are Drones Changing War?*

[47] Zafra and McClure, *Four Factors that Stalled Ukraine's Counteroffensive.*

References

Biddle, S. 2006. *Military Power.* Princeton University Press.

Bonds, T.M., et al. 2017. *What Role Can Land-Based, Multi-Domain Anti-Access/Area Denial Forces Play in Deterring or Defeating Aggression?* RAND Corp.

Bruijstens, Hessel M. 2023. *Waterlinie, stroming tegen doorstroming. Een analyse en vertaling van het ontwerp van de IJssellinie naar de huidige tijd in Litouwen.* BSc Thesis, Den Helder: Netherlands Defence Academy (NLDA).

Creveld, Martin van. 1975. *Military Lessons of the Yom Kippur War: Historical Perspectives.* Beverly Hills, CA: Sage Publications.

Dalen, Hans van LKol (Ret.). 2025. "River Crossing Operation." *LinkedIn.* 12 23, 2025. https://www.linkedin.com/pulse/river-crossing-operation-hans-van-dalen-ndome/?trackingId=4QY5bgFgQyG-D3p%2FwP3Ip7Q%3D%3D.

de Kraker, A.M.J. 2015. "Flooding in River Mouths: Human Caused or Natural Events? Five Centuries of Flooding Events in the SW Netherlands, 1500–2000." *Hydrology and Earth System Sciences,* 6 ed.

Demerre, Guido, and Johan Termote. 2013. "The Flooding of the Yser Plain." *De Grote Rede.*

Department of the Army. 1994. "Planning and Design of Roads, Airfields, and Heliports in the Theater of Operations, Road Design." FM 5-430-00-1, Headquarters, Washington, DC.

Department of the Army. 2002 *FM 3-34.2 Combined Arms Breaching Operations.* Washington, D.C.

Department of the Army. 2014. *Geospatial Engineering.* ATP 3-34.80, Headquarters, Washington, DC.

Department of the Army. 2016. *Combined Arms Mobility and Survivability Operations.* ATP 3-90.4, Headquarters, Washington, DC.

Department of the Army Headquarters. 2020. "Engineer Operations." FM 3-34, Washington, DC.

Dunlop, Storm. 2008. "Dictionary Lemma for 'Rasputitsa'." In *A Dictionary of Weather,* by Storm Dunlop. Oxford University Press.

Fortuin, Liam. 2025. *Gebiedsontzegging in het landdomein: Het kwantificeren van de hinderniswaarde van een tactisch mijnenveld bestaande uit ATMF-62 mijnen ten opzichte van de MR Brigade van de Russische Federatie.* BSc Thesis, Den Helder: Netherlands Defence Academy (NLDA).

Gaszewski, Shaheen. 2025. *Poland's East Shield, A Comprehensive Approach to Secure NATO's Eastern Flank.* Berlin: Konrad Adenauer Stifung.

Grech-Madin, Charlotte. 2021. “Water and Warfare: The Evolution and Operation of the Water Taboo.” *International Security*. 4 ed., 84–125.

Grech-Madin, Charlotte. 2025. “Murky Waters of War: Toward a Clearer Conceptualization of Water as a Weapon.” *Environment and Security*, 457–481.

Hambling, David. 2022. “Mud Season in Ukraine Leaves Russian Tanks Stuck in Mire.” *The Guardian*, April 12.

Headquarters, Department of the Army. 1993. “Field Manual 5-71-100.” *Division Engineer Combat Operations*. Washington, D.C., April 22.

Joint Chiefs of Staff. 2009. *Joint Intelligence Preparation of the Operational Environment*. Washington, D.C.: Department of Defence, United States of America.

Joint Chiefs of Staff. 2018. “Introduction.” In *Joint Publication, 3-15: Barriers, Obstacles, and Mine Warfare for Joint Operations*, I-1 – I-6. Department of Defense, United States of America.

Kentucky Geological Survey. 2026. *Peat and Peatification*. January 21. https://www.uky.edu/KGS/coal/coal-peat.php.

Kindy, Kyle. 2025. “Russian River Crossing Failure During the Battle of the Siverskyi Donets.” *Infantry Journal*, May 8: 46–50.

King, Marcus DuBois. 2016. “The Weaponization of Water in Syria and Iraq.” *The Washington Quarterly*, 153–169.

Kofman, Michael. 2024. *Assessing Russian Military Adaptation in 2023*. Washington, DC: Carnegie Endowment for International Peace.

Krepinevich, Andrew F., Barry Watts, and Robert Work. 2003. “Meeting the Anti-Access and Area-Denial Challenge.” Center for Strategic and Budgetary Assessments, Washington, DC.

Mekenkamp, Clemens. 2022. *Gecontroleerde bresinitiatie met explosieven als noodmaatregel bij hoogwatercrises*. BSc Thesis, Den Helder: Netherlands Defence Academy (NLDA).

Middel, Mark. 2025. “Meer moeras is goed voor verdediging Europa én het klimaat, zegt Nederlandse veenpaus Hans Joosten.” *NRC Handelsblad*, September 23.

Moriarty, Dylan, and Reis Thebault. 2022. “Satellite Images Show Flooding North of Kyiv in Possible Sign of ‘Hydraulic Warfare’”. *The Washington Post*, March 9.

Mossinkoff, Vincent. 2023. *Controlled Explosive Breach Initiation of Levees. From Real-Scale Testing Towards an Established Emergency Response Measure*. BSc Thesis, Den Helder: Netherlands Defence Academy (NLDA).

NATO Standardization Office. 2025. *AJP-3: Allied Joint Doctrine for the Conduct of Operations (Edition D, Version 1)*. Brussels: NATO Standardization Office.

Newton, Michael. 2025. *How are Drones Changing War? The Future of the Battlefield*. 3 November. https://cepa.org/article/how-are-drones-changing-war-the-future-of-the-battlefield/.

Schadd, Maarten, and Nico de Reus. 2023. “Intelligent Operational Decision Support for the Military Engineer.” Conference: Computer Assisted Analysis, Exercise, Experimentation (CA2X2) Forum 2022.

Schauenberg, Tim. 2025. *How Rewetting Peatlands Could Help Protect EU Borders*. October 30. https://www.dw.com/en/peatlands-as-defense-climate-nato-finland-poland-russia-co2-rewetting/a-74513628 (accessed November 25).

Schieving, Ruben. 2024. *Water als hindernis:Het ontwikkelen van een procesmodel voor het ontwerpen van een ‘waterlinie’*. BSc Thesis, Den Helder: Netherlands Defence Academy (NLDA).

Schilt, C, and J.R. Beekmans. 2013. *Drijvende stuwen voor de landsverdediging*. Stichting Menno van Coehoorn & Walburg Pers.

Schmets, A.J.M., D. Janssen, D. Krabbenborg, S.T. Leertouwer, and E. Dado. 2024. “Novel Emergency Response Interventions for Flood Resilience.” In *Climate Security and the Military: Concepts,*

Strategies and Partnerships, by Frerks, G., R. Geertsma, J. Klomp and T. Middendorp, 231–252. Leiden University Press.

Smeets, L.M.J. 2025. *Het nut van militaire inundaties: Een kwantitatieve monetaire benadering voor de balans tussen militaire waarde en maatschappelijke neveneffecten.* BSc Thesis, Den Helder: Netherlands Defence Academy.

Steneker, Henny. 2022. *Bresinitiatie als noodmaatregel. Effecten van bresinitiatie met behulp van explosieven op de bresgroei-ontwikkeling van een zanddijk.* BSc Thesis, Den Helder: Netherlands Defence Academy (NLDA).

U.S. Department of the Army. 2001. *FM 20-32: Mine/Countermine Operations.* Washinghton, DC.

U.S. Department of the Army. 2023. *Field Manual 90-7: Combined Arms Obstacle Integration.* Headquarters Department of the Army, Washington, DC.

van Wieringen, J. S., and J. de Zee. 1994. "Inundatiewerken." In *Water ter verdediging*, 9–24. 's Gravenhage: Stichting Menno van Coehoorn.

Vandenbohede, A. 2016. "The Hydrogeology of the Military Inundation at the 1914–1918 Yser Front (Belgium)." *Hydrogeol J*: 521–534.

Zabrodskyi, Mykhaylo, Jack Watling, Oleksandr V. Danylyuk, and Nick Reynolds. 2022. *Preliminary Lessons in Conventional Warfighting from Russia's Invasion of Ukraine: February–July.* London: Royal United Services Institute for Defence and Security Studies.

Zafra, Mariano, and Jon McClure. 2023. *Four Factors that Stalled Ukraine's Counteroffensive.* December 21. https://www.reuters.com/graphics/UKRAINE-CRISIS/MAPS/klvygwawavg/2023-12-21/four-factors-that-stalled-ukraines-counteroffensive/.

Zoomers, Christiaan. 2019. *Onderzoek naar het explosief doorsteken van dijken.* BSc Thesis, Den Helder: Netherlands Defence Academy (NLDA).

CHAPTER 12

Geography as an Adaptable Strategic Enabler

Military Engineering within Multi-Domain Operations

Johan Kranenburg and Ruud Moeskops

Abstract

Contemporary conflict is characterised by persistent Intelligence, Surveillance and Reconnaissance (ISR), precision fires, and compressed decision cycles, yet prevailing interpretations of Multi-Domain Operations (MDO) tend to understate the role of the physical environment as an adaptable strategic variable. This chapter addresses that gap by examining how geographic adaptability – operationalised through Military Engineering (MILENG) – conditions the effectiveness of MDO. A layered analytical framework is developed that conceptualises the physical environment across strategic, operational, and tactical levels of war. Using this framework, the chapter analyses how MILENG enables the deliberate adaptation of geography to support endurance, deny decisive advantage to the adversary, and sustain cross-domain action. The analysis is based on qualitative research combining conceptual framework development with structured, focused case analysis. Contemporary cases from Ukraine (2022–present) and Nagorno-Karabakh (2020) are examined as contrasting illustrations to test the explanatory value and boundary conditions of the model. The findings demonstrate that under ISR-dominated conditions, operational success is determined less by tempo than by endurance: the capacity to absorb pressure, preserve functionality, and deny the adversary rapid decision-making. Strategic resilience, operational shaping, and tactical survivability emerge as interdependent functions of adaptive geography rather than isolated effects. MILENG is positioned as a critical enabling mechanism that links these functions across levels of war, translating long-term geographic shaping into operational and tactical advantage. By foregrounding geography as a dynamic and malleable factor, this study offers a corrective to technology-centric interpretations of MDO and provides a coherent analytical framework for doctrine development and military planning.

Keywords: Multi-Domain Operations (MDO); Military Engineering (MILENG); Adaptive geography; A2/AD; Operational endurance; Physical environment

Introduction

Since the mid-2010s, strategic competition with China and Russia has intensified, raising fundamental questions about the stability of the international order.[1] These actors increasingly operate in the grey zone, employing sub-threshold and hybrid activities to undermine Alliance cohesion without triggering open armed conflict.[2] A central challenge within this environment is the growing prevalence of Anti-Access/Area Denial (A2/AD) capabilities, aimed at restricting an adversary's ability to enter or operate freely within a theatre of operations.[3]

In response, NATO has shifted from a traditional joint approach towards Multi-Domain Operations (MDO), defined as the orchestration of military activities across all operational domains, synchronised with non-military instruments to generate converging effects at the speed of relevance.[4] While much of the MDO literature emphasises technological integration and cross-domain coordination, less attention is paid to the role of geography as an adaptable factor in enabling operational effectiveness.

This chapter argues that the adaptability of geography constitutes a critical enabler of MDO and that this adaptability is operationalised through Military Engineering (MILENG). By shaping the physical environment – through infrastructure adaptation, mobility support, and obstacle creation[5] – MILENG transforms geography from a static condition into a dynamic strategic variable that directly supports MDO.

The central research question is therefore: How does the adaptability of geography, through MILENG, contribute to the effectiveness of Multi-Domain Operations?

The chapter develops a layered analytical framework to examine how the deliberate adaptation of geography, operationalised through MILENG, conditions the effectiveness of Multi-Domain Operations in contested operating environments.

Theoretical Framework

Contemporary conflict is increasingly characterised by hybrid forms of competition in which military and non-military instruments are integrated to achieve strategic objectives below the threshold of open war.[6] Within this environment, A2/AD provides a concrete analytical lens for understanding how adversaries restrict freedom of movement and action by combining military capabilities with the exploitation of physical geography.[7]

A2/AD is commonly defined as the use of layered physical and operational measures to prevent an adversary from entering a theatre of operations and to constrain manoeuvre within it.[8] These measures integrate long-range fires, ISR, air and missile defence, cyber and electronic warfare with terrain, infrastructure, and natural obstacles. The effectiveness of A2/AD lies precisely in this integration: geography is not incidental, but foundational to the creation of operational barriers.

Multi-Domain Operations (MDO) emerged as the Western response to this challenge. MDO seeks to synchronise actions across land, air, maritime, space, and cyber domains in order to penetrate, disintegrate, and exploit adversary systems, thereby overcoming the standoff created by A2/AD.[9] While MDO doctrine emphasises tempo, convergence, and cross-domain integration, it tends to treat geography as a static backdrop rather than as an adaptable strategic variable.

This omission is increasingly problematic. Geography encompasses not only terrain and natural features, but also infrastructure, population distribution, and socio-technical systems. The arrangement of these elements shapes how effects, sensors, decision-making, and sustainment interact; geography thus functions as an organising logic for MDO rather than as a mere constraint. Moreover, geography is inherently dynamic: infrastructure can be adapted, routes created or denied, and terrain modified to influence operational outcomes over time. Understood in this way, geography structures the relationships between domains, actors, and capabilities, conditioning how forces move, where effects can be generated, how systems are sustained, and which vulnerabilities emerge within the operational system. Rather than passively limiting action, the physical environment actively shapes the configuration of military power by enabling certain interactions while constraining others. Geography therefore provides the spatial logic through which multi-domain effects are integrated, sequenced, and sustained (see figure 12.1).

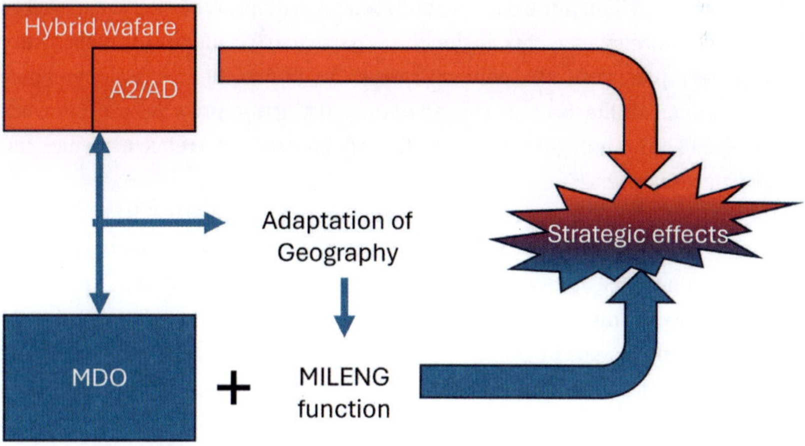

Figure 12.1: Theoretical framework

Within this broader field, engineering contributes through four doctrinal roles – mobility, counter-mobility, survivability, and general engineering – which together shape the physical environment across all levels of war.[10] Through these roles, MILENG transforms geographic conditions into deliberate instruments of operational and strategic effect.

Military Engineering (MILENG) constitutes the primary mechanism through which geographic adaptability can be operationalised. As a functional field, MILENG shapes the physical environment for military ends and encompasses multiple areas of expertise such as engineering, infrastructure management, environmental protection, explosive ordnance disposal, and military search.[11] Within this broader field, engineering contributes through four doctrinal roles – mobility, counter mobility, survivability, and general engineering[12] – which together shape the physical environment across all levels of war. Through these roles, MILENG enables the deliberate shaping of geographic conditions that underpin the generation of operational and strategic effects.

At the same time, geographic adaptation functions as one of several mutually reinforcing enablers within MDO. Through MILENG, it shapes the physical environment in ways that amplify the effectiveness of command and control, logistics, force posture, political decision-making, and the integration of non-military actors. Together, these factors translate multi-domain potential into sustained operational advantage.

At the strategic level, MILENG can be analytically differentiated between overt forms – characterised by explicitly military and visible interventions in the physical environment – and embedded forms, in which latent defensive capacity is integrated into civilian infrastructure and societal systems. While both approaches have historical relevance, this distinction is particularly salient in contemporary MDO, where persistent ISR, political sensitivity, and societal interdependence constrain overt militarisation. This analytical distinction provides a framework for assessing how strategic-level MILENG contributes to deterrence by denial and long-term resilience.

While geography functions as an organising logic for MDO, its effects are neither uniform nor static. The ways in which spatial arrangements structure military action vary across time horizons, scales, and levels of war. Long-term infrastructural resilience, campaign-level manoeuvre, and immediate battlefield survivability are shaped by geography in fundamentally different ways. Analysing geographic adaptation therefore requires a framework that can account for these variations in purpose, temporality, and scale.

To analyse this contribution systematically, the physical environment is conceptualised as a layered model (see Figure 12.2) consisting of strategic, operational, and tactical levels. Each layer represents a distinct horizon of geographic adaptation, with specific objectives and forms of intervention. MILENG connects these layers by translating long-term structural shaping into operational resilience and tactical survivability. This layered framework provides the analytical foundation for assessing how adaptable geography enhances the effectiveness of MDO.

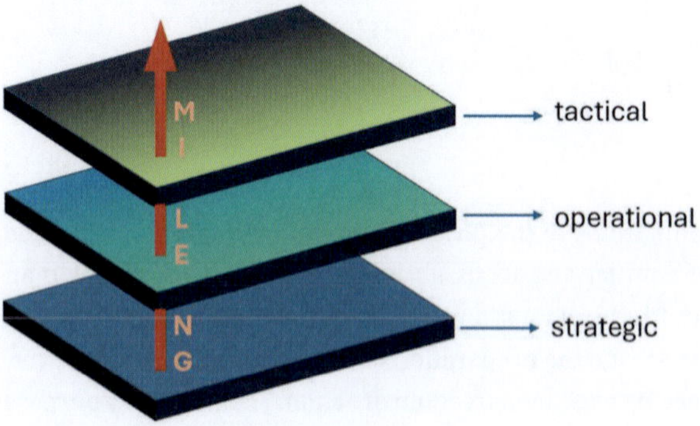

Figure 12.2: Layered model of geographic adaptation

Methodology

This study employs a qualitative analytical research design combining conceptual framework development with structured case analysis. Building on established approaches to theory building and analytical generalisation, a layered strategic – operational – tactical model is developed to examine how geography is adapted through Military Engineering (MILENG) within Multi-Domain Operations (MDO).[13] The framework is derived from systematic analysis of doctrine, policy documents, and scholarly literature on MDO, A2/AD, and MILENG. Historical and contemporary cases are then examined using structured, focused comparison to test and refine the framework rather than to provide exhaustive operational histories.[14] This approach enables analytical generalisation across cases and clarifies how MILENG operationalises adaptable geography across levels of war under contemporary conditions of persistent ISR.

Analysis

In this chapter, the previously introduced layered model is applied to both recent and historical case studies. At each level – strategic, operational, and tactical – the analysis explores how the adaptability of geography, enabled by Military Engineering (MILENG), contributes to the effectiveness of Multi-Domain Operations (MDO).

Application of the Layered Model

To systematically analyse the contribution of MILENG to MDO, this section operationalises the previous introduced layered model by specifying the objectives, actors and forms of interventions associated with each level.

Analysis per Level

Strategic Layer

At the strategic layer, adaptations to the physical environment are implemented well before the onset of conflict and focus on long-term resilience rather than immediate military effect. These adaptations include infrastructure development, redundancy of critical assets, and the preparation of conditions that restrict adversary mobility while preserving one's own freedom of action. Their primary purpose is to establish enduring conditions for deterrence by denial, strategic mobility, and national endurance.

A defining characteristic of strategic-level MILENG is its integration into existing societal and economic functions. Rather than relying on overt military structures, defensive capacity is embedded within civilian infrastructure and the broader landscape. Dual-use transport networks, concealed hardened facilities, and prepared demolition points exemplify how geography can be shaped without visible militarisation. This integration preserves societal functionality in peacetime while quietly shaping the battlespace in advance.[15]

Historically, strategic MILENG has produced decisive effects through overt interventions, such as the Mulberry Harbours in Normandy, the Ho Chi Minh Trail, or railway gauge differences during Operation Barbarossa. While effective in their contexts, such examples relied on visibility and explicit militarisation, limiting their applicability in contemporary environments characterised by persistent ISR, political sensitivity, and societal interdependence.

In contrast, embedded strategic engineering is increasingly relevant for MDO. During the Cold War, states such as Switzerland,[16] Finland,[17] and West Germany[18] systematically incorporated latent defensive features into civilian infrastructure (such as concealed fortifications, prepared demolition points and hardened transport nodes), transforming geography into an instrument of deterrence by denial. Such measures complicate adversary planning, reduce the feasibility of rapid manoeuvre, and preserve national resilience.

Strategic-level MILENG therefore supports MDO by contributing to the material and spatial conditions under which endurance can be sustained, rapid collapse avoided, and operational and tactical freedom of manoeuvre preserved. By embedding resilience, redundancy, and recoverability within the physical environment, strategic geographic adaptation establishes essential preconditions for deterrence by denial and sustained resistance.

At the strategic level, endurance emerges from the interaction of multiple reinforcing factors – political decision-making, command and control, logistics, societal cohesion, and the integration of military and non-military actors. Within this system, strategic MILENG provides the physical foundation for endurance by embedding resilience, redundancy, and recoverability into the national infrastructure. By shaping geography proactively, it creates the favourable starting conditions upon which multi-domain action depends.

Operational Layer

The operational layer concerns the deliberate shaping of the physical environment to enable campaigns and major operations within a theatre. In contrast to the strategic layer, which focuses on long-term, society-wide resilience, operational-level adaptations are explicitly military in nature and oriented toward creating favourable conditions for the conduct of operations. These measures are prepared before combat but only fully realised once a campaign is underway, when time and freedom of action are limited.

At this level, MILENG structures the battlespace to support manoeuvre, sustainment, and force protection across depth and time. By pre-configuring terrain, infrastructure, and movement corridors, operational MILENG restricts adversary options and expands the defender's ability to shape the fight. Typical activities include the establishment of forward-positioned logistical stocks, the preparation of dispersed shelters and staging areas, the construction of defensive positions in depth, and the development of above-and below-ground infrastructure for command, control, and sustainment. Unlike strategic adaptations, these measures are overtly military and prioritise operational effectiveness over other landscape uses.

Operational-level adaptations are critical in defensive operations. Through deliberate pre-shaping of the theatre, MILENG enhances survivability, constrains enemy manoeuvre and enables the concentration and redistribution of forces at decisive points. These preparations allow tactical units to operate from positions of relative advantage, even under conditions of adversary ISR dominance and precision fires.

Crucially, operational MILENG is anticipatory rather than reactive. Once a campaign begins, these adaptations are exploited rather than created, providing the structural conditions that allow tactical actions to succeed. In this sense, the operational layer forms the bridge between strategic shaping and tactical execution: it translates long-term geographic adaptation into concrete military advantage within a specific theatre.

This logic is particularly evident in contemporary defensive preparations along NATO's eastern flank. Poland's *East Shield* (*Tarcza Wschód*) program, launched in 2024, strengthens the country's eastern and northern borders with fortifications, obstacles, surveillance, and military infrastructure.[19] Although often framed in strategic terms due to its national scope, it is best understood as operational-level adaptation: overtly military, theatre-specific, and designed to shape manoeuvre and attrition during a campaign rather than embed long-term societal resilience

Consistent with the anticipatory logic of operational-level MILENG, these adaptations are constructed before conflict but intended for activation once hostilities begin. By pre-structuring avenues of approach, delaying adversary movement, and enabling force concentration at decisive points, the Polish defence line exemplifies how operational MILENG creates the conditions for tactical success and operational endurance within a defined theatre.

Tactical Layer

The tactical layer represents the immediate battlespace and is characterised by continuous, real-time adaptation under conditions of direct contact with the adversary. Unlike strategic and operational adaptations, tactical-level MILENG is inherently transient and reactive, shaped by the dynamics of combat and the constraints imposed by enemy fires, ISR, and manoeuvre. Its primary purpose is to enhance survivability, enable combined-arms action, and preserve combat effectiveness at the point of contact.

At this level, MILENG activities focus on short-term interventions that directly support tactical units. These include barrier planning and emplacement – such as minefields, anti-tank ditches, and expedient obstacles – alongside signature management measures, decoys, and the construction of temporary fighting positions. Rapid repair of damaged infrastructure, including routes, bridges, and field fortifications, is also a defining feature of tactical engineering, enabling units to maintain mobility and cohesion despite sustained enemy pressure.

Tactical MILENG delivers rapid, adaptive interventions that preserve combat effectiveness under continuous observation and precision strike. Operating in compressed temporal and spatial windows, it provides immediate effects – disrupting enemy targeting cycles, reducing vulnerability, restoring mobility, and enabling units to maintain cohesion. In doing so, tactical engineering reinforces both physical and cognitive resilience, allowing forces to operate effectively in highly contested environments.

Within the broader MDO framework, tactical MILENG exploits the conditions created by strategic and operational shaping. While it cannot compensate for the absence of higher-level geographic adaptation, it translates those preconditions into concrete battlefield advantages at the point of contact. Tactical engineering thus forms the final link in the layered adaptation of geography, enabling forces to continue operating effectively within a highly contested battlespace.

By analysis per layer, the model can be expanded as visualised in Figure 12.3.

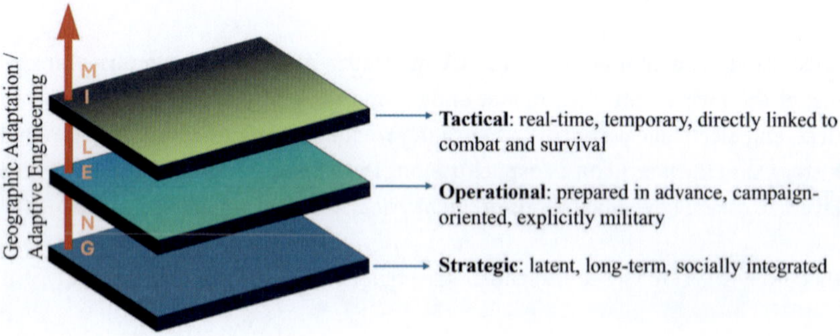

Figure 12.3: Expanded layered model of geographic adaptation

Illustrative Case Studies

The following case studies serve as analytical illustrations that apply and contrast the layered MILENG model developed in this chapter. They are deliberately selected to reflect different operational outcomes under conditions of persistent ISR, precision fires, and multi-domain contestation.

The war in Ukraine (2022–present) is examined as a contemporary case in which adaptations of the physical environment across strategic, operational, and tactical layers have contributed to endurance, resilience, and deterrence by denial. It serves as a positive application of the layered model, illustrating how embedded strategic adaptations, operational shaping, and tactical expedients can reinforce one another under sustained pressure.

By contrast, the war in Nagorno Karabakh (2020) is analysed as a contrasting case that highlights the limitations of structural terrain shaping when geographic adaptation lacks speed, flexibility, and integration across levels. Rather than invalidating the layered framework, this case serves to delineate its boundary conditions by demonstrating how static or poorly integrated adaptations can become liabilities in an ISR-dominated environment.

Historical examples introduced earlier inform the conceptual framework, while the contemporary cases apply it empirically. Together, they enable a comparative assessment of how relative adaptability of geography – rather than absolute preparation or technological superiority – shapes outcomes in MDO. The cases examine how geographic adaptation, enabled through MILENG, shapes operational outcomes in interaction with political, organisational, and military factors under contemporary conditions.

Case Study: Ukraine (2022–Present)

This case study examines the war in Ukraine (2022–present) as an empirical stress-test of the layered MILENG model under conditions of persistent ISR, precision fires, and electronic warfare. Rather than providing a comprehensive operational history, the case is structured explicitly along the strategic, operational, and tactical layers to assess how adaptations of the physical environment contribute to MDO.

Ukraine Strategic Layer: Embedded Resilience and Deterrence by Denial

The Ukrainian experience demonstrates how embedded strategic MILENG, as conceptualised in the previous section, manifests under contemporary MDO conditions. At the strategic level, the war in Ukraine confirms that the physical environment functions as a decisive variable for national endurance rather than a source of operational tempo. National endurance refers to the ability of the state to sustain essential societal functions, governance, and military operations under sustained attack. As Watling and Reynolds note, the ubiquity of ISR and long-range precision fires severely constrains large-scale manoeuvre, shifting the strategic centre of gravity from movement to sustainment and resilience.[20] Russian strike campaigns against energy and transport infrastructure were explicitly intended to erode this endurance by imposing systemic disruption rather than achieving immediate battlefield decision.[21]

MILENG activities at this level therefore centred on the protection, dispersion, and rapid restoration of critical national infrastructure, including energy systems, transport corridors, bridges, and logistical hubs. These measures illustrate embedded strategic engineering: defensive value was integrated into civilian infrastructure that had to remain fully functional for society at large, rather than expressed through overt militarisation. This embedded character emerged from necessity rather than prior doctrinal design, reflecting the constraints imposed by persistent ISR and the need to preserve societal continuity. Redundant power grids, dispersed logistics nodes, and rapidly repairable transport links provided latent defensive capacity without altering the visible character of the landscape, thereby reducing the strategic payoff of Russian precision strikes.[22]

These adaptations did not generate tempo or decisive manoeuvre. Instead, they contributed directly to deterrence by denial by undermining the adversary's ability to impose lasting disruption or achieve rapid control. Geography was thus transformed into a latent strategic instrument, enabling sustained resistance and recovery under continuous attack.[23]

Ukraine Operational Layer: Pre-structured Battlespace and Defensive Advantage

At the operational level, MILENG translated strategic resilience into concrete military advantage within the theatre. Operational adaptations were prepared before the onset of major combat but activated and exploited during campaigns, reflecting the reality that time and freedom of action are severely constrained once execution begins under persistent ISR and precision-fire conditions.

Key measures included large-scale field fortifications, trench systems, dispersed shelters with overhead cover, and hardened command and logistics nodes designed to survive ISR-guided strikes. Limited but critical bridging and route-repair capabilities enabled the restoration of mobility under fire, while camouflage, concealment, and deception were adapted specifically to counter drone-based observation and targeting. Together, these measures created a pre-structured battlespace that shaped how forces could manoeuvre, survive, and sustain operations.

These adaptations were particularly significant in defensive operations. By shaping the operational environment in depth, MILENG constrained Russian manoeuvre, increased Ukrainian survivability, and enabled forces to absorb attacks without operational collapse. Their value did not lie in enabling breakthrough, but in sustaining operational coherence and denying the adversary decisive advantage within the theatre.

Ukraine Tactical Layer: Real-time Adaptation under Fire

At the tactical level, MILENG operated under conditions of continuous contact and rapid targeting cycles. Engineering measures were inherently transient and reactive, focused on immediate survivability and combat effectiveness rather than endurance, as units faced compressed decision-making timelines imposed by drone-enabled ISR and precision fires.

Typical interventions included expedient field fortifications, rapid construction of overhead cover against artillery and loitering munitions, temporary route reinforcement, obstacle emplacement integrated with fires, and the use of decoys and signature-management techniques to disrupt enemy targeting. Rapid repair of damaged routes and fighting positions enabled units to maintain cohesion despite sustained pressure and frequent displacement. As Watling and Reynolds observe, these adaptations were essential for preserving combat power under persistent ISR-driven threat conditions.

These measures did not compensate for the absence of higher-level shaping, but instead exploited the conditions created at the strategic and operational layers. Tactical MILENG converted prepared terrain and infrastructure into immediate battlefield advantage, buying time, preserving force integrity, and enabling continued resistance within a highly contested environment.

Effects on MDO objectives:

- Across all three layers, the Ukrainian case demonstrates that MILENG does not generate tempo in contemporary conflict; it systematically denies the adversary decisive advantage.
- Freedom of manoeuvre: Severely constrained by ISR and precision fires, but locally restored through adaptive infrastructure, rapid repair, and pre-structured routes.
- Endurance: Significantly enhanced through protection, dispersion, redundancy, and rapid restoration of critical infrastructure.
- Convergence: Indirectly supported by preserving C2, logistics, and mobility networks under sustained attack.
- Deterrence by denial: Strengthened by preventing the adversary from exploiting terrain, infrastructure, or operational opportunities at scale.

The war in Ukraine validates the layered MILENG model under contemporary MDO conditions. Strategic-level embedded adaptations generate national endurance; operational-level shaping creates defensive advantage within the theatre; and tactical-level expedients preserve survivability at the point of contact. The decisive factor is not the scale or permanence of terrain modification, but the ability to adapt the physical environment faster than the adversary can detect, target, and exploit it. MILENG thus emerges as a central enabler of deterrence by denial and sustained resistance, achieved through the integrated adaptation of geography across all levels of war.

Analytical Reflection: Limits and Ambiguities

The Ukrainian case also highlights important interactions between geographic adaptation and other drivers of operational outcome. While MILENG consistently shaped the physical conditions for endurance and resistance, its effects were mediated by factors such as force availability, ammunition supply, political decision-making, and external support. In phases where these complementary elements were aligned, geographic adaptation amplified their effectiveness; where they were constrained or delayed, its impact was correspondingly moderated. MILENG therefore functioned not as an independent determinant of success, but as a conditioning mechanism that enabled other military and political factors to operate under sustained pressure.

Contrasting Case: Nagorno-Karabakh (2020)

The war in Nagorno-Karabakh (2020) provides a contrasting case that delineates the boundary conditions of the layered MILENG model. Whereas the Ukrainian case illustrates how adaptive geographic shaping can enable endurance and deterrence by denial under ISR dominance, the Armenian defence in Nagorno-Karabakh demonstrates how static terrain control becomes a liability when adaptation of the physical environment lacks speed, flexibility, and integration across levels of war.[24] Armenian forces relied heavily on pre-constructed defensive works – trenches, bunkers, and fixed artillery positions – reflecting a terrain-centric doctrine oriented toward permanent control rather than adaptive resilience.[25] This approach proved ill-suited to an operational environment dominated by Azerbaijani ISR, precision fires, and persistent drone surveillance.[26]

Nagorno-Karabakh Strategic Layer: Absence of Endurance and Deterrence by Denial

At the strategic level, Armenian MILENG focused on extensive permanent fortifications intended to deny terrain rather than on embedding resilience within broader national infrastructure. While these measures aimed to impose costs on an attacker, they failed to generate national endurance or deterrence by denial. Fixed defensive systems were rapidly detected and systematically neutralised by ISR-enabled strike complexes, accelerating rather than preventing strategic degradation.[27] Azerbaijan, by contrast, invested minimally in long-term structural shaping at the strategic level, instead relying on reach, mobility, and ISR-enabled precision strike to impose cumulative attrition without requiring sustained territorial control. As a result, geography did not function as a latent strategic instrument for Armenia, but as a fixed vulnerability once exposed.

Nagorno-Karabakh Operational Layer: Failure of Anticipatory Shaping

At the operational level, the decisive factor was relative speed of adaptation rather than depth of preparation. Armenian forces remained tied to fixed routes, logistics, and firing positions, limiting their ability to reconfigure the battlespace once combat commenced.[28] Operational-level MILENG lacked the anticipatory flexibility required to shape the theatre under conditions of persistent ISR and precision fires.

Azerbaijani forces, by contrast, exploited temporary logistical hubs, frequent displacement of command and firing positions, and rapid exploitation of terrain without permanent modification.[29] Here, MILENG functioned less as construction and more as an enabling mechanism for tempo and operational flexibility, allowing Azerbaijani forces to remain inside the Armenian targeting cycle and dictate the sequencing of operations.

Nagorno-Karabakh Tactical Layer: Survivability without Structural Support

At the tactical level, static Armenian positions lacked effective concealment against drones and loitering munitions, resulting in rapid attrition of personnel, armour, and air-defence assets.[30] Azerbaijani units relied on expedient positions, mobile launch sites, and minimal terrain modification, prioritising survivability through continuous displacement rather than physical control of terrain.[31]

Tactical engineering thus supported immediate combat effectiveness but remained dependent on operational-level mobility and ISR integration. In the absence of effective strategic and operational shaping, Armenian tactical engineering measures could not offset systemic vulnerability, illustrating the dependency of tactical survivability on higher-level geographic adaptation.

Effects on MDO Objectives

- Across all three layers, the Nagorno-Karabakh case demonstrates that structural terrain shaping alone is insufficient in an environment dominated by integrated ISR and precision fires.
- Freedom of manoeuvre: Severely constrained by fixed routes and positions.[32]
- Endurance: Rapidly eroded due to the absence of embedded resilience.[33]
- Convergence: Undermined by the inability to protect C2 and sustainment nodes under precision strike.[34]
- Deterrence by denial: Absent, as fixed defences failed to impose lasting costs or uncertainty on the adversary.[35]

The conflict in Nagorno-Karabakh further validates the layered MILENG model under contemporary MDO conditions by illustrating its negative case. Strategic-level adaptations failed to generate enduring resilience; operational-level shaping proved insufficiently flexible within the theatre; and tactical-level expedients could not compensate for the absence of adaptive depth at higher levels. The decisive factor was not the scale or permanence of terrain modification, but the inability to adapt the physical environment faster than the adversary could detect, target, and exploit it. MILENG thus emerges not as a guarantor of advantage through fortification alone, but as an enabling function whose effectiveness depends on the balanced integration of structural preparation and rapid adaptation across levels of war.

Analytical Reflection: Limits and Ambiguities

The Nagorno-Karabakh case illustrates how the effectiveness of MILENG is shaped by the degree of integration and adaptability across levels of war. Geographic adaptation was present, but its predominantly static and compartmentalised character interacted unfavourably with an operational environment defined by ISR dominance, precision fires, and high operational tempo. Azerbaijani advantages in ISR integration, mobility, and sequencing of operations were therefore amplified by the physical environment rather than constrained by it. This case demonstrates that MILENG contributes most effectively when structural preparation is combined with the capacity for rapid reconfiguration across strategic, operational, and tactical layers.

Synthesis

The preceding case studies validate the layered MILENG model by demonstrating how the adaptability of the physical environment conditions the effectiveness of MDO. Rather than serving as descriptive illustrations, the cases function as analytical tests of the model under persistent conditions of ISR dominance and precision fires.

The Ukrainian case confirms that coherent adaptation across strategic, operational, and tactical layers enables endurance and deterrence by denial under sustained pressure. By contrast, the Nagorno-Karabakh case delineates the model's boundary conditions, showing that static terrain control and insufficient cross-level integration render geographic shaping ineffective or counterproductive.

Taken together, the cases demonstrate that structural shaping and rapid adaptation are not alternatives but mutually reinforcing components of effective geographic adaptation. MILENG emerges as the integrating function that links long-term structural preparation to operational shaping and tactical survivability, translating adaptable geography into operational advantage.

Discussion

The preceding analysis validated the layered Military Engineering (MILENG) model through contrasting contemporary case studies, demonstrating how the adaptability of the physical environment conditions the effectiveness of Multi-Domain Operations (MDO). This section interprets these findings, reflects on the explanatory value and boundaries of the layered model, and situates its contribution within contemporary debates on MDO.

Interpretation of Findings

The findings challenge interpretations of MDO that privilege speed, technological integration, or cross-domain synchronisation as decisive factors in isolation. While these elements remain important, the analysis demonstrates that their effectiveness is conditioned by the adaptability of the physical environment. In both Ukraine and Nagorno-Karabakh, ISR dominance and precision fires constrained large-scale manoeuvre, compressing decision cycles and limiting the feasibility of rapid operational breakthroughs.

Geographic Adaptation as a Conditioning Variable

This reframes MDO not as a linear progression toward convergence and decision, but as a competitive process in which actors seek to shape, exploit, and deny the operating environment simultaneously. Rather than implying doctrinal symmetry, contemporary conflict reflects an interaction between hybrid environment-shaping strategies – such as A2/AD – and adaptive multi-domain responses. In this context, MDO, when combined with deliberate geographic adaptation, functions as a coherent counter-logic to hybrid warfare rather than as its doctrinal equivalent. The layered MILENG model contributes to this debate by demonstrating that control over the physical environment – through anticipatory shaping, resilience, and adaptation – is a prerequisite for effective cross-domain action rather than a secondary consideration.

Model Boundaries, Ambiguities, and Scope Conditions

While the layered MILENG model demonstrates explanatory value, its applicability is subject to several limitations. The analysis focuses primarily on defensive contexts under conditions of ISR dominance and precision fires; its relevance to offensive operations or low-intensity conflict requires further examination. Additionally, the case studies are limited in number and scope, and future research could test the model across a broader range of conflicts and geographic settings.

The model also assumes a degree of state capacity and institutional coherence that may not be present in all contexts. Where political legitimacy, civil-military coordination, or infrastructural control are weak, the feasibility of strategic-level MILENG may be constrained. These limitations do not invalidate the model but define the conditions under which it is most applicable.

Taken together, the discussion demonstrates that the adaptability of geography contributes to the effectiveness of Multi-Domain Operations by enabling endurance, denying decisive advantage to the adversary, and sustaining cross-domain action under conditions of persistent ISR – effects that are operationalised through MILENG.

Future research could further examine how adaptable geography can be formally embedded within alliance doctrine and planning processes, and how MILENG interacts with hybrid threats across different institutional contexts.

Practical, Policy, and Doctrinal Implications

Building on the analytical findings discussed above, this section outlines the practical, policy, and doctrinal implications of the layered MILENG model.

Implications for Multi-Domain Operations Theory

The findings imply that Multi-Domain Operations theory should place greater emphasis on the conditioning role of the physical environment. Rather than treating geography as a static backdrop, MDO concepts must account for how geographic adaptation shapes endurance, constrains manoeuvre, and enables sustained cross-domain action under conditions of ISR dominance and precision fires. This reframes MDO effectiveness as a function of adaptive interaction with the operating environment, rather than as the product of tempo or technological integration alone.

MILENG as a Distinct Enabling Function within MDO

A central implication of the analysis is the clarification of MILENG's role within the MDO framework. Rather than functioning as a generator of tempo or a purely supporting capability, MILENG emerges as the enabling function that operationalises adaptable geography across levels of war. The cases demonstrate that without coherent geographic adaptation, technological superiority and cross-domain integration fail to translate into sustained advantage.

At the strategic level, MILENG enables endurance by embedding resilience, redundancy, and recoverability within critical infrastructure. At the operational level, it structures the battlespace in advance, translating strategic preparation into campaign-level coherence. At the tactical level, it enables survivability and continued combat effectiveness under conditions of continuous targeting. Crucially, these functions are interdependent: tactical expedients cannot compensate for the absence of higher-level shaping, while strategic adaptations remain inert without operational and tactical exploitation. MILENG thus provides the connective tissue that links long-term preparation to immediate battlefield effect. In this sense, MILENG constitutes the primary mechanism through which geographic adaptability is translated into operationally relevant effects within MDO.

Doctrinal and Planning Implications

The layered model has direct implications for military doctrine and planning. First, it suggests that geographic adaptation should be treated as a core component of operational design rather than as an enabling annex. Strategic-level MILENG requires early integration into national planning processes, including civil-military coordination, infrastructure policy, and resilience planning. Such adaptations demand time, political legitimacy, and procedural alignment, and cannot be improvised once hostilities begin.

Second, operational-level MILENG must be understood as anticipatory rather than reactive. The cases demonstrate that shaping the battlespace under fire is largely infeasible in ISR-dominated environments. Effective operational engineering therefore depends on pre-conflict preparation, informed by realistic assumptions about detection, targeting and attrition. Finally, tactical engineering doctrine must explicitly acknowledge its dependency on higher-level shaping, emphasising rapid adaptation and survivability rather than permanent terrain control.

Implications for Military Engineering Practice

For practitioners, the findings imply a shift in how MILENG is conceptualised and employed. MILENG must be understood not primarily as construction, but as adaptive problem-solving under temporal and informational constraints. This requires a mindset oriented toward flexibility, dispersion, and recoverability rather than permanence and control.

The analysis also highlights organisational implications. Effective MILENG demands coordination across strategic, operational, and tactical echelons, as well as integration with non-military actors responsible for civilian infrastructure. Training, force design, and professional education must therefore emphasise cross-level coherence and the ability to operate within contested socio-technical systems. Without such integration, engineering efforts risk becoming isolated, reactive, or counterproductive.

Conclusion

This study addressed the central research question of how the adaptability of geography, operationalised through Military Engineering (MILENG), contributes to the effectiveness of Multi-Domain Operations (MDO). The findings demonstrate that under conditions of persistent ISR and precision fires, MDO effectiveness depends fundamentally on endurance – the ability to sustain operations and deny the adversary decisive advantage – rather than on tempo or technological integration alone. This endurance is generated through the deliberate, layered adaptation of the physical environment across strategic, operational, and tactical levels.

By applying a layered analytical framework, the study shows that geographic adaptation performs distinct but interdependent functions across levels of war. Strategic-level adaptations embed resilience and deterrence by denial within infrastructure and terrain; operational-level shaping translates this resilience into campaign-level coherence; and tactical-level expedients convert higher-level preparations into immediate survivability and combat effectiveness. Across historical and contemporary cases, the decisive factor is not absolute speed, but relative adaptability: the capacity to shape, exploit, and restore the physical environment faster than the adversary can detect, target, or counter.

The central contribution of this study lies in explicitly positioning geography as an adaptive strategic variable within MDO and identifying MILENG as the primary function through which this adaptability is operationalised. Rather than generating tempo, MILENG integrates adaptations across levels of war, providing the structural linkage that creates the conditions for endurance, resilience, and sustained cross-domain action. In doing so, the study advances MDO theory beyond its predominant focus on technological integration and provides a coherent analytical framework for military planners and practitioners to exploit the malleability of the physical environment in contemporary conflict.

Notes

1 Gilli, Gilli, and Grgić, "NATO, Multi-domain Operations".

2 North Atlantic Treaty Organization, *Allied Joint Doctrine (AJP-01)*.

3 Schmidt, "Countering Anti-access / Area Denial; Ministerie van Defensie, *Beleidsvisie Informatiegestuurd Optreden*.

4 Cannon, "The Alliance's Transition to Multi-domain Operations".

5 North Atlantic Treaty Organization, *AJP-3.12*.

6 Hoffman, *Conflict in the 21st Century*; NATO, *Hybrid Threats and Hybrid Warfare*.

7 Tangredi, *Anti-Access Warfare*; RAND Corporation, *The Foundations of Operational Resilience*.

8 Tangredi, *Anti-Access Warfare*.

9 U.S. Army Training and Doctrine Command (TRADOC), *The U.S. Army*.
10 North Atlantic Treaty Organization, *MC 0560/2*.
11 NATO, AJP-3.12.
12 Ibid.
13 Jabareen, "Building a Conceptual Framework"; Yin, *Case Study Research and Applications*.
14 George and Bennett, *Case Studies and Theory Development*.
15 Weizman, *Hollow Land*.
16 Halbrook, *Target Switzerland*.
17 Limnéll, "Comprehensive Security in Finland".
18 Gaddis, *Strategies of Containment*.
19 Zyguła and Piskorska, "East Shield – A Mechanism".
20 Watling and Reynolds, *Stormbreak*.
21 Prokip, *Russia's War on Ukraine's Energy System*.
22 Sukhodolia, *Energy Security of Ukraine*; Prokip, *Russia's War*.
23 Watling and Reynolds, *Stormbreak*.
24 Shaikh and Rumbaugh, "The Air and Missile War"; Watling and Kaushal, "The Democratization of Precision Strike".
25 International Institute for Strategic Studies, *The Military Balance 2021*.
26 Shaikh and Rumbaugh, "The Air and Missile War".
27 Ibid; Watling and Kaushal, "The Democratisation of Precision Strike".
28 International Institute for Strategic Studies, *The Military Balance*.
29 Kucera, "Azerbaijan's Drone War".
30 Shaikh and Rumbaugh, "The Air and Missile War".
31 Ibid.
32 International Institute for Strategic Studies, *The Military Balance 2021*.
33 Watling and Kaushal, "The Democratisation of Precision Strike".
34 Shaikh and Rumbaugh, "The Air and Missile War".
35 Watling and Kaushal, "The Democratisation of Precision Strike".

References

Cannon, Shaun. 2024. "The Alliance's Transition to Multi-Domain Operations." *Joint Air Power Competence Centre*, May. https://www.japcc.org/articles/the-alliances-transition-to-multi-domain-operations/.

Gaddis, John Lewis. 2005. *Strategies of Containment: A Critical Appraisal of American National Security Policy during the Cold War*. Oxford: Oxford University Press.

George, Alexander L., and Andrew Bennett 2005. *Case Studies and Theory Development in the Social Sciences*. Cambridge, MA: MIT Press.

Gilli, Andrea, Mauro Gilli, and Goran Grgić. 2025. "NATO, Multi-Domain Operations and the Future of the Atlantic Alliance." *Comparative Strategy*, advance online publication. https://doi.org/10.1080/01495933.2024.2445491.

Halbrook, Stephen P. 1998. *Target Switzerland: Swiss Armed Neutrality in World War II*. Cambridge: Da Capo Press.

Hoffman, Frank G. 2007. *Conflict in the 21st Century: The Rise of Hybrid Wars*. Arlington, VA: Potomac Institute for Policy Studies.

Jabareen, Yosef. 2009. "Building a Conceptual Framework: Philosophy, Definitions, and Procedure." *International Journal of Qualitative Methods* 8 (4), 49–62.

Kucera, Joshua. 2020. "Azerbaijan's Drone War in Nagorno-Karabakh," *Eurasianet*, October.

Limnéll, Jarmo. 2015. "Comprehensive Security in Finland." *Journal of Strategic Security* 8 (3), 72–85.

MC 0560/2. n.d. *Military Committee Policy for Military Engineering*. NATO.

Ministerie van Defensie. 2023. *Beleidsvisie Informatiegestuurd Optreden: Sneller Slimmer Sterker*. Versie 1.3. June 30. https://open.overheid.nl/documenten/8c47d888-ecae-4f0f-967c-71e48b947f9e/file.

North Atlantic Treaty Organization. 2020. *AJP 3-12 Allied Joint Doctrine for Military Engineering*, NATO Standardization Office.

North Atlantic Treaty Organization. 2022. *Allied Joint Doctrine (AJP-01)*. Edition F, Version 1 (STANAG 2437). NATO Standardization Office, December.

North Atlantic Treaty Organization. 2024. *Hybrid Threats and Hybrid Warfare: Reference Curriculum*, NATO Defence Education Enhancement Programma.

Prokip, Maksym. 2023. *Russia's War on Ukraine's Energy System*. Washington, DC: Wilson Center.

RAND Corporation. 2016. *The Foundations of Operational Resilience: Assessing the Ability to Operate in an A2/AD Environment*. Santa Monica, CA.

Schmidt, Alexander. 2017. "Countering Anti-Access / Area Denial: Future Capability Requirements in NATO." Journal Edition 23. Joint Air Power Competence Centre, January.

Shaikh, Shaan and Wes Rumbaugh. 2021. *The Air and Missile War in Nagorno-Karabakh: Lessons for NATO*. London: RUSI.

Sukhodolia, Oleksandr. 2022. *Energy Security of Ukraine under Conditions of Military Aggression*. Kyiv: National Institute for Strategic Studies (NISS).

Tangredi, Sam J. 2013. *Anti-Access Warfare: Countering A2/AD Strategies*. Annapolis, MD: Naval Institute Press.

TRADOC (U.S. Army Training and Doctrine Command) 2018. *The U.S. Army in Multi-Domain Operations 2028*. TRADOC Pamphlet 525-3-1. December 6.

Watling, Jack, and Nick Reynolds. 2023. *Stormbreak: Fighting Through Russian Defences in Ukraine's 2023 Offensive*. London: Royal United Services Institute (RUSI).

Watling, Jack, and Sidharth Kaushal. 2020. "The Democratisation of Precision Strike in the Nagorno-Karabakh Conflict." *Royal United Services Institute*, October 22.

Weizman, Eyal. 2007. *Hollow Land: Israel's Architecture of Occupation*. London: Verso.

Yin, Robert K. 2018. *Case Study Research and Applications: Design and Methods*. 6th ed. Thousand Oaks, CA: Sage Publications.

Zyguła, Alicja, and Beata Piskorska. 2024. "East Shield – A Mechanism for Strengthening the Security and Resilience of Central and Eastern European Countries Using the Example of Poland's Eastern Border." *Rocznik Instytutu Europy Środkowo-Wschodniej* 22 (2), 183–197.

CHAPTER 13

The Critical State of Cyberspace

Hans-Dieter Hiep

Abstract

NATO is currently in a 'grey zone':[1] increasingly being challenged in cyberspace while unable to attribute attacks, thus failing to deter. Simultaneously, many governments, businesses, and populations depend on the Internet and the cloud to function properly. This is a dangerous situation, since global Internet connectivity could get disrupted and destabilised.[2] This chapter explains the basic principles from an overall technological perspective, and discusses the challenge of anti-access and area denial (A2/AD) in the context of global Internet disruptions. What is the contested area, what are possible effects, and can those effects be limited? Using a layered approach, we define the cyber 'high ground' as maintenance of and control over critical infrastructure, from core to edge, from hardware to software. To ensure each cyber 'area' can be protected in isolation but still allow global cooperation, we argue it is important to control which paths data takes in the network, similar to air traffic control. This requires a new approach to organisation: we discuss a decentralised architecture for cyberspace, and how the Internet can be structured using federated networks, featuring geofencing and effect isolation, that has the potential to radically change the future of cyber warfare.

Keywords: Internet; Cloud; Control plane; Disruption; Reset; Decentralisation

Introduction

Background

The Internet (originally called the ARPAnet) is a fruit of DARPA. J.C.R. Licklider, DARPA's 1961 director of the Command and Control Research Office (later the Information Processing Techniques Office), envisioned how 'man-computer symbiosis'[3] could form networks of intellectual communities where artificially intelligent agents help humans process information, to support military data acquisition, internal communications, and decision making. Technically, unlike earlier telephone circuit-switched networks, ARPAnet deployed the first wide-area packet-switched network in 1969.[4] It economically supported 'bursty' data exchange, where at unpredictable moments lots of data is sent in a short time span. In the older telephone network this pattern was costly or slow: it required allocating expensive trunk lines that stayed idle most of the time,[5] or setting up a circuit only when enough data accumulated thus increasing latency. ARPAnet also aimed for interoperability: connecting computers from many vendors at many sites, varying in computer architectures and operating systems.

One application was the Worldwide Military Command and Control System (WWMCCS).[6] The WWMCCS Intercomputer Network (WIN) initially used the telephone network, but there were worries about disruption when the central telephone offices could come under coordinated attack.[7] ARPAnet provided a backup network and selected the TCP/IP technology. Related technologies, such as packet radio, were further developed by DARPA. Development cost was about 150 million USD (in 1990 currency).

Eventually ARPAnet split into two parts. One part, NSFnet, grew to connect worldwide research and academic institutes. It was decommissioned and fully commercialised in 1995, forming today's Internet. The other part grew into the U.S. military network MILNET. In 1990 the security architecture for MILNET was updated, and it was noted that "the Internet is international and is only loosely managed through cooperation of its users. No central authority exists to make or enforce security rules."[8] Nonetheless, the Internet community made best-effort improvements, e.g. to support e-commerce and banking, but the scope was limited. Since 1986 the Internet Engineering Task Force (IETF) has acted as an international standards body based on 'rough consensus,' joining the Internet Society (ISOC) in 1993. Both organisations rely on volunteers and lack formal accountability for ensuring privacy and security, so it is unreasonable to expect that the commercialised Internet's underlying technologies were designed to protect against adversaries in cyber or cognitive warfare.

That the Internet aimed for interoperability facilitated its global spread. Early global applications were information systems, such as the World Wide Web developed at CERN, used by librarians and academics at connected universities and laboratories to maintain a global intellectual community. Over three decades, many applications have been built on top, ranging from e-commerce and banking to multimedia streaming, entertainment, and government services, all dependent on a functioning Internet. While the Internet community pursued permissionless innovation, numerous downsides have emerged from connecting mass populations. Harmful applications,[9] possibly more addictive than illegal substances and hard drugs,[10] illustrate why we must speak of 'users.'

The Internet is not only an addictive playground: in decades it has become a new battleground. Humans can be micro-targeted during psychological and cognitive manipulation operations,[11] and societies face viral misinformation[12] that fuels discord and polarisation.[13] Without mass populations connected, these effects would be smaller and more manageable. Individual profiling and fine-grained tracking enable targeted cognitive effects, essential to fifth-generation warfare and grey-zone conflict.[14] The current Internet architecture lets non-state actors gain substantial control over cyberspace, allowing them to influence foreign states by manipulating politicians and civil servants:[15] some view this as a threat to democracy.[16] Authoritarian regimes also use the Internet for social control, exemplified by Iranian shutdowns,[17] while European democracies employ shutdowns to suppress independence movements.[18] Emerging risks include disruptions by digital Luddites.[19] All this serves as preparation for future sixth-generation warfare, 'which targets technical systems to possibly collapse a society through strategic sabotage.'[20]

Motivation & Plan of the Chapter

Since there is no central Internet authority and cyber-attacks are difficult to attribute,[21] disruptions can occur without anyone being held responsible. Adversaries can coordinate attacks that destabilise and disrupt the Internet, locally or globally. This risk defines the 'critical state of cyberspace,' motivating us to address current shortcomings and explore technologies and methods to improve the status quo.

A lack of control in critical networks and a lack of high-assurance computer technology exposes Europe to societal instability crises (cf. sixth-generation warfare).[22] Gartner and NLnet Foundation highlighted this early in their 2018 Next Generation Internet report:[23] "Every technical primitive from the current Internet must be re-evaluated against the full threat catalogue, as we cannot afford our multi-trillion euro economies and societies to not be able to trust the Internet as we move forward." Economies and societies want to trust the Internet: Can this trust be justified, and how can we design computers and networks to ensure true quality and justifiable trust?

Disruption attacks pose a significant national security threat. Adversaries, from lone 'hacktivists' to well-organised state actors, can jeopardise sovereignty by forcing governments to act, under the threat of a nation-wide or even global Internet kill switch that disturbs routing information so severely that many sites and services become unreachable.[24] One could call this attack 'ransomware of the higher-order.' These attacks can be viewed as an anti-access/area-denial challenge, as discussed in the second section.

The complexity of cyberspace is vast: the global state space includes all electromagnetic waves in transmission lines and electronic/photonic circuitry[25] used to exchange or process information. To get a sense of scale, the Internet currently transports zettabytes per year, and is projected to move about one yottabyte by 2030 via roughly 200 billion connections.[26] Constructing a single system to control and secure the entire Internet seems therefore impossible and undesirable. Because sensors and actuators connect cyberspace with the physical world, many hybrid interactions arise: cyber-physical,[27] cyber-biological,[28] cyber-psychological,[29] and cyber-neural systems.[30] First-order effects in cyberspace produce indirect second-order cybernetic effects on these hybrid systems, showing the importance to maintain the 'high ground'.[31] Without control, arbitrary (malicious) actors could manipulate such hybrid systems. In a military context, such second-order effects include weapon system manipulation and even remote firing. We discuss the layered approach of control, and the role of formal methods in high-assurance computing and networking, in the third section.

More concretely, we can illustrate the critical state of cyberspace with several hacks. The Iranian hack of DigiNotar compromised a cryptographic authority, jeopardising the security of large parts of the Internet:[32] a single hack has global impact. Another example involves Europe's Internet backbone.[33] Internet eXchange Points (IXPs) and transit providers increasingly automate maintenance centrally via Software-Defined Networking (SDN).[34] An adversary controlling such a system could deploy configuration changes to hundreds of border routers simultaneously. In 2017 the author discovered a critical vulnerability in one of the world's largest Internet exchanges that could grant this level of control. Due to vulnerabilities in the Border Gateway Protocol (BGP), such access enables dissemination of false routing information globally, thereby disrupting Internet service providers worldwide. In the military context, these disruptions must be handled with resilience, what prof. K. Guttieri calls "fighting through disruption."[35] In the fourth section the BGP-controlled Internet and its weaknesses will be discussed.

A remedy is to connect vital sectors (finance, healthcare, energy, transportation, industry, defence, and education) to secondary 'backup' networks with a more secure architecture featuring failure isolation and geofencing. Technically, SCION (Scalability, Control, and Isolation On Next-generation networks)[36] is a clean-slate, high-assurance networking architecture with formally verified standards, protocols, and open-source reference implementation. It has reached production-grade software quality and is now entering the special-purpose hardware development phase. By adopting this federated architecture, keeping it separate from the current Internet while deploying Isolation Domains (ISDs), the attack surface can be significantly reduced, creating a 'galvanically isolated' higher ground – a protected network accessible only to verifiably trusted parties. The practical, policy, and doctrinal implications are discussed in the fifth section.

Anti-Access and Area Denial

We first examine the Internet from an anti-access and area-denial (A2/AD) perspective.[37] Anti-access refers to long-range prevention of forces entering a theatre, while area-denial refers to short-range restriction of freedom of action, confining operations to areas under adversarial control.[38] We focus on the physical and virtual layers of Grant's model of cyberspace,[39] omitting the cognitive layer. Grant's model is geographical, since cyber overlaps with all physical military domains: an 'area' of cyberspace encompasses all computing and communication technologies within its geographic bounds.

Existing Internet infrastructure enables rapid military deployment and connection to remote sensors/actuators: dense coverage lets forces operate outside established bases without building new communication networks (e.g. for command and control). In peace time those same networks serve civilian purposes. However, adversaries can disrupt this dual-use infrastructure to create anti-access effects: while forces can still physically enter a theatre, they cannot use disrupted communication networks and remote cloud computation. Consequently, commanders lose the ability to coordinate forces and must resort to other, more restricted means of communication. Disruption can also isolate already deployed forces, cutting off communication back home. When this occurs, secondary ad-hoc networks need to be deployed, since repairing the primary networks can be time consuming without the guarantee of restoring full control.

Internet infrastructure ranges from the core to the edge. The core consists of Internet eXchange Points (IXPs) and the backbone networks linking them. The edge includes local Internet Service Providers (ISPs) delivering last-mile connectivity (including 4G/5G/6G[40]), data centres operated by Managed Service Providers (MSPs) hosting services such as video conferencing, e-mail, file exchange, calendars, et cetera. Today, top-ranking IXPs can process 20 terabit per second,[41] and state-of-the-art backbone links reach 10 terabit per second.[42] This traffic traverses the *data plane*, where regular data flows occur. IXPs enable reachability across most of the Internet, a phenomenon known as the 'flat Internet'.[43] On top we have the *control plane*: the structures, including remote access and control systems, that maintain and support core infrastructure.

In flat networks, disruptions near the core have a larger impact. Consequently, core infrastructure is a high-value target,[44] making targeted reconnaissance and surveillance operations strategically attractive to adversaries. The control plane is indirectly a high-value target.[45] Jackpot is a centralised control plane that remotely manages many core infrastructure sites. When vital sectors and military operations rely on Internet connectivity, disruptions can cause second-order effects, such as energy-distribution problems or disrupted logistics of goods. In flat networks, a single successful cyber operation against such a centralised control plane not only can produce a large first-order impact, but the impact is also amplified if cascading second-order failures are coordinated with other effects in the physical military domains.

Summarising the anti-access/area-denial challenge, the 2020s Internet mirrors the 1960s telephone network before ARPAnet was deployed as a backup: today IXPs take the place of the former central offices in the telephone network. Coordinated attacks on IXPs could result in loss of communication with new and existing force deployments, and can have significant economic, social, and political consequences.

Thus arises the question: When is an area of cyberspace 'red' (in control of an adversary), 'grey' (unknown or neutral), or 'blue' (under control by an ally)? What kind of effects can be generated when in control over IXPs? We follow the offensive cyber operation actions, as part of Grant's model,[46] referring to the (unclassified) JP 3-12 from the U.S. Department of Defence:[47]

- **Deny**: To prevent access to, operation of, or availability of a target function by a specified level for a specified time. In our case, the target function is Internet connectivity. Deny is sub-divided into:
  - **Degrade**: To deny access to, or operation of, a target to a level represented as a percentage of capacity for a period of time. Level of degradation is specified. In our case, this means less traffic passes, or specifically only some filtered traffic passes.
  - **Disruption**: To completely but temporarily deny access to, or operation of, a target for a period of time. Can be considered a special case of Degrade, which denies everything. In our case, this amounts to traffic black-holing: all traffic that passes through the IXP is to be dropped, effectively reducing the capacity of the overall network.
  - **Destroy**: To completely and irreparably deny access to, or operation of, a target. Destruction maximises the time and amount of denial. In our case, configurations could be changed to cause physical damage to the underlying supporting hardware.
- **Manipulate**: To control or change information, information systems, and/or networks. It uses an adversary's information resources to create effects not immediately apparent. The targeted network may appear to operate normally until secondary or tertiary effects, including physical effects, reveal evidence. In our case, this means changing route information and information flows (more technical details in the fourth section).

As soon as an adversary can achieve the above effects, the infrastructure is no longer under control of an ally: it is 'red'. Since an adversary may be able to achieve the above effects but does not wish to reveal this, we have the 'grey' area: the contested area. We define the area of cyberspace to be 'grey' when there are risks or uncertainties as to whether the target function is and will remain functional at times when demanded. When evidence is found that the area was (likely) under control of an adversary, the 'grey' area may reveal itself to be 'red' after all.

If an ally can achieve the above effects it does not imply the infrastructure is considered under control, and thus 'blue'. An area is only 'blue' when it is reasonably certain it is neither 'grey' nor 'red', thus denying adversaries the possibility to achieve the above effects. This can only be proven by obtaining evidence about the quality of the critical components within a cyber area (publicly auditable, up to reasonable assumptions): this is the main goal of defensive cyber operations.[48]

Infrastructure and networks constantly change, making it hard to detect adversarial control. Adversaries infiltrate critical Internet infrastructure: entering via outer networks and reconfiguring software/firmware for persistence, so-called Advanced Persistent Threats (APTs).[49] Also hardware supply chain attacks, providing components with covert flaws (hardware Trojans), can compromise operating system security.[50] Risks and uncertainties about device quality are difficult to measure quantitatively. When infrastructure in commercial ISPs or IXPs contains APTs, the network appears functional for many years while still being vulnerable to an adversarial strike: attackers can exploit zero-day vulnerabilities, bypassing cyber defences that rely on invalid assumptions about the underlying technology.

As mentioned above, an area is uncontested only when adversarial action is impossible. This requires full knowledge of the hardware/software stack, its functional dependencies, expected behaviour, and various degrees in which the system can be provably correct. Correctness here is not 'perfect', it merely aims to explain complex systems as simply as possible, and prove that the simplified explanation exactly reflects reality. Two guiding principles to achieve this are: the extended Kerckhoff principle, and the elimination-of-complexity principle.

- The extended Kerckhoff principle requires a system to be secure, even if an adversary knows its internal workings. Applying it to critical infrastructure creates a strategic need for open-source software (OSS) and open-source hardware (OSH) for transparency.[51] Although transparency permits inspection, provable security prevents adversarial effects. Transparency also enables education, independent review, and lets non-critical infrastructure adopt these designs while making different trade-offs, e.g. to improve efficiency albeit with less security.
- The principle of eliminating complexity states that security relies on simplicity and explanation.[52] As computers and computer networks grow more complex over decades, maintenance and keeping control becomes more difficult. This 'boiling-frog' effect, where organisations slowly tolerate more glitches while pressured to deliver more and more complex functionality, pushes towards

normalisation of insecure practices. Moreover, vendors drift apart, causing conceptual fragmentation and technical incompatibility. The principle states that complexity must be eliminated: the behaviour of a system must be specified as simply as possible, while the complex implementation underneath is proven correct with respect to such a behavioural specification.

One example where both principles fail is cloud migration. Inspired by C.S. Magee's "Awaiting the cyber 9/11"[53], cloud computing was proposed to make cybersecurity more manageable. While local 'clouds' aid in security and incident management, remote 'clouds' make the network in between a critical dependency beyond the control of both client and vendor. Moreover, to realise a cloud, data centre deployments use largely homogeneous hardware/software that involve large investments, which incentivise vendors to hide fundamental (hardware) security issues by creating opacity that can be mistaken for security: in reality, decades-old architectures have a large technological footprint and face decades-old critical security issues.[54] Moreover, concentrating responsibility in the hands of a few top-level experts is very risky: a single mistake can nowadays have an economic blast radius in the tens to hundreds of billions of euros, cf. the CrowdStrike incident.[55]

Controlling the Layers of Abstraction

To grasp the complexity of cyberspace, we make use of levels of abstraction (see Figure 13.1). In the time domain, one could consider the range of operating frequencies between the petahertz (the frequency of light and cryogenic computing)[56] and the nanohertz (also called a generation, roughly corresponding to 30 years) to be at least 25 orders of magnitude. In physical space, one could consider the fine granularity of nano-scale devices,[57] each about a hundred of $(nm)^2$, all the way up to the global level near the Earth's surface at which the Internet operates, around $500T(m^2)$, to be at least 30 orders of magnitude. At each order of magnitude we recognise different components. From compositions and integrations of billions of nano-scale devices, one obtains larger components such as information processors and memory. In turn, billions of such higher-level components are linked by electrical/optical channels to create the Internet, where the number of nano-devices is approaching 10^{20}: the most complex socio-technological system humanity ever created.[58]

Cyberspace[59] can be defined as a dynamical system, comprising the electromagnetic field and its cybernetic effects. At each level of abstraction, first-order cybernetic effects are logically explainable observations involving information flow, circularity, and feedback.[60] Recursion can explain such phenomena,[61] and

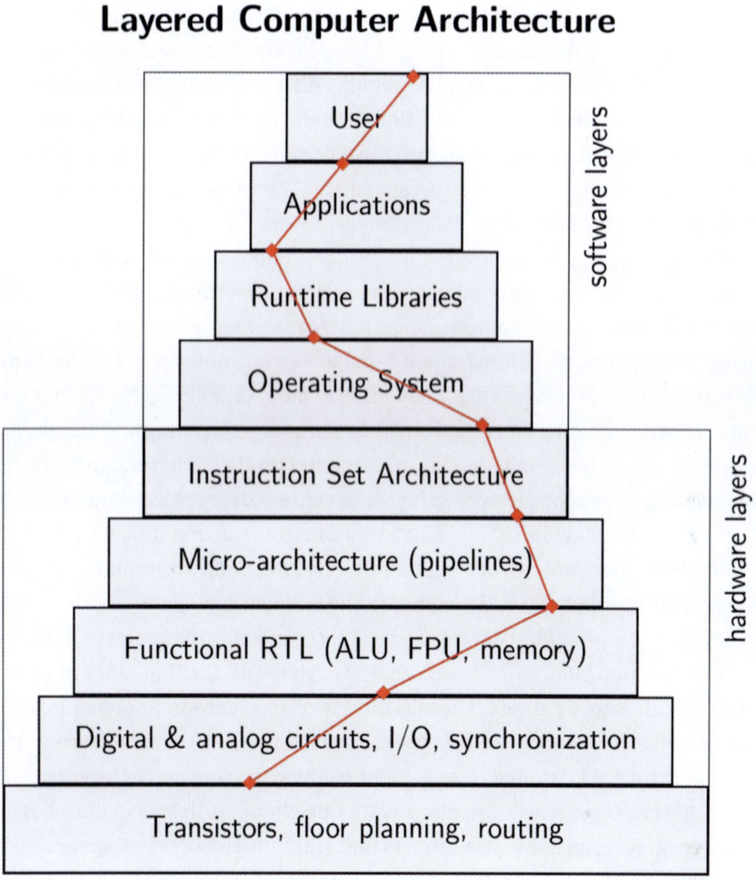

Figure 13.1: Computer architecture, both hardware and software, shown in layers. Instruction Set Architecture layer is the interface between software and hardware. The thin red line shows how flaws in different layers can be used to achieve unexpected effects in higher layers: for example, the RowHammer attack misused timing assumption at the lowest layers to cause bit flips in sensitive locations in memory to bypass defences at higher layers in the stack. It did so because under normal operation of a pipelined execution, such memory accesses would be handled by a cache (part of the micro-architecture), but the attack explicitly avoids that cache, thereby violating a timing assumption in a memory component.

mathematical logic explains relative notions of recursion and the emergence of a-causal yet synchronous effects. Second-order cybernetic effects arise when we also consider the environment, including human behaviour and physical systems, via sensors and actuators.

To raise the level of abstraction, we often ignore the intrinsic complexity of layers below, focusing only on their extrinsic simplicity: observable functionality that support the layers above (Figure 13.1). Ignorance is bliss:[62] only when functionality fails do we recall the inner working of deep layers. Each level of abstraction describes desirable and undesirable functionality differently, creating communication barriers causing 'Babylonian confusion.'[63] While each layer may be correct in isolation, an integrated validation and verification approach across stacked layers is crucial for overall system quality and reliability.[64] Therefore, a common language is essential to prevent misleading abstractions.

Simple descriptions involving recursion can denote systems with complex behaviour. Handling these descriptions requires disciplined thinking as demanded by **formal methods**. Formal methods can be applied at every level of abstraction to define desirable/undesirable functionality, and to validate and verify that systems include/exclude such functionality. They are the most reliable scientific tool for ensuring, from the building blocks upward, the quality of computing and communication systems. Formal methods employ mathematics for the design and integration of systems,[65] and have a long history in high-assurance, safety-critical domains such as railway signalling, avionics, spaceflight, missile control, and nuclear power.[66] The DARPA High-Assurance Cyber Military Systems (HACMS) program has shown how to successfully apply formal methods in the military context.[67]

Zooming in on a single computer, consider modern architectures such as Intel/AMD's x86 and ARM. After the discovery of the FDIV floating point division bug, resulting in a costly product recall, Intel invested more in applying formal methods to catch errors early.[68] Deep design flaws in processors can surface years later, often uncovered through independent study, requiring active **intelligence** gathering. Recent research shows that covert side-channels in speculation-based out-of-order architectures (as used in x86 and ARM processors) lets adversaries exploit timing characteristics of caches to leak data between virtual machines sharing a physical processor.[69] When such fundamental security issues are discovered, higher-level layers should be reassessed. Hence, lower-layer design errors are costlier to fix than those above. Hardware errors are costly, especially when one considers data centre deployments, and the risk of loss of control over many areas of cyberspace, making intelligence regarding their presence highly valuable. If security risks at the theatre-level view are unknown, preparing for impact is difficult. One should not confuse the theatre-level view[70] where rare events still occur, with 'security theatres'[71] that waste resources. Intelligence on existing or potential vulnerabilities enables timely off-loading of low-quality components to less critical markets (e.g. entertainment systems), compartmentalising effects, and lowering cyber insecurity in the military context.

Adversaries penetrate systems by breaking abstraction layers in unexpected ways, generating undesirable effects at higher layers. For example, memory upscaling led to the RowHammer family of attacks,[72] exposing upper layers to new vulnerabilities[73] stemming from neglect of robust software/hardware engineering standards (e.g. exhaustive time-sensitive validation and verification). These issues should be detected early in design, although they are sometimes tolerated for efficiency reasons. Interface mismatches between abstraction layers, causing assumption/guarantee violations, should be identified early via subject matter expert **surveillance**. Proper incident-reporting and pre-mortem processes are essential.[74]

New vulnerabilities and zero-day exploits in legacy systems are increasingly discovered automatically by both allies and adversaries. This activity constitutes **reconnaissance**. AI agents can scan vast networks and our complex technology stack, called the 'tower of Babel,'[75] more efficiently than most humans. While AI can aid vulnerability detection, retaining a human in the hacking loop is necessary to allow human intuition to steer the discovery process. We are not at the singularity where humans cannot keep pace:[76] nevertheless, cyber defenders should take a zero-trust approach, continuously studying unknown devices to detect emerging technologies within protected infrastructure.

Finally, Europe is considered to be lagging in independently developing computer technology. One can easily get the impression there was an implicit division of labour in NATO, where the US focused on hardware while Europe focused on software. This implicit division can partially explain Europe's talent acquisition and retention problems[77] and lack of technology transfer at universities,[78] thus remaining stuck at developing only software prototypes. As Fiott explains,[79] it is Washington's concern that Europe lags so far behind the U.S. in what is called the 'technology gap' in the "areas such as command, control, communications, computers, intelligence, surveillance and reconnaissance (C4ISR)" that it may affect interoperability. Europe now has the second-mover advantage: a unique opportunity to design high-assurance computers and networks according to its own values, taking into account in system design all lessons learnt of the past decades, complying with the highest levels of common criteria for system quality.[80]

Internet Disruptions

The Internet uses a layered architecture of protocols (Figure 13.2). A protocol is an agreement that lets computers communicate. The layers are: application (e.g. HTTP), transport (e.g. TLS, TCP), network (e.g. IPv4/IPv6, OSPF, BGP), link (e.g. Ethernet), and physical. The Internet Engineering Task Force (IETF) and the Institute of Electrical and Electronics Engineers (IEEE) publish the technical specifications for these protocols. Similar to the hardware/software split in computer architecture, network architecture is also divided into two parts:[81] IETF defines the network layer and above, while IEEE defines the link layer and below. This section focuses on the most critical layer: the network layer.

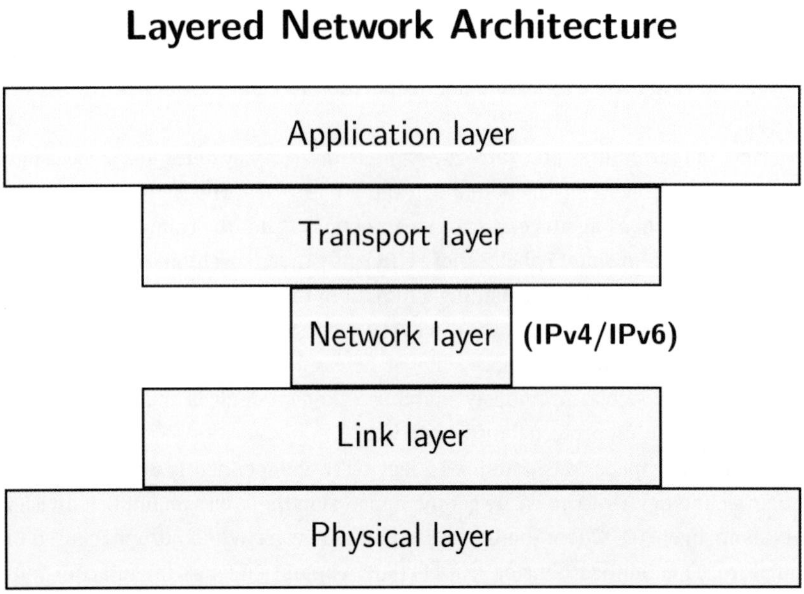

Figure 13.2: Internet networking architecture, shown in layers. The most critical layer with respect to denial and manipulation is the network layer. The figure is shaped as an hourglass due to the Hourglass Lemma.[82] The most critical layer is the network layer, where its main functionality is connectivity.

The Internet is constructed technically and organisationally. Parts of the network are grouped by an organisational entity called an Autonomous System (AS), encompassing computing devices and their communication links. They are called 'autonomous' because a single administrative entity is responsible for the devices, their security, and their interconnection also with neighbouring autonomous systems. Examples include an office, a ship, ad-hoc field networks, or a satellite constellation. Autonomous systems vary in geographic scale, some cover very wide areas.

For inter-network connections, neighbouring autonomous systems are linked by Border Routers (BRs), which are gateways between the interior and exterior of an autonomous system (Figure 13.3). Traffic entering is *ingress*; traffic leaving is *egress*. Globally, the Internet is a patchwork of ASes connected via these border routers. To reach hosts in an arbitrary AS, traffic may traverse one or more third-party ASes that provide transit, carrying traffic that does not have a host in the transit AS as a destination. These third-party ASes belong to various Internet

Service Provider (ISP) tiers, with the mid-tiers covering wider regions, and the top-tiers providing global coverage.

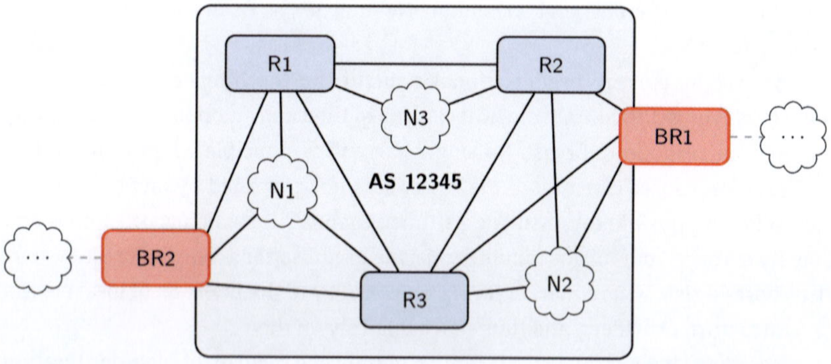

Figure 13.3: Example autonomous system, with internal routers (R1 until R3) and two border routers (BR1, BR2) connected to the outside world. Three subnets (N1 until N3) are owned by this autonomous system, so the border router publicly shares routing information regarding the reachability of these subnets, thus attracting inflow of traffic. Depending on policy, BGP can advertise that this autonomous system allows transit traffic (between BR1 and BR2): note that if the router R3 fails, there is a redundant path via R1 and R2.

The overall system supports dynamic reconfigurations: when links or routers fail, traffic flow adapts automatically. This is also required within an autonomous system, where each organisation uses different techniques for internal redundancy. Monitoring and dynamic reconfiguration tools such as Open Shortest Path First (OSPF) and Intermediate System to Intermediate System (IS-IS) exist, but a global consensus on exterior traffic routing is needed: the *de facto* protocol is the external Border Gateway Protocol (BGP), which specifies route update messages and spreads route information between autonomous systems. Border routers use these update messages to direct and adjust internal traffic flows, ensuring transit traffic flows from border to border.

There are more technical details that can be found in any competent book on computer networking,[83] involving autonomous system numbering (ASN), the global IP address space (IPv4/IPv6), and the international organisations managing these, such as the Internet Assigned Numbers Authority (IANA) and the five Regional Internet Registries (RIRs). Moreover, there are many other critical issues on the Internet, involving the higher layers built on top of the network layer, such as the Domain Name System (DNS) and the Public Key Infrastructure (PKI) used for the Web. We focus on the routing system.

The Internet is not the only way to interconnect. For higher assurance, fixed lines or Multi-Protocol Label Switching (MPLS) can be used, but this requires manual setup, increases reliance on single providers, and could reduce overall network resilience. Fixed lines are not shared and thus entail high costs for a single organisation. Also in the military context, recent conflicts have shown that frontline areas often depend on commodity Internet, as fixed leased lines are not always available.

Within the current organisational structure of the Internet, governing the network layer is difficult. On the BGP-controlled Internet, autonomous systems rely on an implicit notion of trust. No single party is responsible when false routing information causes disruption: the Internet is not governed by any state, nor is any state responsible for ensuring its functionality.[84] Instead, market forces and loosely managed non-profit, volunteer-driven organisations coordinate the system. This poses a risk, where market forces act contrary to the interests of local critical infrastructure providers, and thus states that rely on them.

Moreover, the market lacks incentive to improve security of high-tier Internet service providers, whose autonomous systems are high value targets while they are still trusted by many networks. Spreading false routing information from these systems, especially those employing a centralised control plane, can yield large-scale effects. The core routing protocol BGP could be turned into a 'weapon of mass disruption.'[85] The current trust model enables this behaviour: autonomous systems need not prove the intrinsic security of their control plane; trust is implicit and based solely on extrinsic observed functionality. Lacking intrinsic security, an adversary can strike a well-connected core network to demonstrate insecurity (the 'grey' areas of cyberspace). While attribution remains a difficult problem, these cyber attacks are difficult to trace back to the instigator.

Technically, the BGP-controlled Internet is also vulnerable to route hijacking, route leaking, and address spoofing by compromised or insecure autonomous systems. In a route hijacking attack, adversaries reroute traffic to ASes under their control to capture traffic. Route hijacks may seem legitimate, since many route changes are used as a mechanism to mitigate Distributed Denial of Service (DDoS) attacks. While encryption can improve confidentiality, quantum computing 'harvest now, decrypt later' attacks increase risk, and traffic metadata remains publicly visible for AI to analyse. Route leaking occurs when BGP updates accidentally or maliciously spread to unintended parts of the network, even if these updates originated legitimately. Address spoofing, a major cause of the attribution problem, injects false source and destination addresses, making it difficult to trace global traffic.

More recently discovered vulnerabilities can destabilise global routing altogether,[86] affecting availability when combined with link or router failures. Moreover, BGP was never designed to assign different trust levels to autonomous systems (e.g. NATO-country networks versus external networks). Connecting critical infrastructure and vital sectors, including military networks, to the Internet expands their attack surface, causing an increase in cyber attacks. The same risk applies to the battlefield in the 'Internet of Military Things.'

More recent mechanisms to secure routing information, such as RPKI (Resource Public Key Infrastructure), BGPsec, and ASPA (Autonomous System Provider Authorisation), resulted in a system with only five centralised trust roots, one located at each Regional Internet Registry (RIR). However, controlling only a single trust root is already sufficient to undo the additional security these mechanisms intend to provide. Recent financial troubles at one of the RIRs has raised fears that its root certificate might be sold to an adversary, jeopardising the security of all RIRs. Also deployment is limited: not every part of the network is covered by these mechanisms. Hence, local connectivity is either unprotected or relies on trust roots outside of NATO. Consequently, we lack sovereign control over how the PKIs are used for routing information: a single security failure has global impact. As demonstrated in the DigiNotar breach, this risk is real.

Moreover and currently, it is technically not possible to observe and choose the paths (and thus jurisdictions) where traffic is sent over: this requirement was simply never part of the design.[87] Since Internet routing is destination-based, there is no control over how traffic should travel through different networks. Such lack of control can expose sensitive traffic to hostile or adversarial areas. More technically, BGP-controlled Internet currently offers only one 'best path' per IP address prefix. If the network dynamically reconfigures due to link or router outages, it takes time to converge to a new 'best path'. This process is relatively slow, with convergence times of up to several minutes or even hours in extreme cases, which is problematic for mission-critical applications requiring fast failover. Concurrent usage of different paths is important, especially when planning larger joint operations where one takes into account contingencies in the cyber domain, and needs to anticipate outages and congestion. Moreover, path control can be used as a mechanism to limit the attack surface of systems by tracing the path back to its origin, or (temporarily) filtering traffic based on origin.

Finally, the current architecture lacks mechanisms for ensuring hardware and software supply chain security of devices within an autonomous systems, nor is there an incident reporting mechanism that allows independent reporting of security incidents at Internet service providers – for example when the security of an IXP is breached. Critical network services need to be available locally, such as time synchronisation, root certificate authorities, certificate revocation, and domain naming. Time synchronisation ensures all subsystems share a common clock, which is essential for time-sensitive data such as short-lived certificates. Many operating systems rely on vendor-provided synchronisation services, thus implicitly requiring Internet connectivity. Prolonged outages on these services could prevent desynchronised subsystems to reconnect and recover, where circular dependencies could arise: synchronisation services need valid certificates to authenticate themselves, while certificate validation itself requires prior synchronisation. Similar cyclic dependency arguments also apply to other critical services.

A Potential Internet Crash and Reset: Practical, Policy, and Doctrinal Implications

On the current BGP-controlled Internet, trust relations between autonomous systems are unregulated, lacking transparency and accountability. A zero-trust architecture provides stronger assurance: new and unknown autonomous systems are not trusted by default, but can earn trust over time. Regular verification of trust reduces the risk of control plane compromise and thus large-scale disruptions. Non-compliance with security standards or refusal of security inspections can lead to ejection from a high-assurance network, to incentivise continual security upkeep. In anti-access/area-denial terms, 'blue' areas are proven trustworthy, 'red' zones are demonstrably untrustworthy, and the 'grey' areas contain systems whose trust has not yet been verified, or whose prior verification is now doubtful due to newly discovered fundamental vulnerabilities (see third section).

There is a growing need to connect legacy systems and infrastructures to the Internet, even when they are not IP-based or designed to be exposed to a world-wide network, due to remote-working demands and stay-at-home orders. Society's dependence on commodity Internet connectivity is increasing, and some jurisdictions now legally require governments to ensure Internet connectivity. One mitigation is to use multiple, functionally separated ISPs, but this does not protect higher-tier transit providers and Internet exchanges near the network core, where insecurities can cause cascading effects and affect traffic between nearby autonomous systems. Therefore, any improvement needs to satisfy the *local availability principle*, where connectivity has the smallest possible logical footprint,[88] meaning that, if there exists a path traversing 'blue' areas, no adversary can disrupt it. This ensures distant threats cannot negatively affect local connectivity.

In sixth-generation warfare, critical networking infrastructure and its human support are (automatically acquired) targets of cyber attacks and cognitive and psychological operations, respectively.[89] As attacks intensify, which in the 'flat Internet' has significant impact, first-order disruptions cascade and cause second-order effects. Repeated, intermittent and large-scale disruptions can cause societal collapse due to its large-scale digital dependencies, yielding a surrender to a higher power. This is what is called the 'crash' of the Internet. A well-designed system (including the Internet) must always eventually permit a reset: a return to the initial state. A system satisfies the liveness property of *perpetuity* when, from every reachable state, eventually the initial state can be reached.[90] The overall system can renew and reconfigure itself before the reset completes, by adopting and deploying new technologies to prevent repeating the same kind of failure from reappearing.

The possibility of an Internet crash should be publicly known. Long ignored backbone routing vulnerabilities (colloquially known as *spina bifida*) must be fixed before resetting the system, since otherwise a reset is pointless. To preserve state sovereignty, cyberspace could be structured in a federated and decentralised manner, where each state supports its own cyber area with a domestic military-industrial base supplying high-assurance components. As each area ensures local connectivity, the federated network expands by connecting isolation domains, eventually establishing a secondary global network. Traffic can still flow between isolation domains, now with the full path of each packet known in advance, similar to air traffic control managing aircraft paths. High-assurance software and hardware, as used in avionics architectures where failures are catastrophic,[91] are also needed to protect cyberspace. The new backbone routing system remains backward compatible, to support legacy autonomous systems using IPv4/IPv6 with border routers that only support legacy BGP.

To keep global Internet connectivity stable, states could agree on international laws that prohibit intrusions into other states' protected cyber areas, treating such interference as an act of cyberwar. Each state monitors both outbound traffic to prevent it from reaching foreign protected areas, and inbound traffic to guard its own protected areas. While isolation domains still interconnect globally, due to path control the attribution problem is resolved. Within every isolation domain, essential services (synchronisation, certification, naming, and routing) remain available, and each state regulates, maintains, and verifies these critical functions. Sovereign states can (democratically) regulate their isolation domain, for example by describing basic principles to protect their enclosed humans from cognitive and psychological operations, possibly following the *U.S. Declaration for the Future of the Internet* and the *European Declaration on Digital Rights and Principles*.

This chapter ends with the following bullet points:

- NATO militaries prepare to continue their operations even during global or local Internet disruptions, underscoring Guttieri's vision of 'fighting through disruption.'[92]
- Procurement of cyber within military systems need to use high-assurance operating systems and hypervisors, that are provably secure by employing formal methods, following in the footsteps of the High-Assurance Cyber Military Systems DARPA programme.[93] This class of operating systems can already mitigate many cybersecurity issues in the very short term.

- In the context of NATO's Crossed Swords and Locked Shields cyber exercises, future work builds a more realistic 'simulated Internet' in which vulnerabilities of the BGP-controlled Internet are demonstrated, to show the effects of 'weapons of mass disruption' in a controlled environment.
- NATO should deploy secondary 'backup' networks, functionally separated from the BGP-controlled Internet: 'galvanically isolated', separating low ground from high ground. These high-assurance networks support NATO command and control in case existing commercial Internet infrastructure is disrupted.
- All military domains support cyber intelligence gathering with additional surveillance and reconnaissance activities (C4ISR) covering their hardware/software deployments and supply chains, including critical core network infrastructure of their cyber areas. Progress should be monitored, area by area, resulting in a geographic picture of 'blue', 'grey', and 'red' cyber areas.

Notes

1 Fitton, *Cyber Operations and Gray Zones: Challenges for NATO.*
2 Sampedro, López-Ferrández, and Hidalgo, *Digital Disintermediation*; Jyothi, *Solar Superstorms*; Birge-Lee, Apostolaki, and Rexford, *Global BGP Attacks*; Stöger et al., *BGP Vortex.*
3 Licklider, *Man-Computer Symbiosis.*
4 Reed, Van Atta, and Deitchman, *DARPA Technical Accomplishments.*
5 Roberts, *The Evolution of Packet Switching.*
6 Alberts and Hayes, *Understanding Command and Control.*
7 Andrews, *The Telephone Network of the 1960s.*
8 Shirey, *Defense Data Network Security Architecture.*
9 Livingstone, "Risk and Harm on the Internet"; Carlbring and Andersson, *Commentary: AI Psychosis Is Not a New Threat*; Wong et al., *Rising Cognitive Disability.*
10 Macit, Macit, and Güngör, *A Research on Social Media Addiction.*
11 Lungu, *War.com*; Masakowsi and Blatny, *Mitigating and Responding to Cognitive Warfare*; Kibeom, *A Study on the Concept of Cyber Cognitive Warfare.*
12 Sadoun, *PsyOp and Social Networks.*
13 Ellingsen, *Hybrid Threats as a Threat to Democracy.*
14 Krishnan, *Fifth Generation Warfare.*
15 Aytac, *Digital Domination.*
16 Schaake, *The Tech Coup.*
17 Conduit, *The Political Economy of Digital Authoritarianism.*
18 Sampedro, López-Ferrández, and Hidalgo, *Digital Disintermediation.*
19 Ciriello, Sullivan, and Mindel, *Digital Luddism.*
20 Krishnan, *Fifth Generation Warfare.*
21 Fitton, "Cyber Operations and Gray Zones"; Olmstead, *Into the Gray Zone.*
22 Krishnan, *Fifth Generation Warfare.*
23 Leenaars et al., *Next Generation Internet 2025.*
24 Sampedro, López-Ferrández, and Hidalgo, *Digital Disintermediation.*

25 Marengo and Tu, *Optical Theorem for Transmission Lines*; Westerveld and Urbach, *Silicon Photonics.*
26 Loygue, Al Agha, and Pujolle, *Carbon Footprint of Cloud.*
27 Baheti and Gill, *Cyber-Physical Systems.*
28 Mehrali et al., *Blending Electronics with the Human Body.*
29 Kibeom, *A Study on the Concept of Cyber Cognitive Warfare.*
30 Reed et al., *Fractional Cyber-Neural Systems.*
31 Raymond et al., *Key Terrain in Cyberspace.*
32 Huston, *Hacking Away at the Internet's Security.*
33 Caoui and Steck, *Network Diversity.*
34 Scott-Hayward, Natarajan, and Sezer, *A Survey of Security in Software.*
35 Guttieri, *Fighting Through Disruption.*
36 Chuat et al., *The Complete Guide to SCION.*
37 Russell, *Strategic A2/AD in Cyberspace*; Keithley, *Interoperability.*
38 Krepinevich, Watts, and Work, *Meeting the Anti-Access and Area Denial Challenge.*
39 Grant, *On the Military Geography of Cyberspace*; Grant, *Using C2 Problem Space.*
40 Salahdine, Han, and Zhang, *5G, 6G, and Beyond.*
41 Lallo, *Internet eXchange Points*; Böttger et al., *The Elusive Internet Flattening.*
42 Pires, *On the Capacity.*
43 Livermore et al., *Architecture and Control.*
44 Raymond et al., *Key Terrain in Cyberspace.*
45 Schuchard et al., *Losing Control of the Internet.*
46 Grant, *Using C2 Problem Space.*
47 Joint Chiefs of Staff, *Joint Publication 3-12 – Cyberspace Operations.*
48 Campbell, *Generals in Cyberspace.*
49 Alshamrani et al., *A Survey on Advanced Persistent Threats.*
50 Bhunia et al., *Hardware Trojan Attacks.*
51 Knoll, *Adapting Kerckhoffs's Principle.*
52 Beach et al., *Analysis of Systems Security Engineering.*
53 Magee, *Awaiting the Cyber 9/11.*
54 Rushby, *Partitioning in Avionics Architectures*; Hertogh et al., *Rain: Transiently Leaking Data.*
55 George, "When Trust Fails".
56 Wang et al., *A Wireless Terahertz Cryogenic Interconnect*; Zota et al., *Energy-Efficient Computing.*
57 Cao et al., *The Future Transistors.*
58 Fuchs, *The Internet as a Self-Organizing.*
59 Kuehl, *From Cyberspace to Cyberpower.*
60 Heylighen and Joslyn, *Cybernetics and Second-Order Cybernetics.*
61 Jeon, *Second-Order Recursions*; Kauffman, *Cybernetics, Reflexivity and Second-Order Science.*
62 Kelly, *The Insider Threat to Cybersecurity.*
63 Lipovski, *Hardware Description Languages.*
64 Gleirscher, Foster, and Woodcock, *New Opportunities for Integrated Formal Methods.*
65 Wupper, *Design As the Discovery.*
66 Williams, *Formal Methods in the Development*; Rushby, *Formal Methods and the Certification*; Haxthausen, *An Introduction to Formal Methods*; Dobrovic and Lovrencic, *Application of Formal Methods.*
67 Fisher, "HACMS".
68 O'Leary et al., *Formally Verifying IEEE Compliance*; Gupta, Kiran Kumar, and Ghughal, *Formally Verifying Graphics FPU.*

69 Hertogh et al., *Rain.*

70 Greene, *Cyberwarfare and Our Allies.*

71 Feffer et al., *Red-Teaming for Generative AI.*

72 Mutlu, Olgun, and Yağlıkçı, *Fundamentally Understanding and Solving Rowhammer.*

73 Aslan et al., *A Comprehensive Review of Cyber Security Vulnerabilities.*

74 Eisenbach, Kovner, and Lee, *Cyber Risk and the US Financial System.*

75 Lipovski, *Hardware Description Languages.*

76 Seewald, *A Criticism of the Technological Singularity.*

77 Khan, *European Academic Brain Drain.*

78 Stange et al., *Open Source Hardware in Manufacturing.*

79 Fiott, *A Revolution Too Far?*

80 Rushby, *Formal Methods*, 37; Bowen, *Formal Methods in Safety-Critical Standards*; Fekete, *Common Criteria for the Assessment.*

81 Clark, *Designing an Internet.*

82 Beck, "On the Hourglass Model".

83 Tanenbaum, *Computernetwerken.*

84 Shirey, *Defense Data Network Security Architecture.*

85 Korb, *Weapons of Mass Disruption.*

86 Sampedro, López-Ferrández, and Hidalgo, *Digital Disintermediation*; Birge-Lee, Apostolaki, and Rexford, *Global BGP Attacks*; Stöger et al., *BGP Vortex.*

87 Clark, *Designing an Internet.*

88 de Boer et al., *Footprint Logic for Object-Oriented Components.*

89 Krishnan, *Fifth Generation Warfare.*

90 Hiep, *A Reo Semantics.*

91 Rushby, *Partitioning in Avionics Architectures.*

92 Guttieri, *Fighting Through Disruption.*

93 Fisher, "HACMS".

References

Alberts, D. S., and R. E. Hayes. 2006. *Understanding Command and Control.* Assistant Secretary of Defense: C3I/Command Control Research Program.

Alshamrani, A., S. Myneni, A. Chowdhary, and D. Huang. 2019. "A Survey on Advanced Persistent Threats: Techniques, Solutions, Challenges, and Research Opportunities". *IEEE Communications Surveys & Tutorials* 21 (2), 1851–1877.

Andrews, F. T. 2002. "The Telephone Network of the 1960s". *IEEE Communications Magazine* 40 (7), 49–53.

Aslan, O., S. S. Aktuğ, M. Ozkan-Okay, A. A. Yilmaz, and E. Akin. 2023. "A Comprehensive Review of Cyber Security Vulnerabilities, Threats, Attacks, and Solutions". *MDPI Electronics* 12 (6), 1333.

Aytac, U. 2024. "Digital Domination: Social Media and Contestatory Democracy". *Political Studies* 72 (1), 6–25.

Baheti, R., and H. Gill. 2011. "Cyber-Physical Systems". *The Impact of Control Technology* 12 (1), 161–166.

Beach, P. M., L. O. Mailloux, B. T. Langhals, and R. F. Mills. 2022. "Analysis of Systems Security Engineering Design Principles for the Development of Secure and Resilient Systems". In *Handbook of Scholarly Publications from the Air Force Institute of Technology (AFIT), Volume 1, 2000–2020*, 33–63. CRC Press.

Beck, M. 2019. "On the Hourglass Model". *CACM* 62, (7) (June), 48–57. https://doi.org/10.1145/3274770.

Bhunia, S., M. S. Hsiao, M. Banga, and S. Narasimhan. 2014. "Hardware Trojan Attacks: Threat Analysis and Countermeasures". *Proceedings of the IEEE* 102 (8), 1229–1247.

Birge-Lee, H., M. Apostolaki, and J. Rexford. 2025. "Global BGP Attacks That Evade Route Monitoring". *International Conference on Passive and Active Network Measurement*, 335–357.

Boer, F. S. de, S. de Gouw, H. A. Hiep, and J. Bian. 2022. "Footprint Logic for Object-Oriented Components". *International Conference on Formal Aspects of Component Software*, 141–160.

Böttger, T., G. Antichi, E. L. Fernandes, R. Di Lallo, M. Bruyere, S. Uhlig, and I. Castro. 2019. *The Elusive Internet Flattening: 10 Years of IXP Growth*. RIPE 78 Meeting.

Bowen, J. 1993. "Formal Methods in Safety-Critical Standards". *Proceedings 1993 Software Engineering Standards Symposium*, 168–177.

Campbell, P. 2018. "Generals in Cyberspace: Military Insights for Defending Cyberspace". *Orbis* 62 (2), 262–277.

Cao, W., H. Bu, M. Vinet, M. Cao, S. Takagi, S. Hwang, T. Ghani, and K. Banerjee. 2023. "The Future Transistors". *Nature* 620 (7974), 501–515.

Caoui, E. H., and A. Steck. 2023. *Network Diversity, Market Entry, and the Global Internet Backbone*. https://andrewsteck.github.io/uploads/SSRN-id4478868.pdf.

Carlbring, P., and G. Andersson. 2025. "Commentary: AI Psychosis Is Not a New Threat: Lessons from Media-Induced Delusions". *Internet Interventions*, 100882.

Chuat, L., M. Legner, D. Basin, D. Hausheer, S. Hitz, P. Müller, and A. Perrig. 2022. *The Complete Guide to SCION: From Design Principles to Formal Verification*. Springer.

Ciriello, R., R. Sullivan, and V. Mindel. 2025. "Digital Luddism as Collective Action to Democratise Tech". *Americas Conference on Information Systems (AMCIS 2025)*.

Clark, D. D. 2018. *Designing an Internet*. MIT Press.

Conduit, D. 2026. "The Political Economy of Digital Authoritarianism: Evidence from the Iranian Regime's Implementation of Technology". *Democratization* 33 (1), 62–84.

Dobrovic, Z., and A. Lovrencic. 2013. "Application of Formal Methods in Development of Information Systems". *Central European Conference on Information and Intelligent Systems*, Faculty of organization and informatics varazdin, 268.

Eisenbach, T. M., A. Kovner, and M. J. Lee. 2022. "Cyber Risk and the US Financial System: A Pre-Mortem Analysis". *Journal of Financial Economics* 145 (3), 802–826.

Ellingsen, T. 2024. "Hybrid Threats as a Threat to Democracy: Soft and Smart Tactics and Three Dimensions of Democracy". In *Preparing for Hybrid Threats to Security*, 36–51. Routledge.

Feffer, M., A. Sinha, W. H. Deng, Z. C. Lipton, and H. Heidari. 2024. "Red-Teaming for Generative AI: Silver Bullet or Security Theater?" *Proceedings of the AAAI/ACM Conference on AI, Ethics, and Society* 7, 421–437.

Fekete, A. 2011. "Common Criteria for the Assessment of Critical Infrastructures". *International Journal of Disaster Risk Science* 2 (1), 15–24.

Fiott, D. 2017. "A Revolution Too Far? US Defence Innovation, Europe and NATO's Military-Technological Gap". *Journal of Strategic Studies* 40 (3), 417–437.

Fisher, K. 2012. "HACMS: High Assurance Cyber Military Systems". *Proceedings of the 2012 ACM Conference on High Integrity Language Technology*, 51–52.

Fitton, O. 2016. "Cyber Operations and Gray Zones: Challenges for NATO". *Connections: The Quarterly Journal* 15,109–119.

Fuchs, C. 2005. "The Internet as a Self-Organizing Socio-Technological System". *Cybernetics & Human Knowing* 12 (3), 37–81(45).

George, A. S. 2024. "When Trust Fails: Examining Systemic Risk in the Digital Economy from the 2024 Crowdstrike Outage". *Partners Universal Multidisciplinary Research Journal* 1 (2), 134–152.

Gleirscher, M., S. Foster, and J. Woodcock. 2019. "New Opportunities for Integrated Formal Methods". *ACM Computing Surveys (CSUR)* 52 (6), 1–36.

Grant, T. 2015. "On the Military Geography of Cyberspace". *Leading Issues in Cyber Warfare and Security: Cyber Warfare Security* 2, 119.

Greene, C. V. 2009. *Cyberwarfare and Our Allies: The Importance of Theater Security Cooperation*. Naval War College Newport, Rhode Island.

Gupta, A., M. V. A. Kiran Kumar, and R. Ghughal. 2014. "Formally Verifying Graphics FPU: An Intel(R) Experience". *International Symposium on Formal Methods*, 673–687.

Guttieri, K. 2025. "Fighting Through Disruption". *The Cyber Defense Review* 10 (1), 93–114.

Haxthausen, A. E. 2010. *An Introduction to Formal Methods for the Development of Safety-Critical Applications*. Technical University of Denmark.

Hertogh, M., D. Quakkelaar, T. Raymakers, M. H. Sarma, M. Muench, H. Bos, and van der E. Kouwe. 2026. "Rain: Transiently Leaking Data from Public Clouds Using Old Vulnerabilities". *Symposium on Security and Privacy*.

Heylighen, F., and C. Joslyn. 2001. "Cybernetics and Second-Order Cybernetics". *Encyclopedia of Physical Science & Technology* 4, 155–170.

Hiep, H. A. 2018. *A Reo Semantics for Reasoning about Speculative Execution*. Master's thesis, Universiteit van Amsterdam.

Huston, G. 2011. *Hacking Away at the Internet's Security*. https://www.dotnxdomain.net/ispcol/2011-10/hacking.pdf.

Jeon, W. 2022. "Second-Order Recursions of First-Order Cybernetics: An 'Experimental Epistemology'". *Open Philosophy* 5 (1), 381–395.

Joint Chiefs of Staff. 2018. *Joint Publication 3-12 – Cyberspace Operations*. U.S. Department of Defense.

Jyothi, S. A. 2021. "Solar Superstorms: Planning for an Internet Apocalypse". *SIGCOMM'21: Proceedings of the ACM SIGCOMM 2021 Conference*, 692–704.

Kauffman, L. H. 2016. "Cybernetics, Reflexivity and Second-Order Science". *Constructivist Foundations* 11 (3), 489–497.

Keithley, T. D. 2019. *Interoperability: The Critical Enabler in the Anti-Access Area-Denial Fight?* Marine Corps Command; Staff College.

Kelly, R. F. 2017. *The Insider Threat to Cybersecurity: How Group Process and Ignorance Affect Analyst Accuracy and Promptitude*. Naval Postgraduate School Monterey.

Khan, J. 2021. "European Academic Brain Drain: A Meta-Synthesis". *European Journal of Education* 56 (2), 265–278.

Kibeom, K. 2025. "A Study on the Concept of Cyber Cognitive Warfare and Case Study Methodology From a Psychological Perspective". *Proceedings of the 24th European Conference on Cyber Warfare and Security*, 241–249.

Knoll, T. 2018. "Adapting Kerckhoffs's Principle". *Advanced Microkernel Operating Systems* 93.

Korb, R. A. 2025. *Weapons of Mass Disruption: How Small States Use Cyber to Resist Larger Powers*. PhD thesis, Old Dominion University.

Krepinevich, A. F., B. D. Watts, and R. O. Work. 2003. *Meeting the Anti-Access and Area Denial Challenge*. Center for Strategic; Budgetary Assessments.

Krishnan, A. 2022. "Fifth Generation Warfare, Hybrid Warfare, and Gray Zone Conflict". *Journal of Strategic Security* 15 (4), 14–31.

Kuehl, D. T. 2009. "From Cyberspace to Cyberpower: Defining the Problem". *Cyberpower and National Security* 30.

Lallo, R. Di. 2018. *Internet eXchange Points: Current Challenges and New Opportunities.* PhD thesis, Università degli studi Roma Tre.

Leenaars, M., C. Valayer, C. Peyralbe, K. Billeskov, and M. Gauw. 2018. *Next Generation Internet 2025.* 30-CE-0850616/00-55. European Commission, Directorate-General of Communications Networks, Content & Technology, Publications Office of the European Union.

Licklider, J. C. R. 1960. "Man-Computer Symbiosis". *IRE Transactions on Human Factors in Electronics* HFE, 1, 4–11.

Lipovski, G. J. 2006. "Hardware Description Languages: Voices from the Tower of Babel". *IEEE Computer* 10 (6),14–17.

Livermore, E., R. P. Skillen, M. Beshai, and M. Wernik. 1998. "Architecture and Control of an Adaptive High Capacity Flat Network". *IEEE Communications Magazine* 36 (5), 106–112.

Livingstone, S. 2014. "Risk and Harm on the Internet". *Media and the Well-Being of Children and Adolescents*, 129–146.

Loygue, P., K. Al Agha, and G. Pujolle. 2025. "Carbon Footprint of Cloud, Edge, and Internet of Edges". *Annals of Telecommunications* 80 (1), 153–169.

Lungu, A. M. 2001. *War.com: The Internet and Psychological Operations.* Naval War College Newport, Rhode Island.

Macit, H. B., G. Macit, and O. Güngör. 2018. "A Research on Social Media Addiction and Dopamine Driven Feedback". *Journal of Mehmet Akif Ersoy University Economics and Administrative Sciences Faculty* 5 (3), 882–897.

Magee, C. S. 2012. *Awaiting the Cyber 9/11.* Master's thesis, Command and Staff College, U.S. Marine Corps.

Marengo, E. A., and J. Tu. 2014. "Optical Theorem for Transmission Lines". *Progress In Electromagnetics Research B* 61, 253–268.

Masakowsi, Y. R., and J. M. Blatny. 2023. *Mitigating and Responding to Cognitive Warfare.* NATO Science and Technology Organization.

Mehrali, M., S. Bagherifard, M. Akbari, A. Thakur, B. Mirani, M. Mehrali, M. Hasany, et al. 2018. "Blending Electronics with the Human Body: A Pathway Toward a Cybernetic Future". *Advanced Science* 5 (10), 1700931.

Mutlu, O., A. Olgun, and A. G. Yağlıkçı. 2023. "Fundamentally Understanding and Solving Rowhammer". *Proceedings of the 28th Asia and South Pacific Design Automation Conference*, 461–468.

O'Leary, J., X. Zhao, R. Gerth, and C.-J.H. Seger. 1999. "Formally Verifying IEEE Compliance of Floating-Point Hardware". *Intel Technology Journal.*

Olmstead, M. B. 2024. *Into the Gray Zone: NATO Deterrence in a Graying and Multipolar World.* Università degli Studi di Padova.

Pires, J. J. O. 2024. "On the Capacity of Optical Backbone Networks". *MDPI Network* 4 (1), 114–132.

Raymond, D., T. Cross, G. Conti, and M. Nowatkowski. 2014. "Key Terrain in Cyberspace: Seeking the High Ground". *2014 6th International Conference On Cyber Conflict (CyCon 2014)*, 287–300.

Reed, E., S. Chatterjee, G. Ramos, P. Bogdan, and S. Pequito. 2022. "Fractional Cyber-Neural Systems –A Brief Survey". *Annual Reviews in Control* 54, 386–408.

Reed, S. G., R. H. Van Atta, and S. J. Deitchman. 1990. *DARPA Technical Accomplishments: An Historical Review of Selected DARPA Projects, Volume 1.* Institute for Defense Analyses.

Roberts, L. G. 2005. "The Evolution of Packet Switching". *Proceedings of the IEEE* 66 (11), 1307–1313.

Rushby, J. 1993. *Formal Methods and the Certification of Critical Systems.* 37. SRI International, Computer Science Laboratory.

Russell, A. L. 2017. *Strategic A2/AD in Cyberspace*. Cambridge University Press.

Sadoun, A. A. 2018. *PsyOp and Social Networks*. Naval Postgraduate School Monterey.

Salahdine, F., T. Han, and N. Zhang. 2023. "5G, 6G, and Beyond: Recent Advances and Future Challenges". *Annals of Telecommunications* 78 (9), 525–549.

Sampedro, V., F. J. López-Ferrández, and P. Hidalgo. 2022. "Digital Disintermediation, Technical and National Sovereignty: The Internet Shutdown of Catalonia's 'Independence Referendum'". *European Journal of Communication* 37 (2), 127–144.

Schaake, M. 2024. *The Tech Coup: How to Save Democracy from Silicon Valley*. Princeton University Press.

Schuchard, M., A. Mohaisen, D. Foo Kune, N. Hopper, Y. Kim, and E. Y. Vasserman. 2010. "Losing Control of the Internet: Using the Data Plane to Attack the Control Plane". *Proceedings of the 17th ACM Conference on Computer and Communications Security*, 726–728.

Scott-Hayward, S., S. Natarajan, and S. Sezer. 2015. "A Survey of Security in Software Defined Networks". *IEEE Communications Surveys & Tutorials* 18 (1), 623–654.

Seewald, A. K. 2021. "A Criticism of the Technological Singularity". *The International Conference on Intelligent Systems & Networks*, 91–119.

Shirey, R. W. 1990. "Defense Data Network Security Architecture". *ACM SIGCOMM Computer Communication Review* 20 (2), 66–72.

Stange, M., N. Jäpel, D. Reichelt, and S. Ihlenfeldt. 2023. "Open Source Hardware in Manufacturing–Opportunities and Challenges". *Proceedings of the Changeable, Agile, Reconfigurable and Virtual Production Conference and the World Mass Customization & Personalization Conference*, 709–716.

Stöger, F., H. Birge-Lee, G. Giuliari, J. Subira-Nieto, and A. Perrig. 2025. "BGP Vortex: Update Message Floods Can Create Internet Instabilities". *34th USENIX Security Symposium*, 3613–3629.

Tanenbaum, A. S. 2003. *Computernetwerken*. Pearson Education.

Wang, J., I. Harris, M. Ibrahim, D. Englund, and R. Han. 2025. "A Wireless Terahertz Cryogenic Interconnect That Minimizes Heat-to-Information Transfer". *Nature Electronics*, 1–11.

Westerveld, W. J., and H. P. Urbach. 2017. *Silicon Photonics: Electromagnetic Theory*. IOP Publishing.

Williams, L. G. 1991. *Formal Methods in the Development of Safety Critical Software Systems*. Lawrence Livermore National Lab.

Wong, K.-H., C. D. Anderson, C. Peterson, E. Bouldin, L. Littig, N. Krothapalli, T. Francis, et al. 2025. "Rising Cognitive Disability as a Public Health Concern Among US Adults: Trends from the Behavioral Risk Factor Surveillance System, 2013–2023". *Neurology* 105, (8): e214226(1)–e214226(13).

Wupper, H. 2000. "Design As The Discovery Of A Mathematical Theorem What Designers Should Know About The Art Of Mathematics". *Journal of Integrated Design and Process Science* 4 (2), 1–13.

Zota, C., A. Ferraris, E. Cha, M. Prathapan, P. Mueller, and E. Leobandung. 2024. "Energy-Efficient Computing at Cryogenic Temperatures". *Nature Electronics* 7 (11), 966–974.

CHAPTER 14

Decision Making for Resilient Logistics in Anti-Access/Area Denial (A2/AD)

Herman Monsuur, Paul C. van Fenema and Kerry Malone

Abstract

Logistics for forces encountering an A2/AD theatre face new demands and new 'Service Level Agreements' with the fighting force. Assumptions that it is relatively safe from the rear supply centre to the frontline no longer hold. We consider the operational-tactical level and logistics for forward deployed operations. We propose both formal and informal decision making to face A2/AD dynamics in the logistics arena, where we focus on the concept of resilience. A simple, basic formal network model for thinking about resilience for military logistics is proposed, serving as a platform for discussing investments in, for example, force protection or camouflage. We conclude with practical implications for researchers and professionals to face the challenges of contested logistics.

Keywords: Contested logistics; A2/AD; Resilience; Informal and formal decision making; War against Ukraine; Lessons learned

Introduction

> *'Russian forces have been increasingly targeting Ukrainian logistics routes and positions in the near rear using mothership unmanned aerial vehicles (UAVs), particularly motherships based on variants of the Orlan and Molniya fixed-wing drones, since at least August 2025.'*[1]

This news item marks the evolution of Russian strategy in terms of weapons and technology, such as heavy use of ISR-technology, drones and motherships, but also in terms of tactics and doctrines, namely the targeting of logistics routes, positions and convoys in areas that are supposed to be under control of the defender, far deeper into the defender's territory than in previous conflicts. As a result, it degrades the military supply chain of ammunition, food etc., to the friendly forces at the front. It makes it hard to satisfy the desired service level agreement in terms of delivery to the right place, at the right time, and in the right quantity and quality. This is, of course, exactly the purpose of the targeted disruption or annihilation of the military supply chain. It weakens the strength and mobility of forces at the front. The challenge to operate the logistics for 'forward deployment'[2] i.e., processes connecting resources in staging areas to the fighting force, is in many aspects related to the operational challenge of anti-access/area denial (A2/AD) strategies.

A2/AD as an Operational Challenge

The NATO Defence College proposes the following definition of A2/AD: 'The objective of an anti-access or area-denial strategy is to prevent the attacker from bringing its forces into the contested region (A2) or to prevent the attacker from freely operating within the region and maximising its combat power (AD).'[3] A2/AD capabilities might be defensive, but could also be employed in conducting offensive operations too, something one may envisage for successfully operating the logistics chain. From a NATO-EU perspective, Chinese and Russian A2/AD tend to be perceived as a major challenge, requiring substantial national investments in capabilities and coordination across nations.[4] An important concept for dealing with A2/AD is Multi Domain Operations (MDO)[5] relying for instance on 'pulse' force for coordinated effects across domains.[6] Multidomain is an approach intended to overcome A2/AD challenges introduced by opposing forces. "It revolves around penetrating layered defences, such as anti-access/ area denial (A2/AD) networks, to disrupt an adversary's depth and create exploitable corridors."[7] Service commands are (re)positioning themselves to play a role in this environment. USMC Col. Arlon Smith: "As Marines, our ability to aggregate, orchestrate, analyse, and share fused data at machine speeds is a warfighting imperative ... It is central to our value proposition."[8] This is likely to imply "a short, high-intensity conflict, hitting the A2/AD system's centre of gravity with the factor of surprise and breaking open the obstacles hindering the advance of friendly forces."[9] To deal with A2/AD, military organisations are developing concepts and combining technologies – including AI[10] – to generate effects across multiple domains. They envision new architectures for the set of technologies,[11] and thereby must reconsider their logistics.

As such, A2/AD presents a complex system of systems effort of one party to limit chances of another party to gain control by means of layered defence. Professionals and scholars are debating the strategic-operational-tactical role of A2/AD within a broader set of strategies and capabilities – such as hybrid or grey zone warfare.[12] A2/AD is emerging as an important concept, but this has not always been the case. For instance, "Russia's military has always seen defence to be cost prohibitive, and therefore focused on an offensive damage limitation strategy + functional defeat of the adversary. So whether you think Russian A2/AD systems are amazing, or overrated, just remember – there is no such doctrine or term in the Russian military and the conversation misses the point on how Russian forces actually organise at the operational-strategic level."[13] Overall, however, A2/AD presents an important object of study and military practice for NATO/EU.

Military Logistics for Forward Deployment

We concentrate on and limit ourselves to logistics for 'forward deployment'[14] i.e., processes connecting resources in line 3 at the rear of the warfighting area to the fighting forces at the forward line in line 1. Different logistics segments (lines of support) can be distinguished, each with their own performance criteria and KPIs. These lines include:

- "First line. Support capabilities that are organic or allocated to a ship, unit or squadron;
- Second line. Support capabilities that are organic or allocated to a formation;
- Third line. Support capabilities provided to a military force at the operational level or at installations established along the strategic LOC; and
- Fourth line. Support capabilities provided by strategic-level resources such as national depots and contractors and industry."[15]

Demands for logistics services are likely to become (very) complex, unpredictable and time-pressed when operations start. This applies to virtually all classes of goods handled by the military and required for warfighting.[16] Room to manoeuvre will be limited, with the adversary able to observe, and threats from for instance drones as a menacing third dimension.[17] This extends earlier experiences with IEDs in, for instance, Afghanistan.[18] Currently, the expected threat is from a near-peer adversary. This undermines common assumptions in military logistics considering an area safe to stock close to the battlefront, and even from there to initial sections of the front.[19] Furthermore, with increased adversarial capabilities, the accuracy of and distance covered by weapons systems increases the distance from the front to larger storage areas such as a National Support Area or Land Support Area where line of support 3 requirements are stored. This lengthening of the logistics chain increases the vulnerability of the supply chain.

Current research into supply chains encompasses both military and non-military logistics. On the military side, the notion of contestation comes to the fore, meaning that logistical processes can be militarily disrupted.[20] Various internal and external threats can be encountered.[21] On the non-military side, scholars explore notions like resilience to study how organisations can bounce back when crises hit (mostly non-military induced).[22] Professionals and researchers are starting to explore how logistics can support forward deployment in an A2/AD setting. From a logistics performance perspective, a complex balancing act will supply and repair resources depending on volatile demands of the fighting force and constant threat to logistics operations.[23] The balancing concerns avoidance of presenting targets and ensuring an effectiveness-efficiency mix.[24]

Overall, military organisations demand resilience of military logistics, which calls for new strategies and tactics (passive and active, maybe even offensive). For instance, force protection for logistics appears to become more urgent to ensure logistics performance. This depends on the way a commander prioritises responsibilities in his/ her area of responsibility. Figure 14.1[25] shows opposing forces acting in three ways: A2 and AD bubbles, as well as long range, hybrid and spider-web type threats to logistics in the rear area. The four NATO lines are depicted, supporting 'own forces.' These lines are managed by means of logistics performance criteria serving strategic-operational objectives.

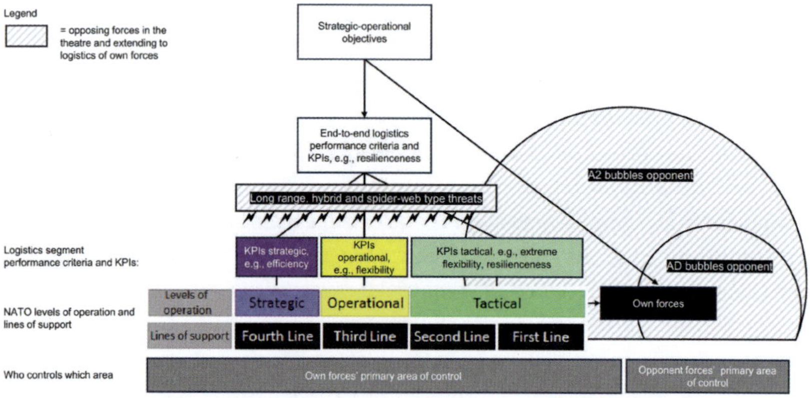

Figure 14.1: A2/AD and Military Logistics Robustness (NATO content from AJP-4[26])

Beyond the military academic realm, research on logistics and supply chain management increasingly pays attention to resilience as a business continuity problem: "the ability to return to a position of equilibrium after experiencing an event that causes operational results to deviate from expectations."[27] Related concepts include responsiveness and robustness.[28] Researchers have examined how logistics networks can become resilient.[29] The extreme setting of logistics supporting operations aimed at overcoming A2/AD presents new opportunities for research.

Research Questions and Setup

Logistics for forces encountering an A2/AD theatre face new demands and new 'Service Level Agreements' with the fighting force. Assumptions that it is relatively safe from the rear supply centre to the frontline no longer hold, as Figure 14.1 illustrates. Therefore, our main question is:

How to ensure resilience of forward deployed logistics despite major threats posed by opposing forces?

This conceptual chapter's main contribution is the development of conceptual logistics concepts. It provides a literature study, where we focus on the operational-tactical level and logistics for forward deployed operations. We also unpack concepts like robustness and resilience. In addition, it proposes both formal and informal decision making to face A2/AD dynamics in the logistics arena. A simple, basic formal network model for thinking about resilience for military logistics is proposed; it can be used as a platform for discussing investments in, for example, force protection or camouflage. We conclude with practical implications for researchers and professionals to face the challenges of contested logistics.

Resilience of Forward Deployed Logistics

Strategies for Resilience and Robustness

We will distinguish between reactive and proactive strategies. A *reactive* strategy, sometimes referred to as *agile*, tries to quickly adjust tactics and operations in case of a disruption. In contrast, a *proactive strategy* does not need to adapt in case of interruptions. Supply chains adopting this strategy are usually referred to as *robust* supply chains.[30] Examples of these strategies are to incorporate redundancy and diversity, or large reserves in their supply chains. Sani et al. discuss the need for robust strategies to mitigate supply chain network risks.[31] One example is Xu et al.[32] who investigated pre-disruption strategies against endogenous attacker efforts in a defender-attacker game. Their findings showed that these strategies can help the defender in counter-terrorism games. These kinds of mathematical approaches with two players, the defender and the attacker, may have added value in identifying and assessing risks and finding robust mitigation strategies.

A proactive or robust strategy may be designed in advance; however, it may become overly specialised leading to even more disruption in case of an unexpected, targeted attack by an adversary. On the other hand, reactive or agile strategies may be most expensive when disruption occurs, due to expensive ad-hoc solutions and excessive allocation of resources to guarantee delivery of goods to the customer. This may be different for military supply chains due to the ever-present 'fog of war,' something the military are used to. Both approaches, reactive and proactive, try to reduce the loss of service levels and, in case of commercial supply chains, focus on the concept of equilibrium in terms of means, modes of transport, etc.

Although common sense dictates that prevention is better than cure, one will have to take a balanced approach to adversarial, targeted attacks in an A2/AD theatre. Therefore, we use a third term, *supply chain resilience*, that corresponds to this balancing of both reactive and proactive strategies.[33] Extending the introduction's definition, resilience is "the ability to prepare for and adapt to changing conditions and withstand and recover rapidly from disruptions. It includes the ability to withstand and recover from deliberate attacks, accidents, or naturally occurring threats or incidents".[34] Kress distinguishes two types of properties of operational logistics:[35] cognitive, and structural and operational properties. Cognitive properties are intangible: synchronisation, foresight, and improvisation – aspects to strive for. Structural and operational properties, on the other hand, are concrete and can be measured: flexibility, attainability, continuity, tempo, simplicity, survivability (unique to military logistics), and efficiency.

We will elaborate on these aspects of resilience with literature on supply chains and A2/AD, for example lessons from Ukraine, where the prevalence of drones and improvements in intelligence, surveillance and reconnaissance play a key role in targeted attacks on the logistics chain.

We remark that the meaning of resilience and the measures that can be taken to assure this desired situation, obtain additional or different dimensions in case of military supply chains, where not only the goods that flow through the chain, but also the means of transport and human capital are at stake. In his publication, Grala introduces *supply chain security* as a specific and important aspect of supply chain resilience.[36] Supply chain security includes the security of transport processes, management of supply chains and modern IT systems. The goal is protecting against external targeted attacks. It therefore not only is about protecting resources, humans, and goods, but also the environment in which threats occur. Elements of the military logistics chain that are most sensitive are, of course, the logistic nodes which are often static, modes of transport, and perhaps most important logistic personnel.

An important distinction between resilience of commercial supply chains and military chains is the role of an equilibrium. Indeed, resilience for commercial supply chains is focused on re-establishing an equilibrium. In contrast, the military logistics does not focus primarily on returning to an equilibrium or some sort of steady state. For military logistics the issue is not bouncing back to previous equilibrium but rather pushing through to maintain the service level. Resilience in the military logistics context balances robustness, which is ex-ante considerations to hedge against worst-case disruptions, and agility, which is an ex-post ability to respond effectively to disruptions. The fog of war introduces conditions which create unstable conditions, for elements in the network are uncertain or in motion. The logistics chain in the contested area focusses on completing its mission, while keeping its personnel, supplies, and transport means as safe and secure as possible. For military supply chains where there is the possibility of targeted attacks, we also have a robust strategy 'deterrence by denial' (DBD) at our disposal. DBD takes place before a potential act of aggression. DBD seeks to "deter an action by making it infeasible or unlikely to succeed [and] deny a potential aggressor confidence to attain its objectives"; in other words, raising the costs of access, a proactive strategy.[37] It is related to A2/AD. In addition, we will need proactive and reactive strategies not only in defensive terms, but also in offensive ones, as already noted by Nelissen:[38] disrupting and disabling an opponent's supply chains is becoming increasingly important in deciding wars. In his paper he asserts that knowledge of an opponent's logistical chains is crucial when disrupting their sustainability, making it more difficult for the opponent to conduct military operations, reducing the enemy's A2/AD capacity.

What are Current Threats to Logistics?

In his paper, Grala presents a whole spectrum of threats to military supply chains.[39] These threats are divided into three categories. First, internal threats, like implementation of IT systems. Secondly, he considers external, unintentional threats, like disasters, flooding etc. The third category is the most interesting to our current study, the external threats due to targeted attacks. Targeted attacks include attacks on warehouses, supply centres, convoys, etc.

The war in Ukraine clearly demonstrates the state-of-the-art threats to logistics. Russia's drone revolution heaps pressure on Ukrainian defences and its logistics. The Russian Federation has demonstrated its capability to destroy targets in 'near real-time'.[40] The use of unmanned aerial systems (drones) as both a weapon and as a sensor is crucial in the Russian Federation's near-real-time targeting.

When used as a sensor, a drone provides information to the adversary. Logistics transports, and the locations of supply nodes and personnel, are immediately available and can be targeted. The drone can provide the coordinates of detected locations for attack, or the drone can itself have a weapon that can be dropped onto the logistics elements. Either way, the time between detection and attack is often a matter of seconds to minutes.

The most important lesson identified in the Russian Ukrainian war is that the area in which friendly forces fight, provide logistics, provide medical care, etc. is transparent. This transparency, combined with targeting, is the threat that needs to be addressed to provide logistics under such circumstances.

Decision Making for Resilient A2/AD Logistics Support

Formal, Model-based Decision Making

Formal models try to connect theory and reality. They aid decision-making by visualising the problem in time and space. To take the example of Operations Research models, one uses decision variables, objective functions in terms of these variables, and constraints on the particular value assigned to variables. Phenomena that occur within an Operations Research model can provide insight into, for example, good tactics and deployment of assets, or insight regarding specifications for new equipment. Models always are incomplete and should be considered as an essay (essay is derived from the French word `essai`, and means experiment, an attempt or a test of thought). One reason for that is the fact that a model always must find a balance between accuracy, generality and simplicity.[41] This means that one must always consider a model as the starting point for discussion, certainly not the endpoint. A model provides a platform for discussions, an instrument for investigation, understanding and learning together with military subject matter experts, meaning that personal experience is of utmost importance to develop, but also interact with the model. Interacting with the model in this way will show that the model's deductive power moves beyond description to inferences from the assumptions. Then, observing similarity in effects, one may infer a similarity in causes.[42] Without involvement into this learning process, finding out what are the important aspects and parameters, one will continue doing things the same way they always have been done, failing to gain insight that may sharpen intuition. This learning may be compared to the metaphor of the blind man with a cane.[43] As he cannot perceive (visual) information directly, he becomes familiar with the process of interpreting the sensations the cane provides him with. Once he becomes accustomed to using the cane, he no longer uses it consciously; it becomes an extension of himself. For logistics officers, this corresponds to learning and training capabilities to effectively use the (output of) models, thereby improving their decision making capabilities. In addition to this, Artificial Intelligence (AI) and Data Science (DS) can detect patterns in complex data which leads to better threat assessments and situational awareness. This means that AI and DS may serve as a lubricator between these formal models and (operational) reality. In the following section, we discuss network science and network models that provide a powerful framework to assess resilience logistics in A2/AD. We also introduce a basic, simple formal model that may be of use if the appreciation of a logistics A2/AD setting points towards survivability being the single objective. But first, we discuss the usefulness of complementary, informal models.

Informal Decision Making

Achieving resilience logistics support for forces entering an A2/AD setting may rely on formal models. But the fluid nature of this setting – and military operations in general[44] – suggests decision-making capabilities that include alternatives to model-based decision making. Figure 14.2 elaborates on Figure 14.1: strategic-operational objectives translate into end-to-end logistics performance criteria and KPIs. Moreover, logistics increasingly needs operational intelligence. From there, we distinguish formal (model-based) and informal modes of logistics decision making, as shown in the grey box in Figure 14.2.[45] Both rely ultimately on a generic decision-making model (illustrated in the centre of Figure 14.2). With A2/AD, reflection on the type of decision making is important. Conditions for relying on formal models may change. Conversely, informal decision making may not suffice in a high-speed high-risk setting.

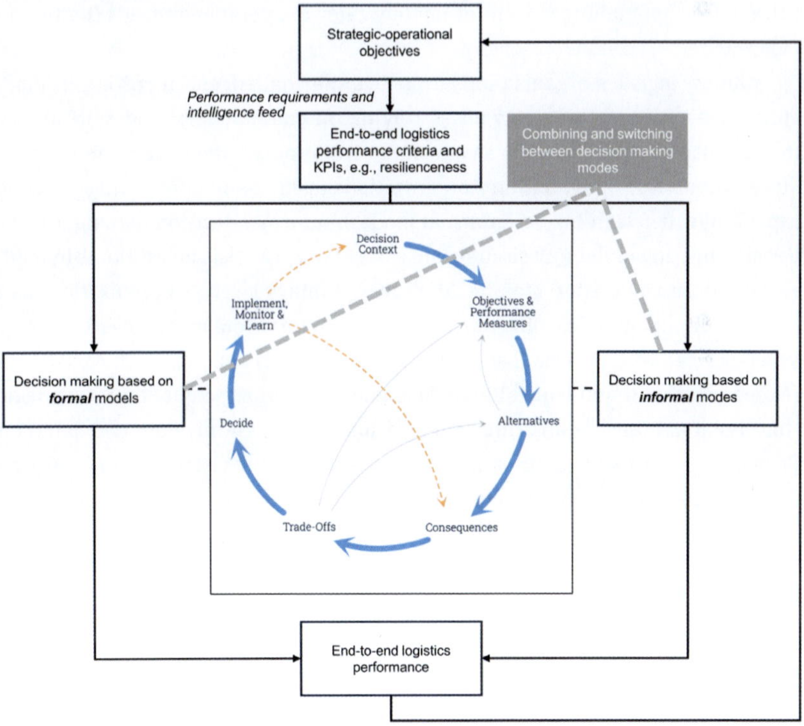

Figure 14.2: A2/AD and variation in military logistics decision making (core cycle visual from https://www.structureddecisionmaking.org/the-steps/)

Military decision-making in a broader sense has been defined as "an iterative process of identification, continuous assessment of threats, available resources, operational objectives, and selection of the best course of action" … "serving as a bridge to translate the strategic vision into tangible actions."[46] Logistics decision making is downstream linked to operational decision making and upstream to supply chain processes.[47] Operational decision making orchestrates the related decision-making processes. So, operational dynamics may influence the extent to which logistics decision making can rely on models. We focus on downstream operational-tactical logistics.

Models, as simplified representations of reality, may vary in their situational applicability. Unimagined or chaotic operational-logistics situations may occur in part as the flip side of the military notion of element of surprise.[48] Literature refers to, for instance, varying levels of structuredness.[49] Varying decision making implies shifting between different practices[50] to achieve a complex and dynamic mix of related operational-logistics goals.[51] The logistics achievements resulting from a chosen decision-making variant ultimately cater to operational needs (bottom of Figure 14.2).

Military logistics organisations thus face the challenge of investing in model-based decision making as well as relying on commanders' and staff ability to make use of model outputs as one of several inputs to their decision making. More broadly speaking, military organisations need combinatory and switching capabilities (between formal-informal modes). Informal decision making affords decision making under any circumstance. It entails pre-crisis investments in models and intuitive decision making. Moreover, military logistics organisations can prepare for enhanced learning during a crisis as human and modelling exercises.[52] Resilience can rely on formal and/ or informal decision making. Since A2/ AD and related command and control literature tend to emphasise qualitative strategic conceptual approaches and informal decision making,[53] we choose to expand on formal decision making. To be sure, formal-informal decision making must be distinguished from performance criteria or logistics outputs, which could entail efficiency and/ or effectiveness (Figure 14.2).

Model-Based Decision Making for Resilience for Logistics in A2/AD

The quantitative language of networks provides a powerful framework and tool for investigation. It may be used to discuss Command and Control,[54] but also to investigate resilience. In the Operations Research literature one may find several methods that determine which nodes and arcs of a network are most critical in the functioning of the network as a whole, as for example illustrated by Xiong et al.[55] They remark that it is difficult to build a metric that reflects military needs comprehensively, which often means that the focus is on economic cost. Instead, the military should use metrics like node degree and average path length in a network. But, as networks dynamically rearrange based on demand and availability along the arcs of the network, these measures miss dynamic performance. On the other hand, the higher the node degree, the more alternative routes toward the demand node exist. Xiong et al. propose military supply capability as the most important metric and apply efficiency metrics when the military need is satisfied. They introduced military supply *incapability* as the percentage of the total period examined that the inventory levels in all the relevant supply nodes are below a certain threshold during wartime. That is, the lower the supply capability, the longer the inventory levels are below the threshold during wartime. Note that this can be redefined to be above certain levels. The economic cost is the total distance covered by the transport movements during the period analysed. Through application of a simulation model, the evolution of the network is allowed to take place. Thus, links are removed and added, like the (un)availability of links in a network in a warfighting situation. Finally, although not explicit in their paper, the supply levels at specific nodes can also be calculated to determine whether the mission-critical demand nodes have the supplies they need during the mission.

In practical terms, JSEC coordinates the 'Hinterland' of the fight in negotiation with nations; it "coordinates and supports the enablement of SACEUR's AOR, through its key functions, to develop, manage and coordinate the RSN. As the functional coordinating authority for enablement at the operational level, JSEC contributes to the ability of Nations to project and sustain combat power."[56] For the Netherlands, for instance, the National Movement Coordination Centre (NMCC), part of DVVO,[57] would have to coordinate with JSEC on the Dutch National Detailed Deployment Plans. Ultimately, JSEC determines modalities, timeslots and (route) deployments between The Netherlands and the JOA.

In Alderson, it is shown that a metric like node or arc degree does not capture the importance of an arc in the network as a whole, as an enemy may reroute the transport because of redundancy in the network, avoiding the specific node or arc.[58] For example, they studied the Soviet railway system that can transport military material from the East to the West front. Through approaches using max-min calculations (using minimal resources for destroying railway tracks trying to maximise damage), the truly critical arc could be determined. This set does not coincide with the arcs that appeared most vital from a more intuitive approach. Given a logistics network, one may use similar techniques to assess the resilience of a network, thereby making possible decisions regarding investments to strengthen logistic nodes and arcs.

There are different approaches to study resilience, depending on objectives like the percentage of nodes for which the inventory falls below a given level. Then the focus is on delivery of the needed resources, i.e., effectiveness.

The basic model below has a different goal. Given a network, it studies how logistics intelligence, camouflage to reduce the footprint and visibility, in conjunction with force protection as an active means to eliminate threats (drones, mines, etc.), maximise the probability that a path is found and that goods are delivered. Here, choices are made on routes and what to improve via investing in camouflage or force protection. Therefore, we consider a network connecting a source node to a demand node. Suppose that intelligence can provide indications of threats by an enemy for each arc of the network: what is the probability that a convoy that passes through the arc will be attacked, and what are its chances of survival.

We convert this to the probability that a convoy will pass an arc (i,j) unharmed: q_{ij}. This may even be equal to 0 if the enemy is able to destroy the segment of routes. So, both tame effects like high values for this probability, as well complete elimination of a segment or arc can be modelled. Suppose that force protection, camouflage or other measures taken on such an arc make it possible to improve this probability to $\overline{q_{ij}}$, larger than before. Of course, the budget is limited (say, only K arcs can be improved in this way). As in reality, this scarcity of resources turns it into a tactical problem. Let x_{ij} be equal to 1 if the survivability of arc (i,j) is improved, 0 otherwise. Given each set of improvements in survivability probability, there emerges an optimal route from source to destination. This can be expressed in terms of a variable y_{ij}, which is 1 if arc (i,j) is on this route, and 0 otherwise. The (quadratic) problem that solves this is to maximise (over x, y) the objective function $\prod_{ij \in E}\left(q_{ij}^{1-x_{ij}}\, \overline{q}_{ij}^{x_{ij}}\right)^{y_{ij}}$. The constraints are $\sum x_{ij} = K$, and in addition also a constraint ensuring that y_{ij} represents a route from source to destination. By taking the log of the objective function, one obtains a quadratic problem that can be solved. Utilising this kind of model enables controlled improvisation.

The optimal value is the probability that a convoy will be able to reach the demand nodes unharmed. It shows how to dynamically reroute the convoy to its destination node. It may also be used to determine how much excess logistics goods must be transported through this network to satisfy demand (on average). It plans for some convoys to be interdicted, like drugs cartels may do. The solution the model presents is a combination of investments in force protection and camouflage on the one hand, and avoidance of certain routes on the other. Similar models may be developed to determine which nodes have to be hardened to survive attacks. Also possible is the addition of human lives lost in case an arc is attacked. This can be used to determine the added value of unmanned convoys.

In the literature one may also find game-theoretic models in which one player (the logistics team) wants to use the network at the lowest possible cost, while the other player, the Breaker, and in our case the enemy tries to maximise the cost of operating the network by means of targeted attacks. The example discussed in Washburn and Ewing[59] concerns the routing of a logistics teams over a network of possible routes, while the enemy can place IEDs on road segments at a certain rate. The logistics team is equipped with a team that can remove IEDs at a certain rate.

Practical, Policy, and Doctrinal Implications

Research Implications

The work in this chapter focuses on the transparency of the warfighting area and what the military logistics chain can do to increase resilience. The dynamic nature of the warfighting situation indicates that although the network structure plays an important part, equally important is how the network is used to determine whether the supplies get to the demand nodes. In addition, this chapter elaborates on formal and informal decision making.

The NATO Joint Analysis & Lessons Learned Center (JALLC) Ukraine Initiative is analysing what NATO can learn from Russia's war against Ukraine.[60] The lessons learned are grouped into Military Strategic Considerations and Operational & Tactical Observations. JALLC studied Contested Logistics as a topic in the latter area. Observations include mobile logistics and lack of fixed depots, making use of 'unpredictable' routing and delivery moments, and using decoys. Although unpredictable routing requires that the logistics network has a choice of routes, there must be multiple routes available to get to the destination. Unpredictability manifests itself also in the moment of delivery as well as the mode used for delivery. With respect to mode, the use of uncrewed ground or aerial vehicles can enable delivery while minimising threats to logistics personnel. Decoy deliveries introduce extra effort by an adversary to disrupt logistics to the front lines. This approach includes approaches that are also used by criminal organisations such as drug cartels: planning for interdiction by sending more supplies than requested while aiming to provide the requested number of supplies on time and at the right place with the requested quality.

Advances in electro-optical and electromagnetic camouflage either by reducing the logistics footprint or by active measures, or both, can increase logistics effectiveness.

These considerations introduce trade-offs. The questions that need to be answered – at various levels, see Figure 14.1 – and weighed against other options include:

- How much should be sent to be 95% sure that the supplies arrive?
- What is unpredictable routing and how should it be implemented?
- What are the options for reducing the logistics footprint and how does that affect the quality of the communication?
- How many transport vehicles are needed to operate in a dispersed fashion and meet the supply needs of the demand nodes?

To make military logistics more resilient requires more supplies, more and varied transport means, technological innovation to reduce the number of personnel to conduct tasks, to completely change whether and how tasks need to be carried out, and new doctrine. The Royal Netherlands Army specifically and the Royal Netherlands armed forces in general are now receiving more funding to make such informed decisions. While our chapter focuses on A2/AD, logistics in related last-mile settings – e.g., urban warfare, attacks on infrastructures[61] – may benefit as well.

Finally, our chapter has implications for the study of decision-making itself as an interdisciplinary phenomenon. The roles of (digital-enabled) formal (model-based) and informal (experience-based, intuitive) decision making in A2/AD settings deserves in depth research. When do models meaningfully support logistics commanders at which level, and when should they rely on 'art'? Studies could include how organisations build combinatory and switching capabilities between decision modes, supported by network science, AI and data science.

Implications for Professionals

In the A2/AD era, forward-deployed logistics supports critical operations and faces major risks itself. Logistics must be designed, planned, and commanded accordingly, releasing the assumption of 'safety until the front'. Resilience depends on a set of approaches:

- Doctrine and multilevel operational planning should incorporate *contested logistics* as a baseline assumption, making for a 360-degree risk environment. This implies prioritising dispersion, mobility, and deception over concentration and efficiency. Moreover, multiple decision-making modes should be anticipated and trained, used to develop multiple scenarios. It is not unlikely that offensive protective measures are considered to safeguard A2/AD logistics. Thus, logistics and operations become more integrated to cater to dynamics along supply routes and in the core theatre, by requiring the author of an order to consider this and provide guidelines.

- Second, logisticians must face a paradox between speed (rush supplies to the front) and unpredictability (deal with threats). This requires investment in coordination between logistics decision making, command and control pertaining to logistics routes, and theatre command and control. Performance may at times be inefficient to ensure resilience or rather daring with deliberate risk-taking to ensure supplies arrive on time.
- Third, force protection for logistics becomes a multidomain responsibility, beyond physical security. It includes camouflage, signature reduction, cyber and electromagnetic protection, and active measures to disrupt adversary ISR. This may require investments in unmanned transport, autonomous resupply, and forward repair (e.g., additive manufacturing).
- Finally, decision-making capabilities deserve sustained attention, relying on training for multiple modes and scenarios. Formal decision making can inform default force protection priorities and routing options – reducing coordination efforts. Informal decision making could ensure logistics performance with improvisation and social-intuitive processes catering to a dynamically evolving A2/AD setting.

Taken together, these implications suggest a shift in mindset: from optimising logistics under assumed protection, to sustaining operations under permanent contestation.

Acknowledgements

The authors want to thank Armand Soeleman, Sjoerd Res, Tim Sweijs and Gijs Ooms.

Use of AI

The authors used AI for subsection “Implications for Professionals” as a search engine in the early design phase, and in a limited sense as a tool to explore initial ideas.

Notes

1. https://understandingwar.org/research/russia-ukraine/russian-force-generation-technological-adaptations-update-october-9-2025/
2. Kress, "Flexibility in Operational-Level Logistics,"; Kress, *Operational Logistics*.
3. Schmidt. "Countering Anti-Access/Area Denial" 70.
4. Radomyski, "Development of the Russian".
5. TRADOC, "Multi-Domain Battle".
6. https://www.hudson.org/missile-defense/airpower-post-dominance-world-restoring-us-air-superiority-through-new-combined-bryan-clark-dan-patt; https://features.csis.org/war-modern-battlefield
7. https://features.csis.org/war-modern-battlefield
8. https://www.nationaldefensemagazine.org/articles/2025/9/23/marine-corps-gets-serious-about-cjadc2-with-project-dynamis#:~:text=JUST%20IN%3A%20Marine%20Corps%20Gets,About%20CJADC2%20with%20%27Project%20Dynamis%27&text=WASHINGTON%2C%20D.C.%20—%20The%20Marine%20Corps,23%20said.
9. https://sameerjoshi73.medium.com/demystifying-the-anti-access-area-denial-a2-ad-threat-doed26ae8b9e
10. https://www.airuniversity.af.edu/JIPA/Display/Article/2425748/penetrating-artificial-intelligenceenhanced-antiaccessarea-denial-a-challenge-f
11. https://www.congress.gov/crs-product/RL32665
12. Bouwmeester, "Krym Nash: An Analysis of Modern Russian Deception Warfare".
13. https://russianmilitaryanalysis.wordpress.com/2020/01/25/russian-a2-ad-it-is-not-overrated-just-poorly-understood
14. Kress, "Flexibility in Operational-Level Logistics"; Kress, *Operational Logistics*.
15. NATO, "NATO Standard AJP-4."
16. https://www.dau.edu/sites/default/files/Migrated/CopDocuments/Military%20Classes%20of%20Supply.pdf
17. https://cepa.org/article/how-are-drones-changing-war-the-future-of-the-battlefield/
18. https://apps.dtic.mil/sti/citations/ADA456446
19. https://pure.tue.nl/ws/portalfiles/portal/303572792/Master_Thesis_Janoux_van_Riet.pdf
20. Low, "Allies Key to Operating in Contested Logistics Environment,"; Sorenson, "Logistics in a Contested Environment".
21. Grala, "Resilience of Military Supply Chains".
22. Ekström, "Supply Chain Resilience".
23. https://digital.nationaldefensemagazine.org/june-2025/page-9?p=11
24. https://digital-commons.usnwc.edu/cgi/viewcontent.cgi?article=8398&context=nwc-review
25. NATO, "NATO Standard AJP-4."
26. Ibid.
27. https://www.ascm.org/topics/what-is-supply-chain-resilience/
28. Saisridhar, Thuerer, and Avittathur, "Assessing Supply Chain Responsiveness".
29. Maharjan and Kato, "Resilient Supply Chain Network Design".
30. Klibi, Martel, and Guitouni, "The Design of Robust Value-creating Supply Chain Networks".
31. Sani, Schaefer, and Milisavljevic-Syed. "Strategies for Achieving Pre-Emptive Resilience".
32. Xu, Zhuang, and Liu, "Modeling and Mitigating the Effects of Supply Chain".

33 Melnyk et al., "Understanding Supply Chain Resilience"; Sáenz and Revilla, "Creating More Resilient Supply Chains".

34 https://www.cisa.gov/sites/default/files/2023-01/ppd-21-critical-infrastructure-and-resilience-508_0.pdf

35 Defined by Kress as, "A collection of means, resources, organizations, and processes that share the common goal of sustaining campaigns and large-scale military operations. This collection, which is derived from strategic logistics, is utilized as input for the tactical logistics. Operational Logistics is designated to sustain battles that are distributed in time and space" Kress, *Operational Logistics*.

36 Grala, "Resilience of Military Supply Chains".

37 https://hcss.nl/wp-content/uploads/2021/12/02-Raising-the-Costs-of-Access-HCSS-2021-V2.pdf

38 https://militairespectator.nl/artikelen/logistics-intelligence

39 Grala, "Resilience of Military Supply Chains."

40 van Rossum, "De UAS-dreiging in de Russisch–Oekraïense oorlog".

41 Frigg and Hartmann, "Models in Science"; Levins, "The Strategy of Model Building".

42 Zhou et al., "Analysing the Resilience of Military Supply Network", 8; Monsuur and Janssen, "Structural Stability of Coalitions".

43 Polanyi, *The Tacit Dimension.*

44 Jomini, Mendell, and Craighill, *The Art of War.*

45 van Fenema and Keers, "Rediscovering Strategic Content".

46 Choo, *The Knowing Organization.*

47 van Fenema and van Kampen, "Foundational Concepts of Military Logistics"; Kress, *Operational Logistics.*

48 Bousquet, *The Scientific Way of Warfare.*

49 Mintzberg, Raisinghani, and Teoret, "The Structure of "Unstructured".

50 Lounsbury, "A Tale of Two Cities"; Schakel, van Fenema, and Faraj, "Shots Fired!".

51 Gnauck, "Applying Ecological Goal Functions".

52 de Waard et al., "Learning in Complex Public Systems"; Saaty, "Modeling Unstructured Decision Problems".

53 Johnson, "Washington's Perceptions and Misperceptions".

54 Grant, Janssen, and Monsuur, eds. *Network Topology in Command and Control*".

55 Xiong et al. "Modeling and Simulation for Effectiveness".

56 From "Enablement Stakeholders Roles, Responsibilities and Authorities for NATO Headquarters"; see also https://atlantische-commissie-s3-bucket.s3.eu-west-1.amazonaws.com/wp-content/uploads/2021/09/06095757/AP4-2021-Knappe-Boeke_compressed.pdf

57 https://english.defensie.nl/topics/doing-business-with-defence/information-for-suppliers-of-means-of-transport

58 Alderson et al., "Sometimes There is No "Most-Vital".

59 Washburn and Ewing, "Allocation of Clearance Assets in IED Warfare".

60 https://www.jallc.nato.int/JALLC-UKR-Initiative

61 Collins and Spencer, *Understanding Urban Warfare.*

References

Alderson, D.L., G.G. Brown, W.M. Carlyle, and L.A. Cox Jr. 2013. "Sometimes There Is No 'Most-Vital' Arc: Assessing and Improving the Operational Resilience of Systems." *Military Operations Research* 18 (1), 21–37.

Bousquet, A.J. 2022. *The Scientific Way of Warfare: Order and Chaos on the Battlefields of Modernity.* Oxford University Press.

Bouwmeester, A.J.H. 2020. "Krym Nash: An Analysis of Modern Russian Deception Warfare." *PhD thesis, Utrecht University, https://dspace.library.uu.nl/bitstream/handle/1874/400504/proefschrift%20krym%20nash%20ajh%20bouwmeester%20-%205fbcd1b309bc9.pdf?sequence=1.*

Choo, C.W. 2006. *The Knowing Organization: How Organizations Use Information to Construct Meaning, Create Knowledge, and Make Decisions.* Oxford University Press.

Collins, L., and J. Spencer. 2022. *Understanding Urban Warfare.* Howgate Publishing.

de Waard, E.J., S. Rietjens, A.G.L. Romme, and P.C. van Fenema. 2021. "Learning in Complex Public Systems: The Case of Minusma's Intelligence Organization." *Public Management Review* 25 (6), 1039–58.

Ekström, Thomas. 2025. "Supply Chain Resilience – an Empirical Exploration of Barriers and Enablers in Military Settings." *Scandinavian Journal of Military Studies.*

Frigg, R., and S. Hartmann. 2025. "Models in Science." In *The Stanford Encyclopedia of Philosophy.* https://plato.stanford.edu/entrieS/models-science/.

Gnauck, A. 1998. "Applying Ecological Goal Functions: Tools for Orientor Optimization as a Basis for Decision Making Processes." In *Eco Targets, Goal Functions, and Orientors,* edited by F. Müller and M. Leupelt. Springer.

Grala, D. 2024. "Resilience of Military Supply Chains." In *Digital Synergy,* edited by I.M.M. El Emary, A. Brzozowska and P. Maśloch. CRC Press.

Grant, T. J., R. H. P. Janssen, and H. Monsuur, eds. 2014. *Network Topology in Command and Control: Organization, Operation, and Evolution.* IGI Global Scientific Publishing. https://doi.org/10.4018/978-1-4666-6058-8

Johnson, J. 2017. "Washington's Perceptions and Misperceptions of Beijing's Anti-Access Area-Denial (A2-Ad) 'Strategy': Implications for Military Escalation Control and Strategic Stability." *The Pacific Review* 30 (3), 271–88.

Jomini, A.-H., G.H. Mendell, and W.P. Craighill. 2007. *The Art of War (Jomini's Book Was Originally Published in 1862).* Dover.

Klibi, W., A. Martel, and A. Guitouni. 2010. "The Design of Robust Value-Creating Supply Chain Networks: A Critical Review." *European Journal of Operational Research* 203 (2), 283–93.

Kress, M. 2000. "Flexibility in Operational-Level Logistics." *Military Operations Research,* 41–54.

Kress, M. 2016. *Operational Logistics: The Art and Science of Sustaining Military Operations.* Springer.

Levins, R. 1966. "The Strategy of Model Building in Population Biology." *American Scientist* 54 (4), 421–31.

Lounsbury, M. 2007. "A Tale of Two Cities: Competing Logics and Practice Variation in the Professionalizing of Mutual Funds." *Academy of Management Journal* 50 (2), 289-307.

Low, T. 2024. "Allies Key to Operating in Contested Logistics Environment." *https://www.nationaldefensemagazine.org/articles/2024/9/13/policy-points-allies-key-to-operating-in-contested-logistics-environment.*

Maharjan, R., and H. Kato. 2022. "Resilient Supply Chain Network Design: A Systematic Literature Review." *Transport Reviews* 42 (6), 739–61.

Melnyk, S.A., D.J. Closs, S.E. Griffis, C.W. Zobel, and J.R. Macdonald. 2014. "Understanding Supply Chain Resilience." *Supply Chain Management Review* 18 (1), 34–41.

Mintzberg, H., D. Raisinghani, and A. Teoret. 1976. "The Structure of 'Unstructured' Decision Processes." *Administrative Science Quarterly* 21 (2), 246–75.

Monsuur, H., and R.H.P. Janssen. 2022. "Structural Stability of Coalitions: A Formal Model Highlighting the Role of Participants Positioned between Members and Neutral Actors." *Games* 13 (1).

NATO. 2018. "Nato Standard Ajp-4. Allied Joint Doctrine for Logistics." https://www.forsvaret.no/en/organisation/nduc/nodefic/pre-study-packages/NATO%201.06%20-%20Vedlegg%20AJP-4%20Doctrine%20for%20Logistics%20(2018).pdf/_/attachment/inline/74011256-cb4f-490c-8386-607fc-1c27609:2f1481cf4b3319f97dac4d4c86dbbc177b62f5f5/NATO%201.06%20-%20Vedlegg%20AJP-4%20Doctrine%20for%20Logistics%20(2018).pdf#page=23.31.

Polanyi, M. 1967. *The Tacit Dimension.* Routledge.

Radomyski, A. 2021. "Development of the Russian Anti Access/Anti Denial Concept as a Threat to Nato and Poland." *Toruńskie Studia Międzynarodowe* 1 (14), 41–54.

Saaty, T.L. 1978. "Modeling Unstructured Decision Problems – the Theory of Analytical Hierarchies." *Mathematics and Computers in Simulation* 20 (3), 147–58.

Sáenz, M.J., and E. Revilla. 2014. "Creating More Resilient Supply Chains." *Sloan Management Review* 55 (4), 22–24.

Saisridhar, P., M. Thuerer, and B. Avittathur. 2024. "Assessing Supply Chain Responsiveness, Resilience and Robustness (Triple-R) by Computer Simulation: A Systematic Review of the Literature." *International Journal of Production Research* 62 (4), 1458–88.

Sani, S., D. Schaefer, and J. Milisavljevic-Syed. 2022. "Strategies for Achieving Pre-Emptive Resilience in Military Supply Chains." *Procedia CIRP* 107, 1526–32.

Schakel, J.K., P.C. van Fenema, and S. Faraj. 2016. "Shots Fired! Switching between Practices in Police Work." *Organization Science* 27 (2) 391–410

Sorenson, C.A. 2020. "Logistics in a Contested Environment: Network Analysis and Flow Optimization." *Doctoral dissertation, Monterey, CA; Naval Postgraduate School.*

TRADOC. 2017. "Multi-Domain Battle: Evolution of Combined Arms for the 21st Century."

van Fenema, P. C., and A.C.M. van Kampen. 2021. "Foundational Concepts of Military Logistics." In *Handbook of Military Sciences,* edited by A.M. Sookermany. https://link.springer.com/referencework/10.1007/978-3-030-02866-4: Springer.

van Fenema, P.C., and B.B.M. Keers. 2020. "Rediscovering Strategic Content in 'Strong Process' Research on Business Network Innovation." *Industrial Marketing Management* 91 (11), 639–56.

van Rossum, R. 2025. "De Uas-Dreiging in De Russisch–Oekraïense Oorlog." *Militaire Spectator,* November.

Washburn, A., and L. Ewing. 2011. "Allocation of Clearance Assets in IED Warfare." *Naval Research Logistics,* 58 (3), 180–187.

Xiong, B, B. Li, R. Fan, Q. Zhou, and W. Li. 2017. "Modeling and Simulation for Effectiveness Evaluation of Dynamic Discrete Military Supply Chain Networks." *Complexity,* 6052037.

Xu, J. J. Zhuang, and Z. Liu. 2016. "Modeling and Mitigating the Effects of Supply Chain Disruption in a Defender–Attacker Game," *Annals of Operations Research* 236 (1), 255–270.

Zhou, Q., B. Xiong, B. Li, J. Huang, and S. Lu. 2016. "Analysing the Resilience of Military Supply Network and Simulation against Disruptions." 8; Tech. Rep. (3).

CHAPTER 15

Self-Defence Zones in Outer Space under International Law

A Lawful A2/AD Measure?

Louisa Handel-Mazzetti, Arjen Vermeer, and Marten Zwanenburg

Abstract

As outer space becomes increasingly integral to military operations, the protection of satellites has become a key concern. This chapter examines the legality of self-defence zones (SDZs) – designated areas around satellites intended to deter or restrict the approach of other space objects. The chapter argues that SDZs sit at the intersection of A2/AD concepts and core principles of international law, raising significant legal uncertainty.

After outlining the different forms of protective zones, contrasting purely communicative warning zones with zones treating entry as potentially justifying defensive measures, it then analyses the legal constraints imposed by the Outer Space Treaty and the United Nations Charter. The chapter further examines potential legal bases for SDZs, including circumstances precluding wrongfulness, such as necessity or countermeasures. It concludes by highlighting practical modalities that influence lawful operation of SDZs as an A2/AD strategy and offers policy recommendations to prevent destabilising unilateral practices in outer space.

Keywords: Legal basis; Outer space; Satellite; Self-defence; Zone

Introduction

Anti-Access and Area Denial (A2/AD) is a military strategy to control access to and within an operating environment. It is usually discussed in the context of the traditional domains of military operations, i.e., land, air, and sea. Relatively little attention has thus far been paid to its application in the space domain.[1] That domain is now recognised as vital for military operations. Satellites are used for a wide variety of purposes by armed forces. This makes them attractive targets for opponents. In addition to their vulnerability to cyber operations and risks to their up- and downlinks, they may be targeted by kinetic attacks using either direct-ascent anti-satellite weapons or on-orbit attacks by other satellites. It is therefore logical that States are considering how they can protect their space objects. Some States are for example considering placing weapons on their satellites or deploying 'bodyguard satellites'.[2] A potential element in such a defensive strategy could be the establishment of a 'self-defence zone' (SDZ) around a satellite. The declaration of such a zone can be considered a form of A2/AD in outer space, as it is implemented to control access to a satellite and may be used to limit freedom of action in the zone. Although no State has yet established such a zone, there have been calls to do so in the literature.[3] One of the reasons that this has not been implemented yet may be that there is much uncertainty concerning the legality of such zones, as they would arguably violate key rules of space law.

Such a scenario would potentially unfold as follows. A State unilaterally declares a self-defence zone around a critical military satellite. The zone is defined as a fixed volume around the satellite, and the State declares that any entry by another object without its consent will be treated as a hostile act and may justify defensive measures. In this scenario, no armed attack against the State has yet occurred. Other States perceive the declaration as an implicit threat of force and a *de facto* restriction on their freedom to use outer space. Although oversimplified, this scenario highlights the potential legal and policy dilemma further analysed in this chapter: whether the designation of a self-defence zone as a form of A2/AD can be reconciled with the *jus ad bellum* and the foundational principles of space law, and what measures can be taken to ensure legality.

This chapter will focus on the question of what the legal basis under international law for the establishment of such a zone could be, and if such a legal basis can be found, what are the requirements that such a zone must meet. The chapter will be limited to a discussion of the law applicable in peacetime and thus will not consider situations of armed conflict and the application of international humanitarian law. The chapter is structured as follows. First, the concept of an SDZ in space will be clarified (second section). The third section discusses some key aspects of international law that arguably limit the possibility of establishing SDZs in space. This section will focus in particular on space law, as the legal regime that principally regulates activities in outer space. It will also take into account the prohibition of the threat of inter-state force laid down in the United Nations (UN) Charter. Against this background, the fourth section will focus on identifying potential legal bases for the establishment of such zones under international law. To the extent such a legal basis can be found, the fifth section discusses certain limitations and necessary safeguards that flow from the applicable law, and it will look at practical, policy, and doctrinal implications. The chapter concludes with a few final remarks in the sixth section.

Types of Zones

This section provides an overview of the different types of zones that have been suggested in academic and policy discussions on outer space, with particular emphasis on SDZs. The rationale behind the establishment of protective zones in space lies in the desire to safeguard space assets and operations from potential threats – a notion that gained traction during the Cold War. Over time, various terms have been used to describe such areas, including warning zones, keep-out zones, security zones, and self-defence zones.[4] Despite these terminological differences, these concepts generally rest upon one of two principles: either the protection of a specific space object or the protection of a defined area to conduct a particular mission or operation.[5]

While these types of zones share a common rationale, their practical implications can differ significantly. One of the earliest and most frequently discussed concepts is the *keep-out zone*, proposed as a transparency and confidence-building measure within the Conference on Disarmament.[6] Keep-out zones were intended to maintain a minimum distance between States' spacecraft to avoid misperception or miscalculation. Importantly, however, the concept did not inherently attach countermeasures to an unauthorised entry into such a zone.[7]

More recent discussions in academic literature and emerging State practice have focused on *self-defence zones*. Given that this chapter concentrates on the legal basis for SDZs, a working definition is necessary. Although there is no universally accepted definition, two conceptions stand out. These have been put forward in the Woomera Manual and by Chow. The Woomera Manual[8] conceptualises an SDZ as follows:

> a State may designate a warning zone to deter potentially threatening spacecraft from approaching its objects, as an extension of its right of self-defence. A deliberate approach by another space object within a specified distance of an important satellite may raise concern about whether an armed attack is imminent. Such a zone serves as a means of communication – internally to military forces and externally to other States – regarding when a State perceives that an armed attack or unlawful use of force might be forthcoming, and at which point it must consider lawful measures in response. Importantly, the mere entry of another space object into this zone does not in itself trigger the right of self-defence; rather, it marks a threshold at which the declaring State would begin to assess whether the right to self-defence might be invoked.[9]

Accordingly, the SDZ described by the Woomera Manual functions primarily as a communicative mechanism – a way of signalling potential red lines rather than authorising the immediate use of force. In this sense, it closely resembles the earlier concept of the keep-out zone, though framed explicitly within the language of self-defence.

Other scholars, however, have proposed more assertive models. For instance, Chow suggests that, in the context of 'stalker satellites' that closely tailgate other space objects, a State could declare a self-defence zone within which the approach of another satellite would be deemed an imminent armed attack, justifying pre-emptive defensive action. He proposes that:

> if the United States wishes to deter and defend against space stalking, it could declare that any country positioning its space objects within a minimum separation (for example, 148 km) or beyond a specified inclination threshold relative to another country's satellites would be considered the aggressor, thereby entitling the defender to exercise self-defence even before an actual attack occurs.[10]

From these differing approaches, the concept of an SDZ can be understood to rest on two key elements:

1) The communication of a defined minimum distance that other space objects must maintain; and
2) The understanding that unauthorised entry into such a zone may be perceived as a hostile act, and potentially may entail the use of countermeasures up to and including the use of force.

It remains unclear to what extent a State could lawfully resort to force prior to an actual attack by the intruding object. Whereas the Woomera Manual characterises an SDZ as purely declaratory and preventive, other commentators suggest that it could entail an immediate right to take defensive action. As will be discussed later, these distinctions – though seemingly minor – have significant implications for the legality of SDZs as a form of A2/AD.

Self-defence Zones and the Limitations Imposed by the Outer Space Treaty

International (space) law does not specifically address the legality of SDZs. Nonetheless, several key principles of space law, as the principal body of law regulating conduct in space, are relevant in this context. These principles articulate a framework that would bear on the establishment of SDZs. In particular, the freedom to explore and use outer space, the non-appropriation principle, and due regard would severely restrict, if not prohibit, a State from establishing a zone in which it wants to project control over the area and potentially deny entry into the zone to space objects under the jurisdiction and control of other States. This section will discuss these principles. It will also discuss the relationship between an SDZ in space and the prohibition of the threat of force laid down in the UN Charter, which at first sight also seems relevant to the lawfulness of an SDZ.

The international law regulating space activities is first and foremost based on the freedom to explore and use space by all States without discrimination of any kind. This basic principle is expressed in Article I of the Outer Space Treaty (OST).[11] At the time of writing, 118 States were parties to the OST, including all the major spacefaring States. Based on Article I of the treaty, a State may thus not interfere with another State's exploration and use of space, as all States enjoy the same freedom, and Article I requires equal treatment of all States.

A further restriction can be found in Article II of the OST, which articulates that outer space is "not subject to national appropriation by claim of sovereignty, by means of use or occupation, or by any other means."[12] Article II OST is often referred to as expressing the "non-appropriation principle".[13] 'Appropriation' has been interpreted by one observer to mean "the taking of property for one's own or exclusive use with a sense of permanence."[14] According to Trapp, "without such a non-appropriation principle, space would be subject to competing claims from various countries that would only be able to enforce these claims through military force."[15] Ironically, an SDZ might arguably just do that. Moreover, it is unclear what it means to nationally appropriate space "by any other means". It could be that claiming part of space to exercise control over it for the purpose of national security may fall within the notion of "or by any other means".[16] The establishment of an exclusive and permanent SDZ may thus run counter to Article II OST.

Conversely, it has been questioned whether Article II prohibits all claims of authority or control of an area that go beyond the physical presence of the space object itself.[17] As Schwetje opined:

> The international community has generally not objected to the establishment of special zones if the zones under the circumstances were reasonable and did not unduly hamper or interfere with another nation's freedom to navigate the seas or supra-adjacent airspace. Important in the following discussion is the notion that the key to gaining international acceptance of such zones has been that they extend no claim of sovereignty. There is a clear distinction between sovereignty and the right to exercise a preventive, protective, or regulatory jurisdiction.[18]

In a similar fashion, Chow argues in favour of adopting an approach that is based on the maritime law concept of contiguous zones, in which a coastal State is granted a number of means to exercise control in the case of infringement of certain rights. He argues that this approach could be applied to a volume around a space asset, and that a right of innocent passage could be applied to all space assets crossing the zone. According to Chow, such conduct by States would not be in violation of Articles I and II OST.[19] While the authors would caution against using analogies with other domains to the space domain, the point Schwetje and Chow make is that the absence of a (unlawful) claim of sovereignty does not necessarily prevent an (lawful) extension of jurisdiction through the establishment of a zone.

Another principle of international space law that has a bearing on an SDZ is the principle of due regard as stipulated in Article IX OST. This article prescribes that States shall conduct their space activities "with due regard to the corresponding interests of all other States Parties to the Treaty."[20] Article IX clearly places limits on the freedom to explore and use space in relation to the activities of other States, in particular to avoid 'harmful interference'.[21] It should be noted that 'harmful interference' is not defined in the OST, but it is generally broadly defined in the literature to include physical and non-physical means that interfere with the peaceful exploration and use of space.[22] Limiting the freedom of action of other States in an SDZ could arguably be considered as such harmful interference. Concerning the obligation to exercise due regard: while what is required by the obligation to exercise due regard may differ depending on the circumstances,[23] it is a standard of care that has to be met.[24] Obviously, the obligation to declare potential harmful interference within an SDZ comes at the risk of exposing one's own asset. The latter part of Article IX OST obliges a State to start consultations with other States if the former believes its activity would cause potentially harmful interference with the activities of the latter States. The threshold is quite low through the insertion of the word 'potentially'. Establishing an SDZ may indeed cause potentially harmful interference. When the SDZ-establishing State is aware of other States' activities that would potentially come within the scope of the SDZ, it has an obligation to undertake consultations before establishing the zone. As the mechanics of that obligation are not spelled out, it can be argued that a general declaration of a planned SDZ suffices as an invitation to States, other than the establishing State is aware of, to request consultations in compliance with the last sentence of Article IX.[25]

Furthermore, it is unclear once the SDZ is in place whether the establishing State should keep assessing other States' activities and provide consultation opportunities. Alternatively, when other States' activities would include passing through the SDZ, may those States request consultations with the establishing State, and must the latter then demonstrate a willingness to engage in such consultations? Besides, it stands to reason that any change to the SDZ requires the establishing State to reassess the situation and its obligations under the OST. Additionally, Article IX does not clarify the consequences if consultations are concluded without a satisfactory outcome. Should the establishing State suspend or cancel its planned or established SDZ, or should the requesting State cease its activity?[26]

This brings us to the final point of this section: the establishment of an SDZ must adhere to the restrictions on the use of force outlined in Article 2(4) of the UN Charter:[27]

> All Members shall refrain in their international relations from the threat or use of force against the territorial integrity or political independence of any State, or in any other manner inconsistent with the Purposes of the United Nations.[28]

This article prohibits both the threat and use of force. Is the declaration of the establishment of an SDZ in itself a threat of force? Some States have suggested that the placement of weapons in outer space constitutes a threat of force.[29] By analogy, the declaration of an SDZ with the implication that force may be used when it is entered into could arguably be seen as a threat of force. However, it must be recalled that Article 2(4) of the UN Charter only prohibits threats of *unlawful* use of force.[30] Threatening lawful use of force, such as the use of force in self-defence, is not prohibited by Article 2(4) of the UN Charter, as self-defence is a lawful exception to the prohibition of the use of force. Thus, an SDZ would only be problematic in the context of the prohibition of the threat of force if it were related to the potential unlawful use of force.[31] This could include a use of force purportedly based on the right to self-defence but not meeting the requirements for the lawful invocation of that right.

In summary, the rules of international law analysed in this section neither explicitly permit nor forbid the creation of SDZs. The legality of a zone for self-defence purposes, be it object- or mission-specific, may depend on the volume, duration, location, and level of exclusion of the designated area in space. A warning zone that foregoes automatic enforcement in case of intrusion seems most likely to be accepted. As the OST does not specifically address the issue of zones, and State practice has not advanced any such zone as yet, acceptance by the international community is the main indicator of its legality.[32] Nonetheless, the OST continues to apply outside and inside such a zone.

Legal Basis

The previous section made clear that the establishment, maintenance and enforcement of an SDZ as a form of A2/AD in space could violate various rules of space law, as well as the prohibition of the threat of force under Article 2(4) of the UN Charter. This section will focus on identifying potential legal bases for the establishment of such zones under international law. It makes a distinction between zones that do not violate any rules of international law on the one hand, and zones that do violate such rules on the other.

Zones That Do Not Violate Any Rules of International Law

It is submitted that the establishment and maintenance of an SDZ that does not violate any rules of international law does not require a separate legal basis. Although as discussed earlier, space law does not provide for the establishment of zones, it also does not explicitly prohibit it. Reference can be made here to the so-called 'Lotus Principle' under international law. This principle, which is based on the judgment of the Permanent Court of International Justice in the Lotus case of 1927, refers to the theory that States have the right to do whatever is not prohibited by international law.[33] Although this judgment is much discussed and the 'Lotus Principle' has been the subject of much criticism, it nevertheless still plays an important role in the practice of international law.[34]

As the Woomera Manual states, a "State may elect to designate a warning zone in order to deter potentially threatening spacecraft from approaching its objects, as an extension of its right to self-defence".[35] Such a State could declare that it will take into account entry into such a zone in its determination whether an armed attack giving rise to the right to self-defence may be imminent.[36] This in itself does not violate any rule of international law, at least as long as the use of force by the State establishing the SDZ zone would not automatically follow after entry into the zone. Entry into the zone would merely be one factor to be considered in determining whether the criteria for the lawful use of force in self-defence have been met, including the requirement of an (imminent) armed attack. Because the use of force that arguably would be threatened in connection with the zone would be limited to use of force in self-defence, such force would be lawful. Consequently, the threat of such force would not violate Article 2(4) of the UN Charter, which only prohibits the threat of unlawful force. As the zone would also not exclude spacecraft, it would also not violate the principles of non-appropriation and freedom of use in space law.

Despite these potential lawful avenues for establishing an SDZ in outer space, the mere designation of such a zone following these criteria does not in itself ensure its legality under international law when this leads to the use of force. The actual exercise of self-defence in relation to such a zone remains subject to the conditions of necessity and proportionality.[37] Necessity requires that the use of force be a last resort, employed only when no peaceful or less intrusive means are available to avert or respond to an armed attack.[38] Proportionality limits the scale, scope, and intensity of the defensive response to what is strictly required to repel or deter that attack. In the unique environment of outer space, compliance with these conditions raises additional legal and practical challenges. The potential generation of space debris, the difficulty of attribution, and the risk of cascading effects on other space assets all complicate assessments of whether the defensive response can be considered proportionate. For these reasons, several commentators contend that any invocation of self-defence in response to an attack on space assets should, as a rule, be confined to terrestrial measures rather than actions conducted in orbit.[39]

Zones That Do Violate Rules of International Law

There may be zones that would *prima facie* be unlawful under international law. An example would be a zone that prohibits spacecraft from entering it without prior approval from the State that established the zone, as in the scenario set out in the introduction. This would constitute a violation of the freedom of use in space, and arguably also of the principle of non-appropriation as well as the obligation of due regard. Another example would be the situation in which a State that has established such a zone would automatically use force against a space object entering the zone without ensuring that such force was justified under international law.

International law may provide a legal basis for conduct that would otherwise be in violation of it. Generally speaking, such a legal basis may either be in the form of a primary rule of international law or of a circumstance precluding wrongfulness. Circumstances precluding wrongfulness are circumstances accepted in the law of State responsibility that, where applicable, provide a defence for what would otherwise be a violation of international law.[40] The UN International Law Commission (ILC) included six such circumstances in its Articles on the Responsibility of States for Internationally Wrongful Acts (ARSIWA). They are mostly seen as reflective of customary international law.[41]

Whether a primary rule of international law or a circumstance precluding wrongfulness provides a legal basis for the establishment or maintenance of an SDZ must be assessed on the basis of the specific rule of international law that such a zone would otherwise violate, as well as the circumstances of the case. In other words, it requires a case-by-case determination. This notwithstanding, it is possible to draw some general conclusions.

First and perhaps surprisingly, the right to self-defence cannot justify the *establishment* of a zone that would otherwise violate a rule of international law. Self-defence is both a primary rule of international law and a circumstance precluding wrongfulness, set out in Article 20 ARSIWA. The right to self-defence, however, is only triggered when there is an armed attack, and arguably also when such an attack is imminent.[42] Although an SDZ is established for the purpose of self-defence, generally speaking, at the time of its establishment, there will not (yet) be an (imminent) armed attack. Consequently, the right to self-defence cannot provide a legal basis for the *establishment* of a self-defence zone as such, if that zone would violate a rule of international law.

However, once such an (imminent) armed attack takes place, the right to self-defence as set out in Article 51 of the UN Charter provides a justification for the use of force against a space object involved in carrying out such an attack. Such use of force would also need to meet the other requirements for the lawful use of force in self-defence, notably that the measures taken in self-defence are necessary and proportional. In such a situation, self-defence as a circumstance precluding wrongfulness would also justify measures that violate other rules of international law than the prohibition of the use of force vis-à-vis the attacking State. It is debated, however, whether such measures must be related to the use of force in self-defence or whether this is not necessary.[43]

Another subject of controversy is whether self-defence can be invoked as a justification for measures that affect the right of third States, i.e. States other than the aggressor State. Consider the situation in which a State excludes all spacecraft from entering its self-defence zone after it determines that an armed attack from another State is imminent. Because of the difficulty of determining the character of spacecraft and whether they are used by the aggressor State, the defending State considers it necessary to exclude all spacecraft.

Second, there is no doubt that the UN Security Council, acting under Chapter VII of the UN Charter, could authorise an SDZ in which a State may take certain enforcement measures. It is generally accepted that the Security Council may authorise measures, whether they involve the use of force or not, that would otherwise breach a rule of international law. This also extends to outer space.[44] This power is not unlimited and would not extend to rules that constitute peremptory norms of international law, but none of the principles of space law discussed above constitute such norms.

Third, States have sometimes resorted to the establishment of 'exclusion zones' in the maritime or air domain. They have done so mainly in the context of armed conflict, but in some cases also during peacetime. This raises the question whether this had led to the acceptance of such zones in those domains, and if yes, if this acceptance can be transposed to the space domain. If so, this could provide a legal basis for an SDZ that would otherwise violate the freedom of use of outer space, the non-appropriation principle, and/or the obligation of due regard. Regarding the first question, as Klein writes, the establishment of security zones "has not generally been accepted under international law because of the vagueness that surrounds these zones and their susceptibility to abuse."[45] Even if this were otherwise, it is questionable whether this could be transposed to the space domain.[46]

Fourth, circumstances precluding wrongfulness, other than self-defence, may also be relevant. In particular, 'necessity' and 'countermeasures' may provide a justification for measures related to a self-defence zone. Both these concepts are accepted as circumstances that may preclude the wrongfulness of conduct that would otherwise be a breach of international law. Necessity refers to those exceptional cases where the only way a State can "safeguard an essential interest threatened by a grave and imminent peril is, for the time being, not to perform some other international obligation of lesser weight or urgency".[47] A countermeasure is an act taken by a State that would normally violate international law, but that is justified because it is taken to induce another State to cease its wrongful behaviour toward the State that deploys a countermeasure. Countermeasures may only be taken subject to a number of requirements, including that measures taken are proportional.[48] Arguably, necessity could in exceptional cases provide a justification for a State excluding satellites from a zone around its own satellite. Consider the case of a military satellite that is of vital importance to a State's security, for example because it provides crucial early warning for ballistic missile launches, which is approached by a known spy satellite of another State. However, it is likely that only in rare cases will the ability for a national security satellite to continue its activities unimpeded constitute an essential interest.[49] In addition, a countermeasure could form the legal basis for a military space activity that would otherwise be in breach of international law.[50] However, such a measure may not take the form of the use of force, except in self-defence.[51] Consider the case of a satellite used by one State to blind the sensors of a military satellite used by another State. Arguably, this would constitute harmful interference with the latter, in violation of article IX OST.[52] The victim State could then be justified in demanding that the perpetrator State cease its wrongful conduct, and if it does not do so, for example, temporarily jam the uplink to the other State's satellite.

Modalities

This section will discuss several considerations that are important for a State that is considering establishing a self-defence zone as a form of A2/AD around a space object. These considerations largely follow requirements that are linked to the possible legal bases analysed in the previous section.

The first consideration is the volume of the zone. It is submitted that such a zone should not be any larger than is necessary for the purpose for which it is established. For a zone that does not otherwise violate a rule of international law, this does not directly follow from international law.[53] However, it may be the case that even if a zone strictly speaking does not breach the rights of any other State, it may still lead other States and satellite operators to be more hesitant to enter such a zone, as a consequence hindering lawful navigation in that zone to some extent. In such a situation, it can be argued that the obligation for States to conduct space activities with 'due regard' to the corresponding interests of other States requires that the zone be as small as possible. For a zone that a priori breaches a rule of international law, the illegality of which is justified by a circumstance precluding wrongfulness, this flows from the fact that proportionality is generally an element of such circumstances. For example, a countermeasure "must be commensurate with the injury suffered, taking into account the gravity of the internationally wrongful act and the rights in question."[54]

The second consideration concerns the communication of an SDZ. The function of an SDZ would not be achieved if such a zone is not clearly communicated to other States that might pose a threat to the State's space object. The deterrent effect of a zone would be lost if other States were not properly made aware of it. While Article 51 of the UN Charter requires States to notify the UN Security Council when they engage in self-defence measures, the establishment of a zone in which potentially the right to self-defence will be exercised does not require such notification. However, as discussed in the third section, a notification duty can be deduced from Article IX OST.

The third consideration is the need for space domain awareness (SDA) on the part of the State establishing an SDZ.[55] SDA can be defined as "the capability to detect, track, identify and characterize Space objects and the Space environment, aimed at supporting Space activities in terms of safety, security, and sustainability."[56] If a State is not aware when a space object enters an SDZ it has established, much of the usefulness of that zone is negated. It is not only important that the State is aware of where a space object is, but also that it is aware of the strategies of other States and the possible intent behind a space object entering the SDZ. This requires more than the technical tracking and identification of space objects that are potentially a threat.

Concluding Remarks and Policy and Military Recommendations

This chapter took as its starting point that SDZs can form a component of an A2/AD strategy in space. In particular, they can form part of the defence of satellites against malicious actors by inhibiting the approach to such satellites. In this context, the chapter defined such a zone as resting on two key elements:

1) The communication of a defined minimum distance that other space objects must maintain; and
2) The understanding that unauthorised entry into such a zone may be perceived as a hostile act, and potentially may entail the use of countermeasures up to and including the use of force.

The chapter demonstrated that the lawfulness of such a zone depends less on any explicit legal authorisation than on a careful alignment with existing principles of space law and the United Nations Charter. From an operational perspective, the analysis suggests that self-defence zones should not function as mechanisms for pre-programmed counterreactions. Rather, where they are considered at all, SDZs should be embedded within broader frameworks of escalation management or grounded in communication and transparency measures. This is because more restrictive or enforcement-oriented zones risk violating the UN Charter as well as fundamental rules of space law on freedom of use, non-appropriation, and due regard unless justified under narrowly construed circumstances precluding wrongfulness.

From this, several policy recommendations follow. These recommendations follow from the considerations concerning the modalities of an SDZ discussed above. The recommendations have implications for both policymakers and military planners. States considering SDZs should prioritise designing zones that are as limited as possible in scope and duration, and that serve primarily as de-escalatory signalling rather than *de facto* exclusionary claims. To do this, the establishment of SDZs can be rooted in already existing international consultation obligations and ongoing processes of norm development. SDZs should be accompanied by clear, advance communication to the international community, drawing on the consultation mechanisms in Article IX OST to mitigate risks of misperception. States should also invest in robust space domain awareness capabilities – not for the purpose of broadening the scope of SDZs, but to ensure that any response to perceived threats remains grounded in accurate situational understanding and complies with the requirements of necessity and proportionality.

Moreover, States should avoid responding through in-orbit defensive measures and instead embed any SDZ policies within a broader strategy that prioritises terrestrial and/or non-kinetic responses wherever possible, thereby reducing risks of debris creation and collateral harm. Finally, because the legal status of SDZs ultimately depends on the acceptance of such zones by the international community, States should continue multilateral dialogue to clarify expectations, reduce uncertainty, and prevent the emergence of destabilising unilateral practices. Only through such engagement can SDZs avoid becoming destabilising precedents and rather enhance stability in outer space.

Notes

1. For examples of the very limited literature available see Caton, *Impacts of Anti-Access/Area Denial Measures*; Betar, "Anti-Access and Area Denial".
2. See e.g. Chan, "US Ally Plans 'Bodyguard'".
3. See e.g. Chow, "Stalkers in Space: Defeating the Threat".
4. Mallowan, Rapp and Topka, "Reinventing Treaty Compliant 'Safe Zones'"
5. Ibid.
6. Conference on Disarmament, "Report of the Conference on Disarmament".

7 Aoki, "Rules on Safety Zones in International Law".

8 The Woomera Manual provides a non-binding but authoritative restatement of how existing international law applies to military operations in outer space. See Beard and Stephens, *The Woomera Manual on the International Law.*

9 Paraphrased from Beard and Stephens, *Woomera Manual*, 163.

10 Paraphrasing from: Chow, "Stalkers in Space," 96.

11 Treaty on Principles Governing the Activities of States, 610 UNTS 205, Article I(2).

12 Ibid., Article II.

13 See e.g. Paliouras, "The Non-Appropriation Principle".

14 Gorove, "Major Legal Issues Arising from the Use of the Geostationary Orbit", 5. Similarly, Jakhu, "Legal Issues Relating to the Global Public Interest", 44. ("Appropriation in the context of outer space implies the exercise of exclusive control or use and denial of use by others.")

15 Trapp, "Taking Up Space By Any Other Means", 1685.

16 Bernstein et al., "Countering Co-Orbital ASATs", 161–162.

17 A state retains jurisdiction and control over its space object. See OST, Article VIII.

18 Schwetje, "Protecting Space Assets", 134.

19 Chow, "Stalkers in Space", 94–95.

20 OST, Article IX.

21 The authors subscribe to the view that the first sentence of Article IX contains a standalone obligation. See also Beard and Stephens, *Woomera Manual*, 171–172.

22 Ibid., 178–179.

23 See, *Chagos Marine Protected Area Arbitration* (Mauritius v UK), para. 519 ("the ordinary meaning of 'due regard' calls for the United Kingdom to have such regard for the rights of Mauritius as is called for by the circumstances and by the nature of those rights. The Tribunal declines to find in this formulation any universal rule of conduct.").

24 Harrington, "Due Regard as the Prime Directive", 69. However, states that conducted ASAT tests in the last decades, such as China in 2007 and Russia in 2021, did not comply with Article IX OST.

25 *Cf.* the public announcement of the US to destroy one of its satellites to safeguard "human life against the uncontrolled re-entry of a 5,000-pound satellite containing over 1,000 pounds of hazardous hydrazine propellant" can be seen as complying with Article IX of the OST. See US Missile Defense Agency, "One-Time Mission."

26 Takeuchi argues in the context of safety zones that "Based on the principles of general international law, the use of force is also prohibited in outer space, the establishing state cannot physically force the trespassing state to follow their requests. The only way to maintain that area's safety will be to suspend the original potentially harmful activity in case the consultation ends unsatisfactorily." Takeuchi, "Rights and Obligations in International Consultation", 378.

27 For the application of the rules on the use of force in outer space, see OST, Article III. See also, Gartner and Friedl, "Safety Zones", 260.

28 Charter of the United Nations, 1 UNTS XVI (24 October 1945), Article 2(4).

29 See e.g. Statement by Mr. Oleg Postnikov, Head of the Delegation of the Russian Federation, at the General Debate of the First Committee of the 80th session of the UNGA, Oct. 10, 2025, https://russiaun.ru/en/news/101020256.

30 Gill and Tibori-Szabo. *The Use of Force and the International Legal System*, 60. The authors recognize that the scope and application of Article 2(4) in outer space is still debated due to the link to "territorial integrity, political independence or in any other manner inconsistent with the UN Charter". However, the authors consider this beyond the scope of this article and start from the assumption that the condition to breach Article 2(4) has been met in these circumstances.

31 Beard and Stephens, *Woomera Manual*, 163–164.

32 Although the US was concerned with its biggest military space rival, the USSR, the point was already made by the Reagan Administration: "Exploration of a possible tacit understanding or even explicit agreement with the Soviets on self-defense zones around many of the satellites in the key space systems of either side, on the same grounds used to justify the self-defense of ships at sea against intrusions that can have no purpose other than a hostile one. Such an arrangement might permit some entries into the self-defense zones and would not affect normal, non-threatening satellite operations, including perhaps some inspections." US, *Commission on Integrated Long-term Strategy*, 55.

33 *S.S. 'Lotus'* (France v Turkey). For a critique of the Lotus Principle see e.g. An Hertogen, "Letting Lotus Bloom".

34 von Bogdandy and Rau, "The Lotus".

35 Beard and Stephens, *Woomera Manual*, 163.

36 Ibid.

37 *Military and Paramilitary Activities in and against Nicaragua*, 95.

38 Corten, *The Law against War*.

39 See for example, O'Meara, "Self-defense in outer space"; Pobjie, "Cosmic Force"; Brown, "Conflict on the Final Frontier".

40 See generally Paddeu, "Circumstances Precluding Wrongfulness".

41 Ibid.

42 Gill and Tibori-Szabo, *Use of Force*, 130–139.

43 See for a discussion Paddeu, "Self-Defence as a Circumstance Precluding Wrongfulness", 90–132.

44 Stubbs, "The Legality of Keep-Out", 207.

45 Klein, *Maritime Security and the Law of the Sea*, 59.

46 Stubbs, "The Legality of Keep-Out", 217.

47 International Law Commission, *Draft Articles on Responsibility*, 80.

48 2001 Articles on State Responsibility of States for Internationally Wrongful Acts (2001) (ARSIWA), See Arts. 49–53 concerning countermeasures. ARSIWA, Art. 51 sets out the requirement of proportionality.

49 Brown, "Conflict on the Final Frontier", 38–39.

50 Beard and Stephens, *Woomera Manual*, 244.

51 ARSIWA, Art. 50 (1) (a).

52 The Woomera Manual takes a broad approach, see Beard and Stephens, *Woomera Manual*, 178–179.

53 However, in the view of Gartner and Friedl: "an indefinite or unreasonably large or restrictive safety zone may be in tension with the prohibition on appropriation of outer space (art. II OST) or other principles of space law." See Gartner and Friedl, "Safety Zones", 261.

54 ARSIWA, Art. 51.

55 Chow, "Stalkers in Space", 107.

56 Di Mare, "The Role of Space Domain Awareness".

References

Aoki, Setsuko. 2022. "Rules on Safety Zones in International Law Applicable to Space Activities." *Proceedings of the International Institute of Space Law* 65 (2), 333–344.

Beard, Jack, and Dale Stephens, eds. 2024. *The Woomera Manual on the International Law of Military Space Operations*. Oxford University Press.

Bernstein, David, Michael B. Cerny, Brandon W. Kelley, and Raphael J. Piliero. 2021. "Countering Co-Orbital ASATs: Warning Zones in GEO as a Lawful Trigger for Self-Defense." In *Space and Missile Wars: What Awaits*, edited by Henry D. Sokolski. Nonproliferation Policy Education Center.

Betar, Edwin. 2021. "Anti-Access and Area Denial in the Space Domain (Part 1)." May 24, https://cove.army.gov.au/article/anti-access-and-area-denial-space-domain-part-1.

Bogdandy, Armin von and Markus Rau. 2023. "The Lotus." In *Max Planck Encyclopedia of Public International Law*, edited by Anne Peters. Max Planck Institute for Comparative Public Law and International Law. https://opil.ouplaw.com/display/10.1093/law:epil/9780199231690/law-9780199231690-e162?prd=MPIL.

Brown, Ross. 2001. "Conflict on the Final Frontier: Deficiencies in the Law of Space Conflict below Armed Attack, and How to Remedy Them." *Georgetown Journal of International Law* 51, 11–80.

Caton, Jeffrey L. 2018. *Impacts of Anti-Access/Area Denial Measures on Space Systems: Impacts of Anti-Access/Area Denial Measures on Space Systems: Issues and Implications for Army and Joint Forces Issues and Implications for Army and Joint Forces*. US Army War College Press.

Chagos Marine Protected Area Arbitration (Mauritius v. UK). 2015. Case No. 2011-03, PCA Award, 18 March.

Chan, Ryan. 2025. "US Ally Plans 'Bodyguard' Against Russia and China's 'Satellite Killers'", *Newsweek*, September 5. https://www.newsweek.com/japan-bodyguard-against-russia-china-satellite-killers-2125005.

Chow, Brian. 2017. "Stalkers in Space: Defeating the Threat." *Strategic Studies Quarterly* 82, 82–116.

Charter of the United Nations. 1945. October 24, 1 UNTS XVI.

Conference on Disarmament. 2025. "Report of the Conference on Disarmament to the General Assembly of the United Nations." CD/2493. September 15.

Corten, Olivier. 2010. *The Law against War: The Prohibition on the Use of Force in Contemporary International Law*. Hart Publishing.

Di Mare, Alessio. 2021. "The Role of Space Domain Awareness: Space Asset Resilience thru Protection." In *Joint Air and Space Power Conference*, edited by The Editor. Joint Airpower Competence Centre.

Gartner, Maximilian and Michael Friedl. 2025. "Safety Zones." In *Elgar Concise Encyclopedia of Space Law*, edited by Mahulena Hofmann and Percy J. Blount. Edward Elgar Publishing.

Gill, Terry. D. and Kinga Tibori-Szabo. 2024. *The Use of Force and the International Legal System*. Oxford University Press.

Gorove, Stephen. 1984. "Major Legal Issues Arising from the Use of the Geostationary Orbit," *Michigan Journal of International Law* 5 (1), 3–12.

Harrington, Andrea J. 2023. "Due Regard as the Prime Directive for Responsible Behavior in Space." *Loyola University Chicago International Law Review* 20, 57–86.

Hertogen, An. 2015. "Letting Lotus Bloom." *European Journal of International Law* 26 (4), 901–926.

Jakhu, Ram. 2006. "Legal Issues Relating to the Global Public Interest in Outer Space." *Journal of Space Law* 32, 31–110.

Klein, Natalie. 2012. *Maritime Security and the Law of the Sea*. Oxford University Press.

Mallowan, Lucas, Lucien Rapp and Maria Topka. 2021. "Reinventing Treaty Compliant 'Safe Zones' in the Context of Space Sustainability," *Journal of Space Safety Engineering* 8 (2), 155–166.

Military and Paramilitary Activities in and against Nicaragua (Nicaragua v. United States of America). 1984. Merits, Judgment, ICJ Rep. 169.

O'Meara, Chris. 2025. "Self-Defense in Outer Space: Anti-Satellite Weapons and the Jus ad Bellum." *Leiden Journal of International Law* 38 (3), 527–549.

Paddeu, Federica. 2015. "Self-Defence as a Circumstance Precluding Wrongfulness: Understanding Article 21 of the Articles on State Responsibility." *British Yearbook of International Law* 85 (1), 90–132.

Paddeu, Federica. 2023. "Circumstances Precluding Wrongfulness." In *Max Planck Encyclopedia of Public International Law*, edited by Anne Peters. Max Planck Institute for Comparative Public Law and International Law.

Paliouras, Zachos. 2014. "The Non-Appropriation Principle: The Grundnorm of International Space Law." *Leiden Journal of International Law* 27, 37–54.

Pobjie, Erin. 2025. "Cosmic Force: A Framework for Applying the Prohibition of the Use of Force in Outer Space." *International Law Studies* 106, 415–454.

Postnikov, Oleg. 2025. "Statement by the Head of the Delegation of the Russian Federation at the General Debate of the First Committee of the 80th UNGA Session." October 10. https://russiaun.ru/en/news/101020256.

S.S. 'Lotus,' (France v. Turkey). 1927. Judgment No. 9, PCIJ Series A No. 10.

Schwetje, T.K. 1987. "Protecting Space Assets: A Legal Analysis of 'Keep-Out Zones.'" *Journal of Space Law* 15 (1–2), 131–146.

Stubbs, Mathew. 2020. "The Legality of Keep-Out, Operational, and Warning Zones in Outer Space." In *War and Peace in Outer Space: Law, Policy and Ethics*, edited by Cassandra Steer and Matthew Hersch. Oxford University Press.

Takeuchi, Yu. 2022. "Rights and Obligations in International Consultation for Establishing Safety Zones in Outer Space." *Proceedings of the IISL Colloquium on the Law of Outer Space*, 392–396.

Trapp, Timothy J. 2013. "Taking Up Space By Any Other Means: Coming to Terms with the Non-Appropriation Article of the OST." *University of Illinois Law Review*, 1681–1713.

Treaty on Principles Governing the Activities of States in the Exploration and Use of Outer Space, Including the Moon and Other Celestial Bodies. 1967. January 27, 610 U.N.T.S. 205.

US 1988. *Commission on Integrated Long-term Strategy – Discriminate Deterrence: Report of the Commission on Integrated Long-Term Strategy.*

US Missile Defense Agency. One-Time Mission: Operation Burnt Frost, https://web.archive.org/web/20120214031001/http://www.mda.mil/system/aegis_one_time_mission.html

CHAPTER 16

Shooting for the Stars, Falling in the Ratings

Missile Tests and Country Risk

Jeroen Klomp

Abstract

This chapter investigates the causal impact of missile testing activities on country risk, as measured by the Worldwide Governance Indicators (WGI). Using a dynamic panel approach, the findings provide robust evidence that missile testing substantially elevates perceived country risk. The baseline estimates reveal that countries engaging in missile testing experience an average increase of approximately 8 percent in country risk in the year following a test. This impact is not merely short-lived, it persists and accumulates over time. These findings highlight how strategic military activities can undermine governance perceptions and investor confidence, ultimately linking defence behaviour to broader political and economic risk profiles.

Keywords: Missile testing; Country risk; A2/AD policies.

Introduction

The proliferation of missile testing activities across the globe represents one of the most significant developments in contemporary international security dynamics, often in the context of growing regional tensions, unresolved geopolitical disputes, and evolving military doctrines.[1] From North Korea's recurrent ballistic missile launches to the hypersonic weapons trials conducted by major powers such as China, Russia, and the United States, these tests have become an integral component of strategic signalling and deterrence in the 21st century.[2] Missile testing thus operates not merely as a technical exercise, but as a deliberate act of political communication – signalling technological sophistication, national resolve, and readiness to defend or project interests in contested regions. By leveraging missile-based deterrence, states can impose asymmetric costs on potential adversaries without matching them in total military expenditure or global reach.[3] However, missile testing is not without costs. Each missile test sends ripples through the international system, affecting diplomatic relations, triggering economic responses, and fundamentally altering perceptions of risk among foreign investors and credit rating agencies, and international organisations.

Based on these considerations, it is reasonable to expect that missile testing activities are reflected in a state's overall country risk profile. Country risk, broadly defined as the likelihood that changes in the business environment will adversely affect operating profits or asset values in a specific country, encompasses political, economic, and social dimensions.[4] It serves as a critical measure for international investors, multilateral institutions, and policymakers in assessing the stability and predictability of a nation's operating environment. From the perspective of international investors, credit rating agencies, and multilateral institutions, missile tests – especially when accompanied by escalatory rhetoric or conducted in violation of international norms – serve as an indicator of rising geopolitical risk.

Understanding the precise mechanisms through which missile testing shapes country risk is not only an academic endeavour – it is a pressing policy concern with far-reaching implications for national security strategy, economic development, and regional stability. In particular, the economic consequences of heightened country risk triggered by missile tests are both substantial and multifaceted. Higher country risk directly translates into increased sovereign borrowing costs, as investors demand higher yields to compensate for perceived instability.[5] For nations dependent on international capital markets, this translates into millions or billions in additional debt service costs. Foreign direct investment responds negatively to elevated country risk, with multinational corporations incorporating governance indicators into location decisions.[6] Studies consistently show that improvements in governance indicators correlate with increased FDI inflows, while deteriorations have the opposite effect.[7] The causality runs particularly strongly when governance changes result from specific government actions, such as missile tests, rather than gradual institutional evolution.

On a related note, trade relationships suffer as elevated country risk complicates financing, insurance costs rise, and concerns about sanctions or political instability affect partner calculations. For export-dependent economies, these effects can significantly impact growth prospects. Domestic investment likewise responds to country risk, with elevated risk contributing to capital flight as domestic investors seek safer havens abroad.[8] The composition of investment shifts toward shorter time horizons, reducing the capital available for long-term productive investments crucial for development. Exchange rates typically depreciate as country risk rises, affecting import costs, inflation, and real incomes. While depreciation may provide some competitiveness benefits, the instability associated with risk-driven exchange rate movements generally outweighs these gains.[9]

Importantly, the economic repercussions of missile testing extend beyond national borders, generating significant regional spillover effects. Neighbouring countries often face heightened risk premiums, declining cross-border investment, and eroding economic confidence – particularly within highly integrated regional markets. For instance, North Korea's Hwasong-15 ICBM test in November 2017 triggered sharp declines in South Korea's stock market and widened credit default swap spreads across East Asia. Similarly, Iran's Shahab and Fateh missile tests have coincided with spikes in Middle Eastern sovereign bond spreads and capital outflows from nearby states. In South Asia, India's Agni-V and Pakistan's Nasr developments have shaped investor sentiment during cross-border tensions, while China's anti-ship ballistic missile tests in the South China Sea have heightened perceptions of political risk across ASEAN economies.[10]

Against this backdrop, the aim of this chapter is to rigorously examine the impact of missile testing on country risk, quantifying both short-run and longer-run effects while exploring the mechanisms through which these impacts operate. As a dependent variable, a composite index based on the Worldwide Governance Indicators is used as a measure of country risk.[11] Methodologically, the empirical analysis builds the Least Squares Dummy Variable Corrected (LSDVC) estimation technique developed by Bruno (2005), which addresses the dynamic panel bias inherent in models with lagged dependent variables and fixed effects. The LSDVC estimator corrects the bias in standard fixed effects estimation that arises in finite samples with lagged dependent variables,[12] providing consistent estimates of both short-run effects of missile testing and longer-run cumulative impacts. More specifically, the findings indicate that countries conducting missile tests experience a significant increase in country risk in the two years following the test.

The remainder of this chapter is organised as follows. The next section discusses the theoretical framework underpinning the relationship between missile testing and country risk. The subsequent section outlines the research design, including the data used and methodology employed. The empirical findings and conclusion then follow.

Theoretical Background

Missile tests are usually embedded within broader military doctrines emphasising Anti-Access/Area Denial (A2/AD) capabilities. A2/AD strategies are designed to deter superior adversaries and constrain their operational freedom by denying access to critical theatres of engagement. As Posen (2003) and Biddle and Oelrich (2016) note, these doctrines rely on layered defence systems – including ballistic, cruise, and hypersonic missiles – to target high-value assets such as aircraft carriers, airbases, and logistical nodes.

Missile testing serves multiple, interrelated objectives that operate across technical, strategic, and political dimensions. Technically, it ensures the reliability, accuracy, and readiness of missile systems under diverse operational conditions, allowing refinements in guidance, targeting, and warhead integration.[13] Politically, tests function as instruments of strategic communication – demonstrating national resolve, deterring adversaries, and reassuring domestic audiences of regime strength.[14] When embedded within A2/AD strategies, repeated and successful tests heighten adversaries' uncertainty about the costs of intervention, reinforce deterrence, and strengthen the protective buffer such strategies aim to create. They also enhance regional bargaining power and complicate opposing coalition-building efforts.

The diversity of missile systems subject to testing reflects the varied roles these weapons play in modern military arsenals. Ballistic missiles, following parabolic trajectories after powered flight phases, range from short-range systems (under 1,000 km) to intercontinental ballistic missiles (ICBMs) capable of striking targets over 5,500 km. Cruise missiles, maintaining powered flight throughout their trajectories and often flying at low altitudes to evade detection, provide precision strike capabilities against both land and maritime targets. Surface-to-air missiles form the backbone of air defence systems, with continuous testing necessary to counter evolving aerial threats. Anti-ship missiles, designed to target naval vessels, receive particular attention in A2/AD contexts. Most recently, hypersonic weapons – capable of travelling at speeds exceeding Mach 5 while manoeuvring during flight – represent a new frontier in missile technology, with major powers investing heavily in testing programs to develop and counter these systems.[15]

As these missile tests send strong signals to adversary states, neighbouring countries, and, more broadly, the international community, they may lead to a reassessment of a country's perceived security environment. Such changes in threat perceptions are likely to be reflected in adjustments to the country risk profile that the state faces. Country risk is a multidimensional construct that refers to the totality of uncertainties and vulnerabilities that may affect the stability of a nation's political institutions, the performance of its economy, the reliability of its legal and regulatory frameworks, and its security environment.[16] More formally, country risk can be defined as the probability that political, economic, or security-related events within a country will cause losses to international investors, disrupt trade relations, or impair the government's ability or willingness to meet its financial obligations. Unlike firm-specific or market-specific risks, country risk captures systemic characteristics embedded in the structure and governance of a nation-state.[17]

Existing research typically distinguishes several categories of country risk. First, political risk includes factors such as regime instability, civil unrest, expropriation, corruption, and policy unpredictability. Second, economic risk encompasses macroeconomic mismanagement, unsustainable fiscal or external balances, and monetary instability. Third, sovereign credit risk relates to the likelihood of government default or debt restructuring. Fourth, security risk – increasingly salient in a militarised and polarised world – concerns the threat of war, terrorism, internal conflict, or militarised disputes. Lastly, legal and institutional risk reflects the reliability and independence of the judiciary, the enforcement of contracts, and the overall quality of governance.

The relationship between missile testing and country risk operates through multiple causal pathways, some direct and immediate, others indirect and operating over longer time horizons. One primary mechanism is geopolitical signalling. Missile tests are often interpreted by other states as acts of escalation or revisionist intent, especially when conducted in violation of international norms or regional confidence-building agreements. From a realist perspective, such signalling increases the likelihood of military confrontation, arms races, or strategic miscalculation. Countries that engage in frequent or provocative missile testing may be perceived as sources of regional instability, which can in turn erode diplomatic relations, reduce trade partnerships, and invite external balancing. This deterioration in the geopolitical environment contributes directly to the security component of country risk.[18]

A second mechanism concerns economic sanctions and retaliatory measures. States that engage in missile testing – particularly when under existing sanctions regimes – often trigger new waves of punitive actions from the international community. Economic sanctions can include restrictions on trade, technology transfers, capital access, and diplomatic engagement. These sanctions degrade macroeconomic performance, generate inflationary pressures, and undermine investment confidence. The expectation or imposition of sanctions also increases borrowing costs and sovereign default risk, as investors demand higher risk premiums.

A third mechanism relates to internal political dynamics and resource allocation. In many cases, missile tests are conducted by authoritarian regimes seeking to consolidate domestic legitimacy, project strength, or distract from internal weaknesses. This diversionary behaviour often comes at the expense of economic and social development. The reallocation of public funds from welfare programs to defence projects may weaken state capacity, reduce popular support, and increase the risk of internal unrest. This channel is particularly salient in countries with fragile institutions, where militarisation amplifies governance deficits and public sector inefficiencies.[19]

A fourth and subtler mechanism concerns institutional credibility and reputational spillovers. Missile testing that violates international commitments, such as non-proliferation treaties or arms control agreements, signals a lack of institutional commitment to global norms. This behaviour undermines the perceived rule-based reliability of a country's governance structures. From the perspective of institutional theory, countries that fail to comply with international law or that pursue destabilising military behaviour are seen as higher-risk environments by multinational firms, institutional investors, and credit rating agencies. Loss of institutional credibility, especially in conjunction with perceived legal or regulatory opacity, contributes to higher perceived country risk.

However, the impact of missile testing extends beyond the country conducting tests as it is also likely to affect neighbouring states through multiple spillover mechanisms. These externalities create regional risk clusters where one country's military activities elevate country risk throughout its strategic neighbourhood.[20] Understanding spillover effects is crucial for comprehensive assessment of missile testing's implications and for explaining regional patterns in governance indicators and economic outcomes.

The most immediate and well-theorised mechanism stems from the logic of the security dilemma in international relations. Direct threat perceptions arise when neighbours must assess whether missile capabilities demonstrated through tests might be directed toward them. Even if the testing country assures neighbours of peaceful intentions, the existence of capabilities creates potential threats that cannot be ignored in security planning.[21] This is particularly acute for countries with historical conflicts, territorial disputes, or alliance rivalries with the testing state. Countries observing missile tests by neighbours face domestic political pressures to respond, whether through indigenous programs, purchases of defensive systems, or enhanced security partnerships with external powers.[22] This perception triggers reactive military investments, doctrinal shifts, or even reciprocal missile development and testing programs. The resulting regional arms build-up heightens the probability of miscalculation, escalation, or accidental conflict, which in turn raises the risk premium associated with the entire region.

Basically, economic spillovers operate through multiple channels. Trade disruptions occur when missile tests prompt sanctions that affect regional commerce. Countries conducting extensive trade with sanctioned states may face secondary sanction pressures, forcing difficult choices between economic ties and relationships with sanctioning powers.[23] Investment diversion sees capital flows shifting away from regions perceived as unstable. Even if specific countries have not conducted missile tests themselves, their location in turbulent regions affects investor risk assessments. Regional infrastructure projects may be delayed or cancelled if investors worry about stability.

Financial markets and multinational investors increasingly rely on regional risk assessments, rather than strictly country-specific analyses. In an interconnected global economy, shocks that affect one country – especially within a politically fragile region – often trigger contagion effects in nearby economies. Missile testing contributes to these effects by introducing tail-risk events that alter regional stability expectations. When a state conducts a provocative test, investors may anticipate not only sanctions against that state but also collateral damage to trade flows, diplomatic cooperation, and capital mobility in adjacent countries.

For example, missile tests by Iran have in the past affected not only Iran's country risk scores but also increased perceived risk in Gulf Cooperation Council (GCC) states, due to anticipated geopolitical responses, potential blockades in the Strait of Hormuz, or retaliatory cyber operations. Similarly, during periods of intensified missile activity by China in the South China Sea, nearby ASEAN countries have seen risk premia rise due to perceived entrapment in a broader regional conflict. These perceptions feed into sovereign bond pricing and foreign direct investment decisions, amplifying the economic dimensions of risk spillovers.

Diplomatic and reputational spillovers extend security and economic effects. Missile tests may erode the institutional foundations of regional cooperation. Many regions rely on confidence-building measures, arms control agreements, or non-aggression pacts to stabilise relations between neighbours. Missile testing – especially when conducted without prior notification or in violation of agreements – undermines these mechanisms by signalling unilateralism and distrust. The result is a reduction in institutional trust that can spill over into other domains, including trade, migration, and cross-border regulatory cooperation.

Regional reputation suffers when one country's actions affect international perceptions of an entire region. Investors and partners may apply heuristics treating nearby countries similarly, even if specific situations differ substantially.[24] This lumping together of countries based on geographic proximity affects access to finance, trade relationships, and diplomatic engagement.

This erosion of institutional credibility can affect neighbouring states even if they are not direct parties to the missile dispute. For instance, when Russia tests new long-range missile systems near its western borders, not only do the Baltic states and Poland experience heightened risk assessments, but so do adjacent EU members that rely on coordinated security policy. The weakening of institutional trust is particularly harmful in fragile or transitional democracies, where external instability can reverberate into domestic political fragmentation or populist backlash.

Based on the preceding theoretical considerations, this study proposes the following hypotheses:

Hypothesis 1 (Domestic effect): *Missile testing adversely affects country risk for the testing country, with effects visible in the immediate aftermath of tests and persisting over subsequent years.*

Hypothesis 2 (Neighbour effect): *Missile testing heightens country risk for a neighbour country, with effects visible in the immediate aftermath of tests and persisting over subsequent years.*

Research Design

Data on Missile Tests

My primary independent variable is drawn from the comprehensive Missile Testing dataset compiled by Birol (2023). This dataset offers detailed coverage of missile test events across countries and time. To be precise, the dataset offers high-resolution data on ballistic and cruise missile test launches from 1945 through 2015, encompassing more than 7,000 individual launch events across nearly ninety states.

The Missile Testing dataset was constructed through triangulated coding from military reports, declassified government archives, satellite data, defence white papers, news wire services, and specialised think tank repositories such as the Missile Defence Project, SIPRI, and Jane's Defence Weekly. Each missile test entry includes precise information on the date of the launch, the country conducting the test, the classification of the missile system (e.g., SRBM, MRBM, IRBM, ICBM, cruise, or SLBM), the launch platform (land-based, sea-based, air-based, or submarine-launched), and an assessment of whether the launch succeeded or failed. The dataset also codes for the context in which the test occurred, such as peacetime routine testing, military exercises, responses to regional crises, or strategic escalations.

For the purposes of this study, the missile test data have been aggregated to the country-year level, producing a balanced panel structure in which each country is observed annually for its number of missile tests. Figure 16.1 shows the number of tests during my period of analysis.

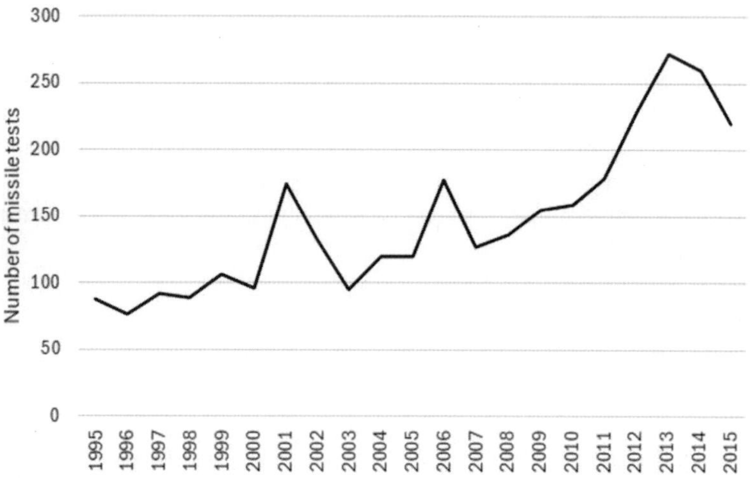

Figure 16.1: The number of missile tests

For my empirical analysis, two primary variables are constructed from the Birol dataset. First, a binary indicator variable equals one if a country conducted at least one missile test in a given year, and zero otherwise. This dummy captures the presence or absence of missile activity in a given time period, and it reflects the political signal associated with engaging in such strategic behaviour. The occurrence of a missile test – regardless of the number or type – may trigger international condemnation, heighten regional tensions, or provoke shifts in investor and diplomatic confidence.

As noted above, my analysis extends beyond domestic effects to also capture potential spillover impacts on the country risk of neighbouring states. To this end, a second dummy variable is included that takes the value one when a neighbouring country has undertaken missile testing activities in a given year, and zero otherwise. The data used to identify neighbouring countries are taken from the CIA fact book.

Country Risk Data

To assess the effect of missile testing on country risk, this study employs the Worldwide Governance Indicators (WGI) as its principal measure of institutional quality. Governance constitutes a core component of country risk, as it shapes the credibility, predictability, and reliability of state behaviour in economic and political interactions. Developed by the World Bank and published annually, the WGI dataset covers more than 200 countries from 1996 onward and provides a comprehensive, multidimensional assessment of governance performance.[25] The WGI synthesise information from over thirty underlying data sources, including firm and household surveys, assessments by commercial risk-rating agencies, evaluations conducted by international organisations, and expert judgments from think tanks and non-governmental organisations.[26]

The quality of governance within the WGI framework is captured across six distinct but closely related dimensions. The first, Voice and Accountability, reflects the extent to which citizens are able to participate in selecting their government, as well as the degree of freedom of expression, association, and media. The second dimension, Political Stability and Absence of Violence/Terrorism, measures perceptions of the likelihood that a government will be destabilised or removed through unconstitutional or violent means, including terrorism and politically motivated violence. Government Effectiveness assesses the quality of public service delivery, the professionalism and independence of the civil service, the quality of policy formulation and implementation, and the credibility of government commitments. Regulatory Quality captures the state's ability to design and enforce policies that support private sector development. Rule of Law reflects confidence in and adherence to societal rules, including contract enforcement, property rights, policing, and judicial effectiveness. Finally, Control of Corruption measures the extent to which public authority is exercised for private gain, encompassing both petty and grand corruption and state capture by elites.

Substantial cross-country variation is observed. Advanced economies such as Norway, Canada, and Germany consistently achieve high scores across all dimensions, while countries such as Venezuela, Syria, and North Korea score poorly. Temporal variation is also meaningful, with sharp deteriorations following political upheavals or security crises and gradual improvements accompanying institutional reform in other cases.

The use of the WGI as proxies for country risk is justified on both theoretical and empirical grounds. Conceptually, the indicators capture core aspects of state capacity, credibility, and institutional reliability that underpin investor confidence and sovereign risk assessments. Empirically, several WGI dimensions – most notably political stability, rule of law, and control of corruption – closely align with how country risk is defined in the academic literature. Moreover, governance quality is known to be a powerful predictor of sovereign credit ratings and borrowing costs. Rating agencies explicitly incorporate institutional weakness as a source of policy uncertainty, fiscal mismanagement, and elevated default risk, implying that countries with low WGI scores tend to face higher financing costs and reduced access to capital.[27]

Although the WGI do not directly incorporate macroeconomic variables such as debt ratios or inflation, they capture the institutional environment within which economic policy is formulated and executed. Importantly, unlike sovereign ratings that may be influenced by market sentiment or discretionary judgment, the WGI are constructed using transparent and replicable methodologies. They thus provide a stable and conceptually grounded barometer of institutional health – an attribute that is particularly relevant in contexts of missile testing and military escalation, where governance capacity may be strained.

Missile testing can affect governance through multiple channels. Government effectiveness may deteriorate as fiscal and administrative resources are diverted toward military priorities at the expense of public services. Regulatory quality can weaken as budgetary pressures constrain enforcement capacity. Political stability may erode when opportunity costs become visible through declining service provision or rising economic insecurity, potentially generating domestic discontent,[28]

The six WGI dimensions are strongly positively correlated, with pairwise correlations ranging between 0.60 and 0.90, reflecting the systemic nature of governance. To exploit this structure, the primary dependent variable is constructed as the first principal component of the six indicators, derived using principal components analysis separately for each year. This synthetic index captures the largest share of common variation – over 85 percent – while loading positively and relatively evenly on all dimensions. The resulting standardised index provides a parsimonious and interpretable measure of country risk rooted in governance quality.

As a preliminary assessment, a nonparametric comparison of country risk levels before and after missile tests reveals a statistically significant decline following testing events (χ^2 test, $p = 0.02$). While suggestive, this result does not account for unobserved heterogeneity or confounding factors and therefore motivates the multivariate analysis that follows.

Empirical Model

To formally examine the impact of missile testing on country risk, the following baseline specification of a dynamic panel model is estimated.

$$CountRisk_{i,t} = \\ \propto + \beta\, CountRisk_{i,t-1} + \sum_{t=0}^{m} \theta_m\, MisTest_{i,t}^{D} + \sum_{t=0}^{m} \gamma_m\, MisTest_{i,t}^{N} + \mathbf{x}'_{i,t-1} + \mu_i + \lambda_t + \varepsilon_{i,t} \qquad (1)$$

Where $CountRisk_{i,t}$ represents country risk for country *i* in year *t*, measured by the factor scores from the first principal component of the World Wide Governance Indicators, as explained above; $MisTest_{it}^{D}$ and $MisTest_{it}^{N}$ are respectively the domestic and neighbouring country missile test indicator (either binary or count); $CountRisk_{it_{-1}}$ is the lagged dependent variable to capture autoregressive tendencies; $\mathbf{x}'_{it_{-1}}$ is a vector of lagged time-varying control variables; μ_i represents country fixed effects; λ_t represents year fixed effects; and, finally, $\varepsilon_{i,t}$ is the error term.

The coefficients of primary interest, θ_m and γ_m, capture the short-run and long-run effect of missile testing on respectively the domestic and neighbouring country risk. A negative coefficient indicates that missile testing increases country risk, consistent with my theoretical predictions. Since the country risk indicator is expressed in natural logarithms, the estimated coefficients can be interpreted as elasticities.

Country fixed effects (μ_i) control for all time-invariant differences across countries that might correlate with both testing and country risk. By including these fixed effects, my identification strategy exploits within-country variation over time rather than cross-sectional differences. Year fixed effects (λ_t) control for global shocks and trends affecting all countries similarly.

My empirical model incorporates a set of control variables to isolate the effect of missile testing on country risk from potential confounding factors and to mitigate omitted variable bias. These controls include real GDP per capita and total real GDP (both taken from the World Development Indicators reported by the World Bank), trade openness (measured as the sum of exports and imports as a share of

GDP), natural resource dependence (proxied by natural resource rents as a share of exports), the level of democracy (measured by the Polity V score), the presence of armed conflict (a binary indicator for whether a country experienced conflict in a given year recorded by UCDP), and military expenditure as a share of GDP. All control variables are included with a one-year lag to address potential simultaneity bias with the missile testing indicators.

The empirical analysis employs the LSDVC (Least Squares Dummy Variable Corrected) estimator, an estimation technique well-suited for dynamic panel models with fixed effects and a short time dimension. Traditional fixed-effects estimation in dynamic models suffers from the Nickell bias, whereby the inclusion of a lagged dependent variable induces a downward bias in the estimated coefficient when the time dimension is small. This bias is particularly problematic in panels with many countries observed over a relatively short period, as is the case with the WGI data (1996–2023).

The LSDVC estimator, developed by Bruno (2005), provides a solution by applying a bias-correction procedure to the least squares dummy variable (LSDV) estimator. It begins by estimating the model using a consistent, but possibly inefficient, estimator such as Arellano-Bond GMM. It then uses analytical approximations of the bias to correct the fixed-effects estimator. The result is a consistent and efficient estimator that is particularly well-suited for panels with a moderate number of time periods and unbalanced structures, as is common in international governance data.

Since the dependent variable is derived using principal component analysis (PCA), which may introduce measurement error, I employ bootstrap estimation with 1,500 replicators to obtain consistent standard errors in the panel regression.

Empirical Findings

Table 16.1 reports the baseline estimates of the impact of missile testing on country risk. Columns (1) and (2) present the results using the binary testing indicator, while Columns (3) and (4) employ the count variable capturing the number of tests conducted within a given year. The results provide strong empirical support that missile testing exacerbates country risk. In Column (1), the coefficient on the binary testing indicator suggests that conducting at least one missile test within a year increases the country risk index by approximately 8 percent, a result that is statistically significant at the 5 percent level. Column (3) further refines this relationship using the count variable, showing that a hundred percent increase in the number of tests is associated with about four percent increase in risk, also significant at the 5 percent level.

In the bottom of Table 16.1, I also assess whether missile testing generates regional spillover effects in country risk. The rationale for this test derives from theories of security interdependence and regional contagion. In geographically proximate regions, missile testing by one state may heighten perceived insecurity, trigger defensive reactions, and increase political risk in neighbouring countries. As explained above, these spillovers can operate through multiple channels, including security dilemmas, trade disruptions, capital flight, or diplomatic contagion. The economic interpretation is clear, missile testing generates substantial negative externalities for neighbouring countries. These effects are strongest among economically integrated or politically rival states, consistent with the idea that interdependence amplifies transmission. Spillovers are also magnified when the testing country deploys longer-range systems, whose perceived threat radius extends beyond immediate borders. The results in the bottom part indicate that missile testing is not only a domestic issue but a regional public bad, undermining governance and stability across entire security complexes.

So far, my analysis has focused on the short-run effects of missile testing. Columns (2) and (4) extend the investigation to longer temporal horizons by incorporating multiple lags of the missile testing variable, allowing us to assess whether the observed impacts persist over time. The results indicate two key patterns. First, missile testing significantly increases country risk within the first two years following a test. Second, beyond this period, country risk remains elevated, showing no indication of reverting to pre-test levels. These findings suggest that the effects of missile testing are persistent and partially cumulative, rather than purely transitory, with enduring adverse implications for a state's risk profile.

The temporal pattern of the estimated coefficients provides further insight into the underlying transmission mechanisms. The pronounced immediate effect, followed by smaller yet statistically significant lagged effects, points to a combination of rapid and gradual adjustment processes. In the short run, missile tests trigger immediate reactions from markets and investors, reflecting heightened geopolitical uncertainty and risk aversion. Over time, the cumulative imposition of international sanctions, trade restrictions, and diplomatic isolation amplifies these effects, as the associated economic and financial costs accumulate.[29] Additionally, domestic resource reallocation toward missile development diverts funding from productive sectors, further undermining macroeconomic stability. Reputational damage from repeated testing reinforces sustained risk elevation, as international partners remain cautious and constrain economic engagement.

Taken together, these findings provide robust evidence that missile testing increases perceived country risk in the short run – both in the testing country and in neighbouring states – and that its adverse effects persist and accumulate over time through a combination of economic, financial, and reputational channels. Based on this evidence, I can accept the two hypotheses outlined above.

Table 16.1: Missile tests and country risk

	(1)		(2)	
Domestic effect				
t = 0	-0.081	**	-0.078	**
	(0.024)		(0.022)	
t = 1			-0.023	*
			(0.013)	
t = 2			-0.008	
			(0.006)	
T = 3			-0.101	
			(0.037)	
Neighbouring effect				
t = 0	-0.022	**	-0.028	**
	(0.006)		(0.014)	
t = 1			-0.009	
			(0.012)	
t = 2			-0.003	
			(0.002)	
T = 3			-0.037	
			(0.011)	
Observations	3198		3198	
Countries	201		201	
Country fixed effects	YES		YES	
Year fixed effects	YES		YES	
Control variables	YES		YES	

Note: Bootstrapped standard errors in parentheses. ** and * denote statistical significance at 5% and 10% level, respectively.

Conclusion

Key Findings

This chapter has examined the relationship between missile testing and country risk using comprehensive global data spanning six decades. The analysis yields several key findings with important implications for understanding the economic costs of military modernisation programs and the strategic calculations underlying missile testing decisions.

First, the short run results indicate substantial short-run effects of missile testing on country risk. Countries conducting tests experience approximately 8 percent deterioration in governance quality measured by the Worldwide Governance Indicators composite index. This effect is statistically significant, robust across specifications, and economically meaningful, affecting sovereign borrowing costs, foreign direct investment, trade relationships, and development assistance.

Second, the longer-run analysis reveals that effects persist and partially accumulate over multiple years rather than dissipating immediately. Cumulative three-year effects reach approximately ten percent. This persistence reflects mechanisms including accumulating sanctions, compounding resource allocation distortions, persistent reputational damage, and gradually evolving regional tensions.

Finally, the spillover analysis reveals significant externalities from testing to neighbouring countries, suggesting that missile testing represents a beggar-thy-neighbour policy externalising costs onto nearby countries.

Practical, Policy, and Doctrinal Implications

Several theoretical and policy-relevant implications emerge from these findings. First, the magnitude of the estimated governance costs indicates that conventional evaluations of military modernisation programs systematically understate their true economic impact. By focusing narrowly on defence expenditures and technical capabilities, standard assessments overlook broader institutional and credibility effects that materially influence sovereign risk, borrowing conditions, foreign investment, trade relations, and access to development finance. Missile testing, in this sense, should be understood not merely as a military or technical act, but as a policy choice with enduring economic and institutional consequences.

Second, the presence of sizeable regional spillovers underscores missile testing as a classic source of negative externalities. Testing activity by one state systematically affects the risk perceptions and economic outlook of neighbouring countries, as illustrated by spillovers from North Korean tests to Japan and South Korea, and from Indian missile developments to Pakistan. These patterns are consistent with missile testing functioning as a beggar-thy-neighbour policy that externalises governance and country-risk costs beyond national borders, thereby amplifying regional instability. From a policy perspective, this strengthens the case for regional arms control agreements, testing moratoria, advance notification regimes, and cooperative monitoring frameworks designed to internalise these externalities.

Despite limitations related to endogeneity, measurement error, and data availability, this chapter makes several contributions. It provides systematic quantitative evidence on the economic consequences of military signalling through missile testing, demonstrates how discrete security actions affect institutional quality, and quantifies governance costs that enrich cost–benefit analyses of military programs. Taken together, the results underscore a clear conclusion – the strategic benefits of missile testing must be weighed against tangible, persistent, and partly externalised costs to governance, economic credibility, and regional stability.

Notes

1 Birol, "Global Missile Testing Database Version 2"; Birol, "Ready to Fire?".
2 Kroenig, *The Logic of American Nuclear Strategy;* Acton, "Escalation through Entanglement".
3 Montgomery, "Extended Deterrence in the Second Nuclear Age".
4 Erb, Harvey, and Viskanta, "Political Risk".
5 Cantor and Packer, "Determinants and Impact".
6 Wei, "How Taxing Is Corruption".
7 Busse and Hefeker, "Political Risk, Institutions".
8 Hermes and Lensink, "Capital Flight".
9 Park and Park, "Rare Disaster Risk and Exchange Rates".
10 Birol, "Global Missile Testing Database Version 2"; Birol, "Ready to Fire?".
11 Kaufmann et al., "The Worldwide Governance Indicators".
12 Nickell, "Biases in Dynamic Models".
13 Birol, "Ready to Fire?"
14 Hymans, *The Psychology of Nuclear Proliferation.*
15 Speier, et al., *Hypersonic Missile Nonproliferation.*
16 Schroeder, "The Underpinnings of Country Risk Assessment".
17 Erb, Harvey, and Viskanta, "Political Risk"; Schroeder, "The Underpinnings of Country Risk Assessment".
18 Gupta et al., "Corruption and Military Spending".
19 Gupta et al., "Corruption and Military Spending"; Knight, Loayza, and Villanueva, "The Peace Dividend".
20 Murdoch and Sandler, "Civil Wars and Economic Growth".
21 Jervis, "Cooperation under the Security Dilemma".
22 Fordham and Walker, "Kantian Liberalism, Regime Type".
23 Hufbauer et al., *Economic Sanctions Reconsidered.*
24 Kono and Montinola, "Does Foreign Aid Support".
25 Kaufmann et al., "The Worldwide Governance Indicators".
26 Ibid..

27 Afonso et al., "Short- and Long-Run Determinants"; Cantor and Packer, "Determinants and Impact".

28 Ross, "Does Oil Hinder Democracy?"

29 Hufbauer et al., *Economic Sanctions Reconsidered.*

References

Acemoglu, D., S. Johnson, and J.A. Robinson. 2001. "The Colonial Origins of Comparative Development: An Empirical Investigation". *American Economic Review*, 91 (5), 1369–1401.

Acton, J. M. 2018. "Escalation through Entanglement: How the Vulnerability of Command-and-Control Systems Raises the Risks of an Inadvertent Nuclear War". *International Security*, 43 (1), 56–99.

Afonso, A., P. Gomes, and P. Rother. 2011. "Short- and Long-Run Determinants of Sovereign Debt Credit Ratings". *International Journal of Finance & Economics*, 16 (1), 1–15.

Biddle, S., and I. Oelrich. 2016. "Future Warfare in the Western Pacific: Chinese Anti Access/area Denial, US Air Sea Battle, and Command of the Commons in East Asia". *International Security*, 41 (1), 7–48.

Biglaiser, G., and K. DeRouen. 2007. "Following the Flag: Troop Deployment and US Foreign Direct Investment". *International Studies Quarterly*, 51 (4), 835–854.

Birol, C. 2023. "Global Missile Testing Database Version 2". *Journal of Conflict Resolution*, 67 (4), 789–812.

Birol, C. 2025. "Ready to Fire? Missile Tests and Crisis Involvement". *Journal of Global Security Studies*, 10 (4), ogaf024.

Bruno, G. S. 2005. "Approximating the Bias of the LSDV Estimator for Dynamic Unbalanced Panel Data Models". *Economics Letters*, 87 (3), 361–366.

Busse, M., and Hefeker, C. 2007. "Political Risk, Institutions and Foreign Direct Investment". *European Journal of Political Economy*, 23 (2), 397–415.

Cantor, R., and F. Packer. 1996. "Determinants and Impact of Sovereign Credit Ratings". *Economic Policy Review*, 2 (2), 37–53.

Cline, W. R. 1984. *International Debt: Systemic Risk and Policy Response.* Washington, DC: Institute for International Economics.

Cross-National Time-Series Data Archive. 2021. *International Status Rankings 1960–2020.* Databanks International.

Collier, P., and A. Hoeffler. 2004. "Greed and Grievance in Civil War". *Oxford Economic Papers*, 56 (4), 563–595.

Cordesman, A. H. 2014. *Iran's Rocket and Missile Forces and Strategic Options.* Washington, DC: Center for Strategic and International Studies.

Dollar, D., and V. Levin. 2006. "The Increasing Selectivity of Foreign Aid, 1984–2003". *World Development*, 34 (12), 2034–2046.

Erb, C. B., C.R. Harvey, and T.E. Viskanta. 1996. "Political Risk, Economic Risk, and Financial Risk". *Financial Analysts Journal*, 52 (6), 29–46.

Fearon, J. D. 1994. "Domestic Political Audiences and the Escalation of International Disputes". *American Political Science Review*, 88 (3), 577–592.

Fordham, B. O. and T.C. Walker. 2005. "Kantian Liberalism, Regime Type, and Military Resource Allocation: Do Democracies Spend Less?" *International Studies Quarterly*, 49 (1), 141–157.

Gleditsch, N. P., P. Wallensteen, M. Eriksson, M., Sollenberg, and H. Strand. 2002. "Armed Conflict 1946–2001: A New Dataset". *Journal of Peace Research*, 39 (5), 615–637.

Gupta, S., L. de Mello, L. and R. Sharan. 2001. "Corruption and Military Spending". *European Journal of Political Economy*, 17 (4), 749–777.

Hermes, N., and R. Lensink. 2001. "Capital Flight and the Uncertainty of Government Policies". *Economics Letters*, 71 (3), 377–381.

Howell, L. D. 2001. *The Handbook of Country and Political Risk Analysis*. PRS Group.

Hufbauer, G. C., J.J. Schott and K.A. Elliott. 2007. *Economic Sanctions Reconsidered* (3rd ed.). Washington, DC: Peterson Institute for International Economics.

Hymans, J. E. C. 2006. *The Psychology of Nuclear Proliferation: Identity, Emotions, and Foreign Policy*. Cambridge: Cambridge University Press.

Jervis, R. 1978. "Cooperation under the Security Dilemma". *World Politics*, 30 (2), 167–214.

Kaufmann, D., and A. Kraay. 2002. "Growth without Governance". *Economía*, 3 (1), 169–229.

Kaufmann, D., A. Kraay, and M. Mastruzzi. 2011. "The Worldwide Governance Indicators: Methodology and Analytical Issues". *Hague Journal on the Rule of Law*, 3 (2), 220–246.

Knack, S. and P. Keefer. 1995. "Institutions and Economic Performance: Cross-country Tests Using Alternative Institutional Measures". *Economics & Politics*, 7 (3), 207–227.

Knight, M., N. Loayza, and D. Villanueva. 1996. "The Peace Dividend: Military Spending Cuts and Economic Growth". *IMF Staff Papers*, 43 (1), 1–37.

Kono, D. Y. and G.R. Montinola. 2009. "Does Foreign Aid Support Autocrats, Democrats, or Both?" *Journal of Politics*, 71 (2), 704–718.

Kroenig, M. 2018. *The Logic of American Nuclear Strategy: Why Strategic Superiority Matters*. Oxford: Oxford University Press.

Montgomery, E. 2016. "Extended Deterrence in the Second Nuclear Age". *CSBA, Washington*, 24–26.

Murdoch, J. C., and T. Sandler. 2004. "Civil Wars and Economic Growth: Spatial Dispersion". *American Journal of Political Science*, 48 (1), 138–151.

Neufeld, M. J. 1995. *The Rocket and the Reich: Peenemünde and the Coming of the Ballistic Missile Era*. Free Press.

Nickell, S. 1981. "Biases in Dynamic Models with Fixed Effects". *Econometrica*, 49 (6), 1417–1426.

North, D. C. 1990. *Institutions, Institutional Change and Economic Performance*. Cambridge University Press.

O'Rourke, R. 2021. *China Naval Modernization: Implications for U.S. Navy Capabilities*. Washington, DC: Congressional Research Service.

Park, C., and S. Park. 2020. "Rare Disaster Risk and Exchange Rates: An Empirical Investigation of South Korean Exchange Rates Under Tension Between the Two Koreas". *Finance Research Letters, 36*, 101314.

Posen, B. R. 2003. "Command of the Commons: The Military Foundation of US Hegemony". *International Security*, 28 (1), 5–46.

Ross, M. L. 2001. "Does Oil Hinder Democracy?" *World Politics*, 53 (3), 325–361.

Sartori, A. E. 2005. *Deterrence by Diplomacy*. Princeton, NJ: Princeton University Press.

Schelling, T. C. 1966. *Arms and Influence*. New Haven, CT: Yale University Press.

Schroeder, S. K. 2008. "The Underpinnings of Country Risk Assessment". *Journal of Economic Surveys*, 22 (3), 498–535.

SIPRI (Stockholm International Peace Research Institute). 2021. *SIPRI Military Expenditure Database*. Available at https://www.sipri.org

Speier, R. H., G. Nacouzi, C. Lee, and R.M. Moore. 2017. *Hypersonic Missile Nonproliferation: Hindering the Spread of a New Class of Weapons*. Santa Monica, CA: RAND Corporation.

Sutyagin, I., and J. Bronk. 2017. *Russia's New Ground Forces: Capabilities, Limitations and Implications for International Security*. London: Royal United Services Institute.

Wei, S. J. 2000. "How Taxing Is Corruption on International Investors?" *Review of Economics and Statistics*, 82 (1), 1–11.

World Bank. 2021. *World Development Indicators*. Washington, DC: World Bank.

Discussion and Conclusions

Edwin Dado, René Janssen, Kerry Malone, Roy Hamans and Marc Gouw

Anti-Access / Area Denial (A2/AD)

According to a 2016 conference report published by the NATO Defence College, which has since been widely cited across NATO-related scholarship and policy debates, Anti-Access/Area Denial (A2/AD) is defined as follows:

> The objective of an anti-access or area-denial strategy is to prevent the attacker from bringing its forces into the contested region (A2) or to prevent the attacker from freely operating within the region and maximizing its combat power (AD).

Within this formulation, a conceptual distinction is drawn between two complementary components. **Anti-Access (A2)** refers to capabilities, strategies, and measures designed to prevent or deter an adversary from entering a specific theatre of operations. These capabilities may include long-range precision strikes, integrated air defence systems, maritime interdiction, or strategic deterrence tools intended to raise the cost of intervention to an unacceptable level. **Area Denial (AD)**, by contrast, focuses on limiting the freedom of movement and operational effectiveness of an adversary once it has entered the contested region. This can involve layered defensive systems, electronic warfare, anti-armour capabilities, and other measures that degrade the attacker's ability to manoeuvre and project power.

In NATO discourse, this definition has frequently been used to emphasise the defensive logic underpinning A2/AD strategies. It frames A2/AD primarily as a means of protecting territory and sovereignty by complicating an adversary's operational planning. However, from a military-analytical perspective, this definition can be interpreted as largely technology-centric. It tends to focus on platforms, weapons systems, and spatial control of defined geographic areas. While such a perspective has analytical clarity, recent developments suggest that it may no longer be sufficient.

The full-scale invasion of Ukraine by the Armed Forces of the Russian Federation in 2022 has profoundly reshaped contemporary understandings of warfare. In light of this conflict, and amid rising geopolitical tensions globally, there is an urgent need to reassess the conceptual foundations of A2/AD. Examining the concept from a Ukrainian perspective is particularly illuminating, as Ukraine has experienced large-scale, high-intensity warfare against a near-peer adversary in real time.

Two Major Lessons from the War in Ukraine Stand Out

First, the notion of a clearly bounded 'contested region' is increasingly obsolete. In Ukraine, the battlefield has not been confined to a specific geographic area such as the eastern regions. Rather, the entire national territory has been treated as a potential battlespace. There is no stable or reliable separation between a combat zone – where soldiers engage in direct hostilities – and a secure rear area where civilians remain insulated from military operations. Missile strikes, drone attacks, cyber operations, and sabotage have targeted infrastructure, logistics hubs, energy facilities, and urban centres across the country. Civilians and military personnel alike have been exposed to risk, and both civilian and military assets have been targeted. In effect, there is no protective 'military shell' surrounding a safe civilian geographic domain. The traditional spatial logic embedded in the A2/AD definition – where a distinct region can be denied or defended – does not adequately capture this reality.

Second, the war has demonstrated the growing importance of Multi-Domain Operations (MDO). MDO is a modern operational concept aimed at integrating and synchronising activities across all domains of warfare to generate strategic advantage against peer or near-peer adversaries. Historically, warfare was conceptualised primarily in terms of three domains: land, sea, and air. Contemporary conflict, however, extends beyond these traditional spheres. MDO incorporates additional functional domains, including space, cyberspace, the electromagnetic spectrum, and the information environment.

The Ukrainian experience illustrates this clearly. A military convoy operating in the land domain, for example, cannot rely solely on ground-based protection. It requires aerial surveillance and drone support to detect and neutralise threats from above. It depends on cyber capabilities to disrupt or deceive incoming precision-guided systems. It may require electronic warfare to interfere with enemy communications and satellite navigation. Information operations shape both domestic morale and international support. Thus, survivability and operational effectiveness are no longer confined to one domain; they depend on synchronised, cross-domain integration.

In this context, A2/AD cannot be understood merely as a set of technological barriers within a geographically defined theatre. It must be reconceptualised as a dynamic, multi-domain strategy that operates simultaneously across physical, digital, and cognitive spaces. Denying access or operational freedom today may involve disrupting satellite communications, degrading cyber networks, manipulating information flows, or contesting electromagnetic signatures – actions that transcend traditional geographic boundaries.

Therefore, the widely cited definition of A2/AD requires refinement. A predominantly technology-centric and territorially bounded perspective is too narrow to capture the complexity of contemporary high-intensity conflict. The lessons drawn from Ukraine underscore the need to expand the conceptual framework to incorporate multi-domain integration, societal resilience, civilian vulnerability, and the blurring of front lines.

The chapters in this book collectively contribute to this necessary revision. Across sixteen contributions, the authors reassess A2/AD from diverse operational, strategic, technological, and doctrinal perspectives. Together, they argue for a broader, more adaptive understanding of A2/AD – one that reflects the realities of modern warfare and provides a more robust foundation for future strategic planning.

Advances in A2/AD

Taken together, the chapters in this book demonstrate that Anti-Access/Area Denial (A2/AD) can no longer be treated as a coherent or singular military concept. Rather, it emerges as a plurality of practices and logics, shaped by domain, scale, and strategic intent. What unites this family of practices and logics is not a shared definition, but a common concern with the control of access, freedom of action, and decision-making under contested conditions.

Several contributions explicitly challenge dominant strategic narratives that portray contemporary A2/AD as a revolutionary or near-decisive new concept. For example, van Maurik's historical and conceptual analysis in Chapter 1 shows that contemporary A2/AD systems remain structurally constrained by the same dynamics that have historically limited denial strategies: predictability, reactivity and vulnerability to adaptation. By analysing denial logics from the Maginot Line through to Soviet integrated air defence systems, his chapter highlights a recurring strategic–operational gap, in which political narratives tend to overstate the durability and effectiveness of denial while downplaying its inherent operational vulnerabilities.

A second cluster of contributions approach A2/AD primarily as a dynamic system-of-systems problem. Hamans and Bertrand in Chapter 2 conceptualise A2/AD as an adaptive hybrid construct that integrates kinetic and non-kinetic actions across domains and time. From this perspective, A2/AD is less about static defensive 'bubbles' and more about shaping adversary behaviour through uncertainty, coercion, and decision-dominance. This interpretation aligns closely with the operational insights derived from NATO war-gaming and kill-chain analysis, which define A2/AD confrontations as contests of cognition, timing, and risk management rather than attritional exchanges, as discussed by Hörnedal and his co-authors in Chapter 3.

A third cluster of contributions adopts a domain-specific, capability-driven perspective. Savelsberg, Kampert and Vermeulen in Chapter 4 demonstrate how hypersonic glide vehicles compress decision time and undermine layered missile defence concepts, thereby reinforcing anti-access effects through uncertainty rather than impermeability. Voskuijl and his co-authors in Chapter 8 show how swarming attacks seek to saturate defensive decision-making capacity rather than defeat individual systems. Complementing these kinetic perspectives, chapters on electromagnetic dominance (Chapter 6), navigation warfare (Chapter 7), and cyberspace (Chapter 13) illustrate how access denial increasingly operates through reversible, often invisible disruption of information and timing dependencies.

A fourth cluster of contributions emphasises endurance, resilience, and the material foundations of access. Monsuur, Van Fenema and Malone in Chapter 14 demonstrate how logistics itself has become a primary target of A2/AD strategies, effectively transforming rear areas into contested operational spaces. Kranenburg and Moeskops in Chapter 12, as well as Chapter 11 on hydraulic warfare, further demonstrate that geography is not merely a backdrop but an adaptable strategic variable, deliberately shaped to deny manoeuvre and impose long-term costs on adversaries.

Finally, legal and political-economic chapters broaden the analytical scope of A2/AD beyond military effectiveness alone. Handel-Mazzetti, Vermeer and Zwanenburg in Chapter 15 demonstrate how space-based A2/AD practices intersect with legal ambiguity and norm contestation, while Klomp in Chapter 16 links missile testing as an A2/AD-related signalling practice to increased country risk and economic costs.

Taken together, the different perspectives point to a broader conceptual shift in how A2/AD is understood across the contributions in this book. The implications of this shift – from earlier, more technology-centric and penetration-focused conceptions toward a view of A2/AD as a condition of ongoing competition – are explored in greater depth in the following section.

Implications for Defence Organisations

The conceptual shift outlined above has direct and actionable consequences for defence organisations. If A2/AD is no longer understood as a static set of denial capabilities to be penetrated or neutralised, but rather as a persistent condition of multidomain strategic competition, then existing assumptions underpinning defence planning, force development, and organisational design require a systematic review. Across the contributions in this book, authors repeatedly point to implications that – when taken together – suggest the need for policy-driven organisational adaptation, rather than incremental technical improvement within individual domains.

Defence Planning and Force Development

From a policy perspective, a first implication concerns the orientation of defence planning and force development. Several chapters caution against planning models that implicitly assume a short phase of contested access followed by a return to operational freedom. Instead, A2/AD emerges as a condition that shapes operations across all domains and throughout the full spectrum of competition, crisis, and conflict. This is illustrated by historical and conceptual analyses that show how denial strategies have repeatedly been overestimated at the strategic level while remaining operationally fragile and adaptive in practice (Chapter 1), as well as by operational studies that emphasise temporality, concentration of effects, and risk acceptance rather than permanent access (Chapter 3).

For defence organisations, this implies that planning assumptions, capability priorities, and readiness models must be aligned with the expectation of sustained multidomain contestation. Policy choices should therefore favour force structures that are resilient under degradation, capable of operating with partial access, and able to adapt across domains. Importantly, several technically oriented chapters illustrate that such adaptability depends critically on the integration of capabilities across technical domains, rather than on platform performance in isolation. Studies on hypersonic systems, swarming attacks, counter-UAS, and directed-energy applications demonstrate that operational effects emerge from the interaction between sensors, shooters, networks, and decision-support systems spanning multiple domains (Chapters 4, 5, 8, 10).

Command, Control, and Governance

A second policy-relevant implication concerns command, control, and governance. The chapters in this book consistently demonstrate that contemporary A2/AD strategies aim to disrupt decision-making processes through saturation, information denial, and temporal compression across domains. This dynamic is explicitly discussed in studies on electromagnetic dominance and navigation warfare, which show how disruption of spectrum access (Chapter 6) and **Positioning, Navigation, and Timing** (PNT) information can degrade command-and-control without direct kinetic engagement (Chapter 7).

As a result, defence organisations must expect degraded situational awareness and intermittent connectivity as a baseline condition rather than an exception. This has implications for governance models and command philosophy. Policy frameworks should explicitly reinforce mission command in multidomain environments, supported by clear delegation of authority and decision rights. At the same time, the technical interdependence of digital networks, sensors, and effectors means that command-and-control resilience increasingly depends on technical interoperability and cross-domain system integration. Contributions examining AI-enabled decision-support systems caution that increased automation introduces new vulnerabilities if robustness under adversarial conditions and integration across technical layers are insufficiently addressed (Chapter 9), underscoring the continued importance of human judgment embedded within technically resilient architecture.

Personnel Policy, Education, and Training

A third implication concerns personnel policy and professional development. Across domains – including cyber operations, electromagnetic warfare, logistics, and military engineering – the chapters highlight the growing importance of personnel who can operate beyond narrow functional or domain-specific boundaries. The need for such cross-domain professionalism is implicit in analyses of cyber contestation and information manipulation (Chapter 13), as well as in studies that emphasise the integration of engineering, geography, and operational planning (Chapter 12).

For defence organisations, this suggests a need to prioritise cross-domain competence in education, career development, and training policy. In addition to conceptual understanding, personnel must increasingly be able to understand technical dependencies between domains, such as the reliance of kinetic operations on space-based PNT, electromagnetic spectrum access, and resilient digital networks. Exercises and training programmes should therefore expose personnel not only to degraded conditions, but also to failures and trade-offs arising from imperfect technical integration across domains.

Sustainment, Resilience, and Organisational Endurance

A fourth implication, with significant policy relevance, concerns sustainment and resilience. Several contributions demonstrate that logistics, infrastructure, and geographic enablers are increasingly targeted by A2/AD strategies across multiple domains. This is particularly evident in analyses of contested logistics, which show how ISR, long-range strike, cyber disruption, and unmanned systems combine to extend denial effects deep into rear areas (Chapter 14). Complementary studies on hydraulic warfare and adaptive geography further illustrate how deliberate shaping of the physical environment can impose enduring constraints on manoeuvre and sustainment (Chapters 6 and 12).

These findings challenge long-standing assumptions about secure rear areas and uninterrupted support. Defence organisations must therefore reconsider logistics and sustainment policies through a resilience-oriented lens, recognising that sustainment itself depends on the integration of technical systems across domains, including communications, navigation, energy supply, and protection.

Legal, Political, and Alliance Considerations

Finally, the chapters highlight that organisational responses to A2/AD are inseparable from legal, political, and alliance considerations. Multidomain A2/AD practices in cyberspace and outer space raise questions of legality, legitimacy, and escalation management, as illustrated by analyses of self-defence zones in space (Chapter 15). In addition, political-economic studies demonstrate that A2/AD-related signalling practices, such as missile testing, can generate significant economic and reputational effects beyond the military domain (Chapter 16).

For defence organisations, this underscores the need for integrated policy coordination between military planners, legal advisors, political leadership, and alliance partners. In practice, such coordination increasingly depends on shared technical standards, interoperable systems, and aligned approaches to cross-domain capability integration, particularly in alliance contexts.

A Future Research Agenda

The analyses in this book collectively suggest that future research on A2/AD should place greater emphasis on area denial as a technologically enabled, multidomain practice, while avoiding a narrow focus on individual systems or domains. The conceptual shift from A2/AD as a static defensive construct to A2/AD as a condition of ongoing competition calls for a research agenda that integrates technology, operations, organisations, and governance.

Area Denial As a Dynamic, Technology-driven Practice

A first priority for future research concerns the evolving nature of area denial itself. Several chapters demonstrate that contemporary area denial is increasingly achieved through technological means that shape uncertainty, timing, and decision-making rather than through permanent physical exclusion. Hypersonic weapons, swarming systems, electromagnetic disruption, navigation warfare, and cyber operations all illustrate how denial effects are often temporary, reversible, and context-dependent.

Future research should therefore move beyond assessing individual technologies in isolation and instead examine how combinations of technologies generate denial effects over time. Key questions include how area denial scales across geographic space, how denial thresholds are established and eroded, and how adversaries adapt technologically and organisationally to denial measures. Particular attention should be paid to the gap – highlighted in historical analyses – between strategic expectations of durable denial and the operational realities of fragility and adaptation.

Technical Integration and Cross-domain Dependencies

A second research line concerns technical integration across domains. Many contributions implicitly demonstrate that area denial effectiveness depends on the interaction between sensors, effectors, networks, and decision-support systems spanning land, air, maritime, space, cyber, and the electromagnetic spectrum. Disruption in one technical layer – such as spectrum access or PNT – can cascade across domains and undermine broader operational effectiveness.

Future research should therefore focus on cross-domain dependencies and failure modes. This includes studying interoperability challenges, cascading effects of technical disruption, and trade-offs between performance, resilience, and complexity. Research on system integration should explicitly consider adversarial interference, degraded connectivity, and contested information environments as baseline conditions rather than edge cases.

Decision-making and Human-technology Interaction under Area Denial

A third priority lies in understanding decision-making under technologically enabled area denial. Several chapters highlight that denial strategies increasingly target cognition and decision processes through saturation, compression of time, and information disruption. At the same time, defence organisations are increasingly reliant on AI-enabled decision-support tools to manage multidomain complexity.

Future research should examine how humans and automated systems interact under these conditions. This includes questions of robustness, trust, and accountability in AI-supported decision-making, as well as the organisational implications of operating with degraded situational awareness and intermittent connectivity. Importantly, research should avoid treating technology as a substitute for human judgment and instead explore how technological systems can support decision-making without amplifying vulnerability.

Area Denial, Endurance, and the Material Foundations of Operations

A fourth research direction concerns the relationship between area denial and endurance. Contributions on logistics, geography, and hydraulic warfare demonstrate that denial increasingly targets sustainment, infrastructure, and mobility, extending contestation deep into rear areas. Area denial thus operates not only at the tactical or operational level, but also through long-term constraints on endurance and resilience.

Future research should further explore how technological area denial interacts with logistics, infrastructure, and geographic adaptation. This includes studying how denial effects accumulate over time, how organisations absorb and recover from disruption, and how material and technical resilience can be designed into force structures and support systems.

Legal, Political, and Strategic Stability Implications of Technological Area Denial

Finally, a comprehensive research agenda from a whole-of-society perspective must address the legal and political implications of technologically enabled area denial. Chapters on space, cyber operations, and missile testing illustrate how denial practices intersect with legal ambiguity, norm contestation, and broader political-economic effects. The dual-use, international and civilian-military characteristics of space and cyberspace are especially challenging. As denial technologies become more precise, reversible, difficult to attribute, and with impacts in a much broader physical and virtual area, questions of legitimacy, escalation control, and strategic stability become increasingly salient.

Future research should therefore integrate legal, political, and economic analysis into the study of area denial technologies, particularly in alliance contexts. This includes examining the broader societal impacts, i.e., how shared technical standards, interoperability requirements, and legal frameworks shape collective approaches to A2/AD, and how misalignment across these dimensions may generate strategic risk.

About the Editors

Edwin Dado is Associate Professor of Military Engineering at the Netherlands Defence Academy. His current research interests include hydraulic warfare, area denial and force protection engineering.

Marc Gouw is a lieutenant-colonel in the Royal Netherlands Army and an Associate Professor at the Faculty of Military Sciences of the Netherlands Defence Academy. He teaches within the field of military operational sciences, focusing on military operations in warfighting scenarios. His research is currently dedicated to understanding modern land warfare, notably the interplay between military intelligence and targeting.

Roy Hamans is Assistant Professor of Defence Economics at the Faculty of Military Sciences, Netherlands Defence Academy. His expertise lies at the intersection of innovation, technology, strategy, and finance, with a focus on challenging environments. Drawing on experience in both the private sector and government, he holds a PhD in Engineering from the University of Cambridge and an MBA from IMD Lausanne, among other qualifications.

René Janssen is Associate Professor of Mathematics and Operations Research at the Faculty of Sciences of the Netherlands Defence Academy. His research interest focusses on network theory and stochastics related to military operations research.

Kerry Malone is a Senior Scientist at TNO, and a Senior Research Fellow at the Netherlands Defence Academy. Her interests include methodologies to assess supply and logistics concepts in a defence context. Kerry is a national scientific representative in the NATO Science and Technology Systems Analysis and Studies Panel.

About the Authors

Colonel **Robert Bertrand** is a colonel in the Royal Netherlands Marechaussee and Associate Professor of Accounting, Control and Defence Economics at the Faculty of Military Sciences, Netherlands Defence Academy. He joined the Royal Netherlands Air Force in 1995 and switched to the Royal Netherlands Marechaussee in 2009. He defended his PhD on contract audits in 2016 at Maastricht University. His current research interests focus on auditing, compliance, internal control, procurement, ethics and leadership.

Stian Betten is a Master of Engineering in Aeronautical Engineering from Imperial College, University of London, and is a Research Manager at the Norwegian Defence Research Establishment (FFI) in operational analysis of offensive air operations. His current focus is distributed simulation of air operations, tactics development for network enabled weapons and advanced collaborative air combat.

Rémon ten Bhömer joined Netherlands Organisation for Applied Scientific Research (TNO) as a researcher addressing socioeconomic problems through systems thinking. Later he started focussing on capability development to bridge the gap between strategic concepts and the operation. Currently in the field of Integrated Electromagnetic Operations, Rémon holds a degree from TU Delft in Engineering, and Policy Analysis.

Edwin Dado is Associate Professor of Military Engineering at the Netherlands Defence Academy. His current research interests include hydraulic warfare, area denial and force protection engineering.

Jochem van Dam is a Master's student in Applied Mathematics at Leiden University. During his thesis internship at the Faculty of Military Sciences in Den Helder, he researched UAV route optimisation under probabilistic radar-detection constraints, focusing on multi-objective decision support in A2/AD environments.

Paul van Fenema is a Professor of Military Logistics at the Netherlands Defence Academy. His research concerns service network innovation, network/organisational governance, reorientation, innovations like digital transformation, and data enabled decision making in value chains. He participated in empirical research projects in the military sector (maritime, land, and urban operations).

Roy Hamans is an Assistant Professor of Defence Economics at the Faculty of Military Sciences of the Netherlands Defence Academy. Drawing on experience in both the public and private sectors, his research interests focus on the intersection of strategy, innovation, technology, and finance, particularly in challenging environments.

Louisa Handel-Mazzetti is Assistant Professor of Military Law at the Netherlands Defence Academy. She is also conducting her PhD research at Leiden University on the application of the prohibition of the use of force to the outer space context for both private and public actors.

Hans-Dieter Hiep is a specialist in formal methods and automated reasoning, with an interest in high-assurance computing and networking. In early 2024 he completed cum laude a PhD program in computer science (Leiden University). He briefly worked at the largest cloud provider in the world, Amazon Web Services, as applied scientist on hypervisor verification, and at the influential organisation where the Internet started in Europe, NLnet, as technology assessor and project manager, before joining the Netherlands Defence Academy as Assistant Professor in late 2025.

Andreas Hörnedal is a Deputy Research Director at the Swedish Defence Research Agency (FOI), Division of Defence Technology. His main field of work is air domain kill chain analysis, with modelling and simulation. He holds a Master of Science in Vehicular Engineering from the Royal Institute of Technology (KTH), Sweden.

Danny Janssen is an Assistant Professor of Military Engineering at the Netherlands Defence Academy. In his research, he investigates the multifunctional application of military equipment during high-water crises, with a specific focus on the BresDefender. He also researches animal burrows in hydrological structures.

Relinde Jurrius works as Assistant Professor of Mathematics & Operations Research at the Netherlands Defence Academy. Her research and teaching focuses on the application of new techniques in mathematics and computer science to relevant military problems, as well as the theory behind it.

Mitchell Kampert is an engineering officer in the Royal Netherlands Navy. He completed his Bachelor's degree at the Netherlands Defence Academy, specialising in Radar- and weapon systems. In his thesis, he researched the manoeuvring capabilities of hypersonic glide vehicles based on open-source information using computer modelling and simulations.

Jeroen Klomp is a full Professor on arms export control at the Faculty of Military Sciences of the Netherlands Defence Academy and Associate Professor of macroeconomics at Wageningen University & Research. His research is mainly concerned with the political economy aspects of the international arms trade & control and the strategic competition among states.

Dennis Krabbenborg is a Lecturer at the Faculty of Military Sciences of the Netherlands Defence Academy. After graduation Dennis worked at a construction company in the Civil Engineering industry on large infrastructural projects. In 2005 he joined the Military Engineering section at the faculty as a Lecturer, supporting and facilitating research projects. For these purposes he is also managing the Military Engineering laboratory in Breda.

Johan Kranenburg (MA OF-4) is a military engineer with operational experience in Bosnia and Afghanistan. He leads the Defence Expertise Centre (DEC) Military Engineering, contributes to NATO MILENG doctrine development, and has extensive experience in inter agency cooperation, crisis response planning and lessons learned. He holds an MA in Military Strategic Studies.

Major **Sebastiaan Leertouwer** works as a Lecturer at the Faculty of Military Systems and Technologies at the Netherlands Defence Academy. Within the faculty he works in the section of Military Engineering, providing classes for the cadets and midshipmen mixing the operational environment and problems with academic education. He is researcher in the field of Military Engineering, focusing on Military Infrastructure.

Roy Lindelauf is a Professor of Data Science in Military Operations (Netherlands Defence Academy) & Safety, Security (Tilburg University). His research interest lies at the interplay between mathematical modelling and (security) intelligence analysis with a focus on mathematical and computer science methods to the analysis of terrorism, political violence and related topics such as network resilience, cyber and network design.

Barend Lubbers is an Assistant Professor of Navigation Technology at the Netherlands Defence Academy. He received both his B.Sc. and M.Sc. degrees in Aerospace Engineering from Delft University of Technology. His research focuses on Global Navigation Satellite Systems (GNSS), such as GPS and Galileo, their vulnerabilities, potential alternatives, and navigation warfare.

Kerry Malone is a Senior Scientist at TNO, and a Senior Research Fellow at the Netherlands Defence Academy. Her interests include methodologies to assess supply and logistics concepts in a defence context. Kerry is a national scientific representative in the NATO Science and Technology Systems Analysis and Studies Panel.

Peter van Maurik is a retired Naval Commander with extensive experience in maritime operations and strategy. He served as a master trainer for warfare officers, specialising in developing operational and tactical expertise in the naval domain. He is the author of two books and many public articles on naval warfare, focusing on modern maritime conflict and strategic doctrine. An avid reader of history, he maintains a particular interest in military history and its relevance to contemporary defence and security challenges.

Ruud Moeskops (MA OF-3) is a military and civil engineer with operational military experience in Afghanistan and operational civilian experience in the field of interagency cooperation. He leads the strategy and innovation section at the DEC MILENG with a focus on Civil Military Cooperation. He holds an MA in Military Strategic Studies.

Denise Mollink is an engineering officer who is currently part of the Ground Based Air Defence section of the Netherlands Material and IT Command. She got her Bachelor's degree in aeronautical engineering at the Netherlands Defence Academy. In her thesis, she researched the capabilities of a specific Chinese loitering Munition, the CH-901 Feihong, based on open-source data and simulation modelling.

Herman Monsuur is Professor of Military Operations Research at the Netherlands Defence Academy. His current research interests include robust logistics, routing of maritime drones for surveillance, game theory for security forces, network theory, search and detection, critical infrastructure security and resilience, and optimisation of tactical UAV formations.

Ronald Nijboer is an Associate Professor at the Netherlands Defence Academy. He holds a Master's degree in Applied Mathematics and a PhD in Plasma Physics. He worked at the Royal Netherlands Aerospace Centre (NLR) and at the Netherlands Institute for Radio Astronomy (ASTRON) before joining the Netherlands Defence Academy. Currently he is working on Sensor Systems and Navigation in a military context.

Alexander van Oers is an Assistant Professor at the Netherlands Defence Academy. His research combines artificial intelligence, operations analysis, and optimisation to support military decision-making in the presence of intelligent and adaptive adversaries, with particular attention on countering AI-enabled systems.

Lieutenant Commander **Joost van Oostrum** is a weapon-engineering officer. He earned his M.Sc. in Electrical Engineering at Delft University of Technology, specialising in radar systems with a particular interest in bistatic and dual-use radars and completed his Master thesis at TNO.

Simon Purton joined NATO HQ Supreme Allied Commander Transformation in 2007, working in the Deputy Operational Analysis Branch. He conducted analytical studies on many topics and received a NATO Meritorious Medal for his service (2024). He was also involved in a number of multinational analysis studies, for which he received the NATO award for Scientific Achievement (2011). In Summer 2024, Simon became NATO HQ SACT's Branch Head for Innovation.

Petteri Puusa has a background as a pilot in the Finnish Air Force. In 2017-2023 he worked in the Finnish Defence Research Agency (FRDA) on Air and Space domains and as a Research Manager for Joint Capabilities. He retired in 2026 as COS of Lapland Air Wing.

Mark Patrick Roeling is Assistant Professor at the Data Science Centre of Excellence of the Netherlands Defence Academy. He holds a PhD in statistics and cybersecurity and has extensive hands-on data science experience in national security and intelligence. Within DSCE, he leads research on cyber security and AI.

Ralph Savelsberg is an applied physicist and is Associate Professor for missile defence at the Faculty of Military Sciences, Netherlands Defence Academy. His research is focused on modelling and simulation of ballistic and hypersonic missile threats, based on open sources, and he contributes to missile defence projects on behalf of the Royal Netherlands Navy.

Martijn Scherpenzeel is an engineering officer in the Royal Netherlands Navy. He is completing his Master's degree at the Netherlands Defence Academy, specialising in radar systems. In his thesis, he researched the smart prioritisation of radar tasks in swarming attacks based on the radar function to which the task belongs using computer modelling and simulations.

Alexander Schmets is an Assistant Professor of Military Engineering at the Netherlands Defence Academy. A physicist by training, he has worked on neutron scattering, novel battery materials, and complex fluids. Selected highlights include co-initiating the international research field of self-healing materials, organising international conferences and developing a minor in Integrated Infrastructure Design. His current research spans explosion-induced seismic effects, protective design, area denial, hydraulic warfare, and emergency response to extreme events such as floods and high-impact CBRN incidents.

Ramon Soesan is a researcher at the Data Science Centre of Excellence of the Netherlands Defence Academy. He got his Master's degrees in Econometrics and Computational Science and has hands-on experience as a data scientist within the Ministry of Defence. His research is focused on the robustness of AI-based decision support for military operations.

Tomas van Summeren is a communication and information systems officer in the Royal Netherlands Air and Space Force. He has completed his Master's degree at the Netherlands Defence Academy, specialising in systems engineering. In his thesis, he researched a high-speed air defence quadcopter for air defence purposes using computer modelling and simulations.

Susan Varenbrink is an Assistant Professor of Sensor Systems at the Netherlands Defence Academy. She holds a Bachelor's degree in electrical and electronic engineering, and a Master's degree in cybersecurity. Within the Ministry of Defence she held various positions related to radar and electromagnetic warfare. With the increased relevance of electromagnetic warfare in current conflicts, she fully focusses her work on research and spreading of knowledge in this field and related subjects.

Jory Veldman completed his Bachelor's degree in Military Systems and Technology at the Netherlands Defence Academy, specialising in Aerospace Engineering. In his thesis, he researched the defence against swarm behaviour of loitering munitions, conducting a case study on the APKWS system and the Feihong 901 using modelling and simulation techniques.

Arjen Vermeer is Assistant Professor of Military Law at the Netherlands Defence Academy. He is also conducting his Ph.D. research at the University of Amsterdam on the legal consequences for states under international law for the provision of satellite data by commercial entities in support of a party to an international armed conflict.

Arthur Vermeulen is Assistant Professor at the Faculty of Military Sciences, Netherlands Defence Academy. He has lectured on Guided Weapons, Electrical Power Systems and Control Engineering. Currently his focus is on Ballistics and Weaponeering.

Mark Voskuijl is Professor of Weapon- and Aviation Systems at the Faculty of Military Sciences at the Netherlands Defence Academy. His research is focused on the performance and flying qualities of fighter aircraft, helicopters and unmanned aerial vehicles.

Marten Zwanenburg is Professor of military law at the Netherlands Defence Academy and the University of Amsterdam. He previously held positions as legal counsel at the Netherlands Ministry of Defence and the Ministry of Foreign Affairs. Professor Zwanenburg is a member of the editorial boards of the Military Law and the Law of War Review and the Dutch-language Militair Rechtelijk Tijdschrift. He has published extensively on issues concerning the use of force, international humanitarian law and human rights law and the application of international law to the military use of technology.